OXFORD
UNIVERSITY PRESS

SUCCEED
PROGRESS

Essential
Biology
for Cambridge IGCSE®

2nd Edition

Gareth Williams
Richard Fosbery
Series Editor: Lawrie Ryan

Oxford excellence for Cambridge IGCSE®

OXFORD

OXFORD
UNIVERSITY PRESS

Acknowledgements

Page make-up by GreenGate Publshing Services, Kent

The authors and publishers wish to thank Julietta Watson for her assistance on the research and commenting on the manuscript and Paul Connell for his work on the Alternative to practical paper section.

® IGCSE is the registered trademark of Cambridge International Examinations.

The questions, example answers, marks awarded and/or comments that appear in this book were witten by the authors. In examination, the way marks are awarded may be different.

The authors and publisher are grateful to the following for permission to reproduce photographs in this book.

Cover image: © Michael & Patricia Fogden/CORBIS; p2: © megasquib/istock; p3: © Markanja/istock; p4: © Rinek/istock; p4: © Shadow_Hawk/istock; p4: © konmesa/istock; p6: DAVID SCHARF/Science Photo Library; p6: GREGORY DIMIJIAN/Science Photo Library; p7: M.I. WALKER/Science Photo Library; p8: Heather Angel / Natural Visions; p8: © Mark Kostich/istock; p8: M.H. SHARP/Science Photo Library; p9: © Juanvi Carrasco / Alamy; p9: Oxford Scientific/Mike Birkhead; p10: TED KINSMAN/Science Photo Library; p10: © Lupos/istock; p10: STUART WILSON/Science Photo Library; p10: DAVID AUBREY/Science Photo Library; p11: M.H. SHARP/Science Photo Library; p11: © johnaudrey/istock; p12: SINCLAIR STAMMERS/Science Photo Library; p12: BJORN SVENSSON/Science Photo Library; p12: ADRIAN BICKER/Science Photo Library; p18: Carolina Biological Supply, Co/Visuals Unlimited, Inc.; p19: POWER AND SYRED/Science Photo Library; p21: DR GOPAL MURTI/Science Photo Library; p21: DAVID AUBREY/Science Photo Library; p21: CNRI/Science Photo Library; p29: THOMAS DEERINCK, NCMIR/Science Photo Library; p33: Mike Samworth.; p33: Mike Samworth.; p43: © GoGo Images Corporation/Alamy; p43: A. BARRINGTON BROWN/Science Photo Library; p46: J.C. REVY, ISM/Science Photo Library; p47: LAGUNA DESIGN/Science Photo Library; p54: BIOPHOTO ASSOCIATES/Science Photo Library; p56: Ecoimages/Fotolia; p56: Science Photo Library/Cordelia Molloy; p58: MARTIN DOHRN/Science Photo Library; p58: © IslandLeigh/istock; p63: ROSENFELD IMAGES LTD/Science Photo Library; p63: © Nigel Cattlin / Alamy; p64: © Roel Smart/istock; p66: GEOFF KIDD/Science Photo Library; p73: © filo/istock; p74: © Maica/istock; p74: © imageBROKER / Alamy; p74: GUSTOIMAGES/Science Photo Library; p77: BIOPHOTO ASSOCIATES/Science Photo Library; p77: BIOPHOTO ASSOCIATES/Science Photo Library; p80: IAN BODDY/Science Photo Library; p85: EYE OF SCIENCE/Science Photo Library; p86: CNRI/Science Photo Library; p87: © Les Stone/Corbis; p87: Eco Images; p90: Alamy; p90: LOUISA PRESTON/Science Photo Library; p91: Noel Sturt; p91: CHUCK BROWN/Science Photo Library; p91: Gene Cox; p92: POWER AND SYRED/Science Photo Library; p92: Gene Cox; p94: DR JEREMY BURGESS/Science Photo Library; p94: BIOPHOTO ASSOCIATES/Science Photo Library; p96: BIOPHOTO ASSOCIATES/Science Photo Library; p97: CLOUDS HILL IMAGING LTD/Science Photo Library; p104: RIA NOVOSTI/Science Photo Library; p105: Photo Researchers, Inc. /Science Photo Library; p108: BIOPHOTO ASSOCIATES/Science Photo Library; p108: CNRI/

Science Photo Library; p110: BIOPHOTO ASSOCIATES/Science Photo Library; p111: © ChristianAnthony/istock; p112: STEVE GSCHMEISSNER/Science Photo Library; p112: BIOLOGY MEDIA/Science Photo Library; p113: BIOPHOTO ASSOCIATES/Science Photo Library; p118: ADAM HART-DAVIS/Science Photo Library; p118: © satit_srihin/istock; p119: © gregepperson/istock; p122: © jo unruh/istock; p123: © Nathan Benn/Ottochrome/Corbis; p123: © RGB Ventures/SuperStock/Alamy; p123: © Eric Carr/Alamy; p124: Blend Images - Ariel Skelley; p124: © Richard Levine/Demotix/Corbis; p125: Charmaine A Harvey; p125: © FAYYAZ AHM epa/Corbis; p128: CNRI/Science Photo Library; p134: istock; p137: © PHOTOTAKE Inc./Alamy; p138: MichaelTaylor/Shutterstock; p140: ANDREW SYRED/Science Photo Library; p140: bikeriderlondon; p141: Peter Muller/CULTURA/Science Photo Library; p147: Richard Fosberry; p147: Richard Fosberry; p150: AJ PHOTO/HAMERICAIN/Science Photo Library; p155: MEHAU KULYK/Science Photo Library; p156: BIOPHOTO ASSOCIATES/Science Photo Library; p156: BIOPHOTO ASSOCIATES/Science Photo Library; p159: PROF S. CINTI/Science Photo Library; p160: Michael Freema/Alamy; p161: Richard Fosberry; p164: © Photofusion Picture Library/Alamy; p167: ASTRID andamp;amp; HANNS-FRIEDER MICHLER/Science Photo Library; p168: © JohnPitcher/istock; p170: JEROME WEXLER/Science Photo Library; p171: NIGEL CATTLIN/Science Photo Library; p174: MAURO FERMARIELLO/Science Photo Library; p175: © Eric Audras/Onoky/Corbis; p175: JOHN DURHAM/Science Photo Library; p176: Vladimir Gjorgiev/Shutterstock; p177: Wellcome UK; p177: ©Janine Wiedel/Photofusion; p179: © inhauscreative/istock; p179: © Richard Levine/Alamy; p180: © Underwood & Underwood/Corbis; p181: MOREDUN ANIMAL HEALTH LTD/Science Photo Library; p181: LIVING ART ENTERPRISES, LLC/Science Photo Library; p185: DR JEREMY BURGESS/Science Photo Library; p186: AMI IMAGES/Science Photo Library; p186: © Frank Walker/Alamy; p187: © blickwinkel/Alamy; p187: EYE OF SCIENCE/Science Photo Library; p1 DARWIN DALE/Science Photo Library; p189: DR JEREMY BURGESS/Science Photo Library; p190: © Peter Arnold, Inc./Alamy; p190: SUSUMU NISHINAGA/Science Photo Library; p193: STEVE GSCHMEISSNER/Science Photo Library; p193: PHILLIPS/Science Photo Library; p195: PROF. P. MOTTA/DEPT. OF ANATOMY/UNIVERSITY; p199: ANATOMICAL TRAVELOGUE/Science Photo Library; p199: ANATOMICAL TRAVELOGUE/Science Photo Library; p199: JAMES STEVENSON/Science Photo Library; p200: DR G. MOSCOSO/Science Photo Library; p2 BlueSkyImage/Shutterstock; p201: MAURO FERMARIELLO/Science Photo Library; p202: istock; p204: DR J. P. ABEILLE/Science Photo Library; p209: © monkeybusinessimages/istock; p210: © sean sprague / Alamy; p210: NIAID/NATIONAL INSTITUTES OF HEALTH/Science Photo Library; p214: © video1/istock; p214: CNRI/Science Photo Library; p218: MICHEL DELARUE, ISM/Science Photo Library; p2 M.I. WALKER/Science Photo Library; p219: Dimarion/Shutterstock; p220: D. PHILLIPS/Science Photo Library; p221: PASCAL GOETGHELUCK/Science Photo Library; p223: © Martin Shields/Alamy; p224: © PeterAustin/istock; p226: REVY, ISM/Science Photo Library; p227: © CuboImages srl/Alamy; p229: DAVID NICHOLLS/Science Photo Library; p229: BIOPHOTO ASSOCIATE/Science Photo Library; p232: Ian Shaw/Alamy; p233: JEROME WEXLER/Science Photo Library; p234: © ChristianAnthony/istock; p234: FRANCIS LEROY, BIOCOSMOS/Science Photo Library; p236: Kim Fenemma/Visuals Unlimited, Inc./Getty; p236: Mirko Zanni/Getty; p236: © karlumbriaco/Fotolia; p237: MAURO FERMARIELLO/Science Photo Library; p237: STEVE ALLEN/Science Photo Library; p238: Science Photo Library; p238: FLETCHER andamp;amp; BAYLIS/Science Photo Library; p238: TONY CAMACHO/Science Photo Library; p240: JOHN MOSS/Science Photo Library; p240: John Giustina; p240: (c) 2008 Michele Westmorland All Rights Reserved; p241: driftlessstudio/istock; p244: © Elliot Nichol/Alamy; p253: HUGH SPENCER/Science Photo Library; p253: ANDREW SYRED/Science Photo Library; p254: © arishwad/istock; p260: Image Source; p260: SM Rafiq Photography/Getty; p2 Adrian Muttitt / Alamy; p261: ROSENFELD IMAGES LTD/Science Photo Library; p262: POWER AND SYRED/Science Photo Library; p264: © Olha Rohulya/istock; p265: MAXIMILIAN STOCK LTD/Science Photo Library; p266: © chas53/istock; p266: © Nigel Cattlin/Visuals Unlimited/Corbis; p270: © CARLOS BARRIA/Reuters/Corbis; p270: Sukpaiboonwat/Shutterstock; p271: © frei-impuls/istock; p272: Thomas Marent/Getty; p272: © osiris59/istock; p272: © jeanro/istock; p274: John Potter/Alamy; p275: © davide chiarito/istock; p276: DAVID NUNUK/Science Photo Library; p277: DR JEREMY BURGESS/Science Photo Library; p278: SOHM/VISIONS OF AMERICA/Science Photo Library; p280: MARTIN BOND/Science Photo Library; p280: DIRK WIERSMA/Science Photo Library; p280: JOHN HESELTINE/Science Photo Library; p281: WILL andamp;amp; DENI MCINTYRE/Science Photo Library; p281: SHEILA TERRY/Science Photo Library; p281: YVES SOULABAILLE/LOOK AT SCIENCES/Science Photo Library; p282: © FABRICE BENSCH/Reuters/Corbis; p282: © Olivier Coret/Corbis; p283: © George Clerk/istock; p283: PETER SCOONES/Science Photo Library; p284: ROSENFELD IMAGES LTD/Science Photo Library; p284: MARTIN BOND/Science Photo Library; p2 Richard Fosberry; p285: © paulprescott72/istock; p285: © Lourens Smak/Alamy; p285: © kozmoat98/istock; p286: © Corbis. All Rights Reserved.; p286: Michael Patricia Fogden/Getty; p286: Richard Fosberry; p287: © Enjoylife2/istock; p287: GoodOlga/istock; p288: Erick Margarita Images/Shutterstock; p288: © mkos83/istock; p288: © Juniors Bildarchiv/Alamy; p289: SCUBAZOO/Science Photo Library; p289: © National Geographic Image Collection/Alamy.

Artwork by Greengate Publishing Services and OUP.

Every effort has been made to contact all copyright holders, but if any have inadvertently been overlooked, the publishers will be happy to make the necessary amendments at the first opportunity.

Contents

Contents

Support website:
www.oxfordsecondary.com/9780198399209

This book is designed specifically for Cambridge IGCSE® Biology 0610. Experienced teachers have been involved in all aspects of the book, including detailed planning to ensure that the content gives the best match possible to the syllabus.

Using this book will ensure that you are well prepared for studies beyond the IGCSE level in pure sciences, in applied sciences or in science-dependent vocational courses. The features of the book outlined below are designed to make learning as interesting and effective as possible:

STUDY TIP...

Experienced teachers give you suggestions on how to avoid common errors or give useful advice on how to tackle questions.

LEARNING OUTCOMES

- These are at the start of each spread and will tell you what you should be able to do at the end of the spread.

Supplement

- Some outcomes will be needed only if you are taking a supplement paper and these are clearly labelled, as is any content in the spread that goes beyond the syllabus.

KEY POINTS

These summarise the most important things to learn from the spread.

SUMMARY QUESTIONS

These questions are included in every topic and allow you to test your understanding of the work covered in the topic.

DID YOU KNOW?

These are not needed in the examination but are found throughout the book to stimulate your interest in biology.

At the end of each unit there is a double page of examination-style questions written by the authors.

At the end of the book you will also find:

Alternative to practical paper

A glossary of many important terms used in the book.

PRACTICAL

These show the opportunities for practical work. The results are included to help you if you do not actually tackle the experiment or are studying at home.

Assessment structure

Paper 1: Multiple Choice (Core)

Paper 2: Multiple Choice (Supplement)

Paper 3: Theory (Core)

Paper 4: Theory (Supplement)

Paper 5: Practical Test

Paper 6: Alternative to Practical

Extra resources, including answers:

www.oxfordsecondary.com/9780198399209

1 Characteristics and classification of living organisms

1.1

Characteristics of living organisms

Biology is the study of living things or living organisms, which are classified into five major groups called **kingdoms**:

animals, plants, fungi, protoctists and prokaryotes (bacteria)

There are seven characteristics of living organisms. These characteristics are often described as life processes:

nutrition, respiration, movement, growth, excretion, sensitivity and reproduction.

Nutrition

Nutrition is the obtaining of food to provide energy and substances needed for growth, development and repair. Nutrients are compounds that may be large and complex (like carbohydrates, proteins and vitamins) or simple (like mineral ions).

Nutrition in green plants involves **photosynthesis**, in which the energy from sunlight is absorbed and used to turn carbon dioxide and water into simple sugars. Plants then convert these simple sugars into complex compounds, such as cellulose and proteins. They need mineral ions from the soil to make proteins.

Salmon provide nutrition for bears.

Animals cannot make their own food like plants, so they have to eat plants or other animals to gain energy and nutrients. The process of taking in food is called **ingestion**. The food is digested, absorbed into the blood and then assimilated by cells for growth and repair. Food which is not digested and absorbed is egested in faeces.

Respiration

All living organisms respire because they all need energy. Respiration involves chemical reactions that occur in cells to break down nutrients, such as glucose, to release energy. Oxygen is usually needed for respiration to happen. This is the word equation for respiration involving oxygen:

glucose + oxygen $\longrightarrow$ carbon dioxide + water + energy released

Living things use this energy for movement, growth, repair and reproduction.

Movement

Organisms move themselves or move parts of themselves into new areas or to change position.

Plants move slowly when they grow. Their roots move down into the soil and their leaves and stems move up towards the light. Leaves can move to face the Sun so they can absorb as much light as possible.

Most animals are able to move their whole bodies. They move to obtain their food or to avoid being caught by predators. Some animals remain fixed to one place throughout their lives, but they are able to move parts, such as the tentacles on a sea anemone.

Sea anemone.

Growth

Growth is a permanent increase in size of an organism. This involves an increase in cell number, cell size or both. It always involves making more complex chemicals, such as proteins, which is why the dry mass increases. Plants carry on growing throughout their lives. Animals stop growing when they reach a certain size.

Excretion

All living organisms produce toxic (poisonous) waste substances as a result of metabolism. **Metabolism** is all the chemical reactions that occur in an organism. Respiration is a major part of metabolism.

Excretion is the removal of these waste materials and substances in excess of requirements from the body. Plants store waste substances in their leaves, so the waste chemicals are removed when the leaves fall off. Animals breathe out carbon dioxide; other waste substances leave the body in the urine.

Sensitivity

Living organisms are able to detect or sense changes in their internal and external environments. A change like this is a **stimulus** (plural: stimuli). Sensitivity is the ability to detect these stimuli and **respond** to them.

Plants respond to movement of the Sun by moving their leaves to face the light. The flowers of some plants open in the morning and close at night. Animals have sensory cells and sense organs for detecting light, sound, touch, pressure and chemicals in the air and in food.

Reproduction

Organisms reproduce to make new individuals. **Asexual reproduction** involves one parent giving rise to offspring that are often identical to each other and to the parent. **Sexual reproduction** involves two parent organisms producing **gametes** (sex cells) which fuse to give rise to the next generation. The offspring show **variation**. They are not identical to each other or to their parents.

KEY POINTS

The mnemonic, 'Mrs Gren' can help you remember the seven characteristics of living organisms:

Movement causes an organism to change its position or place.

Respiration involves chemical reactions that release energy in cells.

Sensitivity is being able to detect and respond to stimuli.

Growth is a permanent increase in size and dry mass.

Reproduction results in the formation of new individuals.

Excretion is the removal of waste chemicals made in the cells during metabolism.

Nutrition involves the use of food for energy and growth.

SUMMARY QUESTIONS

1 A visitor from outer space lands on Earth. The first thing that it sees is a motor car passing by.
   a   Give two reasons why the visitor thinks that the car is alive.
   b   Give two reasons why you think that the visitor is wrong.

2 Plants and animals are two groups of living organisms. Find out and then describe how plants differ from animals in the ways in which they:
   a   feed,   b   move,   c   grow,   d   use their senses.

Classification

LEARNING OUTCOMES

- Define the term *binomial system*
- Describe the importance of classification to evolution

We classify organisms into groups. The largest grouping is the **kingdom** (see Topic 1.1) and the smallest is the **species**. The organisms in a kingdom share some similar features. For example, within the plant kingdom, all plants are green and carry out photosynthesis. Each kingdom is subdivided into groups known as **phyla** (singular: **phylum**). But first we must explain how organisms are named.

The binomial system

Meerkat, suricate and Sun angel are names for a type of mongoose that lives in the Kalahari Desert in southern Africa. To avoid confusion, organisms are given scientific names using the **binomial system** (binomial means 'two names').

A **species** is a group of individuals that look alike. They live in the same habitat and breed together to give offspring which are fertile and can also breed together.

Each species is given two names: the first name is for the **genus** and the second name is the **trivial** name that applies to one species within the genus. The trivial name should never be used on its own.

A genus is a group of species that are closely related, but do not interbreed with each other. Some genera (plural of genus) consist of only one species, as is the case with meerkats. This may be because other species in that genus are extinct.

The binomial system is used by biologists all over the world as it is an international language for naming organisms. The table has some examples.

A meerkat.

common name	scientific name
meerkat	*Suricata suricatta*
human	*Homo sapiens*
baobab tree	*Adansonia digitata*
cholera bacterium	*Vibrio cholerae*
malarial parasite	*Plasmodium falciparum*
oyster mushroom	*Pleurotus ostreatus*

The great white pelican, *Pelecanus onocrotalus*.

The importance of classification

The sorting out of the vast array of living organisms into groups is known as **classification**. Scientists have traditionally looked at the differences and similarities between organisms.

They have looked at the **morphology** and **anatomy** of living organisms. Morphology is the outward appearance of an organism, whereas anatomy is its internal structure.

Similarities between organisms may occur because they have evolved along the same lines. For instance, the limb bones of mammals follow a similar pattern known as the **pentadactyl limb**.

The spot-billed (grey) pelican, *Pelecanus philippensis*.

Front limb bones of three species of mammal are shown in Figure 1.2.1. Each has one upper limb bone (blue), two lower limb bones (yellow) and five digits (green, numbered 1–5). However, we cannot always rely on this method of classification. To obtain evidence from evolutionary relationships, scientists study not only the anatomy of organisms and the fossil record, but also the structure of their proteins and the sequence of bases in their DNA.

There are now more accurate methods to classify organisms. These methods involve studying the sequences of amino acids in proteins (see Topic 4.1) and the sequences of bases in DNA (see Topic 4.3). Haemoglobin is a protein that transports oxygen in many animals. For example, human haemoglobin has 574 amino acids. Each amino acid is coded by three bases, so there are at least 1722 bases that code for the protein. These sequences are compared with sequences for the same protein in different animals. The more similarities there are between sequences for any two animals the more closely related they are likely to be.

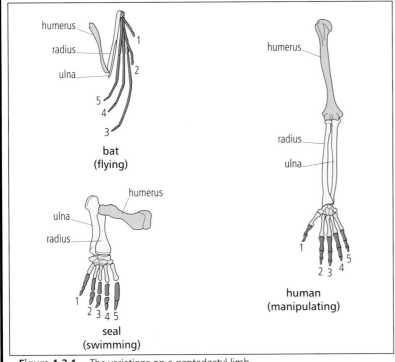

Figure 1.2.1   The variations on a pentadactyl limb.

SUMMARY QUESTIONS

1 Copy and complete the sentences using these words:

   fertile   genus   shared trivial   classification binomial   breed

   The system of putting organisms into groups is called _____. The _____ system is used for naming organisms. Each organism has two names. The first is called the _____ name and the second is the _____ name. A species is a very similar group of individuals that can _____ together and produce _____ offspring. We classify living things on the basis of their _____ features.

2 a Explain the term *binomial system*.

   b Find the scientific names of two plant species and two animal species not mentioned in this topic and write them alongside their common names.

3 In the 18th century, Carl Linnaeus worked out the binomial system of naming living things. Find out about his life and work and then write a brief biography in your own words.

KEY POINTS

1 The binomial system uses two names for each species: the genus name and the trivial name.

2 A species is a group of individuals, living in the same habitat that breed together to produce fertile offspring.

3 Classification is important in the study of evolution.

4 The sequence of bases in DNA is used as a more accurate means of classification.

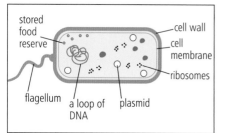

Figure 1.3.1 A typical bacterium.

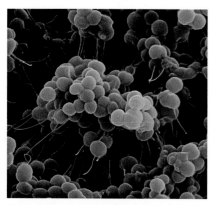

Staphylococcus, a type of bacterium found on the human skin (×4000).

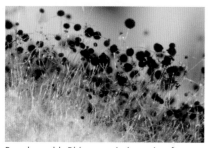

Bread mould, *Rhizopus nigricans*, is a fungus. The black structures contain spores.

The cells of all living organisms are made up of a **cell membrane**, **cytoplasm** and **DNA**. They also have present in the cytoplasm small structures called **ribosomes** that carry out protein synthesis and enzymes that are involved in processes such as respiration.

There are two main types of cell. Bacteria have a simple cell structure that you can see in Figure 1.3.1. Cells of bacteria do not have a nucleus and are known as prokaryotic cells. The cells of all other organisms have nuclei and these cells are known as eukaryotic.

Biologists use the features that organisms share to classify them into five large groups known as **kingdoms**. For example, all bacteria are classified together into the same kingdom because they have the same cell structure. There are five kingdoms, including the plant kingdom and the animal kingdom.

Plant kingdom Plants are multicellular organisms. The bodies of plants are not compact like those of animals. Their roots spread out through the soil to gain water and salts and their stems grow many leaves to absorb light. Growth occurs in special places such as the tips of roots and stems. This kingdom includes mosses and ferns, which do not reproduce using seeds, and the seed-bearing plants, such as conifers and the flowering plants. All plants have chlorophyll and carry out photosynthesis, using light to make their own food. Plant cells have chloroplasts full of chlorophyll, cellulose cell walls and a large sap-filled vacuole.

Animal kingdom Animals are also multicellular organisms. Most animals have much more compact bodies than plants. Growth occurs throughout the body. Animals do not photosynthesise and obtain their food by eating plants and/or other animals. Animal cells do not have cell walls, chloroplasts or large vacuoles. All animals have nervous systems which coordinate their responses to stimuli and their movement.

Prokaryote kingdom Bacteria have a simple cell structure that you can see in Figure 1.3.1. Some bacteria are spherical and some are rod-shaped. Many exist in short chains of cells. Most bacteria are a few micrometres in length and can only be seen with light and electron microscopes. (One micrometre is one-thousandth of a millimetre.)

Bacterial cells are surrounded by cell walls. The cells of some bacterial species are also surrounded by slime capsules.

There is no nucleus, just a loop of DNA within the cytoplasm. Bacteria often have additional loops of DNA inside their cytoplasm called **plasmids** (see page 267). The structure of bacteria differs from animal and plant cells as they do not have chloroplasts or mitochondria. Some bacteria have extensions called flagella (singular: flagellum) for moving through water or other fluids.

Supplement

Protoctists This kingdom contains a diverse group of organisms which are classified in this kingdom as they do not belong in any of the other four kingdoms. All protoctists have cells with nuclei; many are unicellular and some are multicellular. Unicellular protoctists include organisms commonly known as protozoans including *Plasmodium*, which is the parasitic organism that causes malaria. Algae are classified as protoctists and they vary in size from tiny unicellular organisms to the kelps which are giant seaweeds that form underwater 'forests' in parts of the sea where there is plenty of light so they can photosynthesise.

Fungi kingdom Fungi are visible under a light microscope. Many are also visible to the naked eye. Most fungi are multicellular although yeasts are single-celled. Each cell has a nucleus and a cell wall which is made out of chitin, not cellulose as in plants. Fungi do not have chlorophyll and cannot carry out photosynthesis.

The main fungus body is called the **mycelium**. It consists of a branching network of threads or **hyphae** which grow over the surface of its food source, releasing **enzymes** which digest the food outside the fungus. The digested food is then absorbed by the hyphae.

Fungi reproduce by making spores that can be carried by the wind. Most fungi are **saprotrophs**, which means they feed on dead or decaying matter, but some are **parasites**.

Viruses

Viruses are not cells. They are particles made up of genetic material (DNA or RNA) surrounded by a protein coat. The genetic material is composed of a few genes that code for the proteins that form the coat and maybe other proteins that help it reproduce. Viruses are **parasites** that enter the cells of another organism (the **host**) in order to multiply. Viruses take over the host cell and direct it to make new viruses. Viruses are not classified in any of the kingdoms described here. All the organisms classified in the five kingdoms have cells. Viruses do not have cells and so are classified in a different way. Their classification is based on the type of genetic material and protein coat that they have.

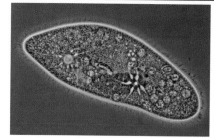

Paramecium, a single-celled protoctist (×206).

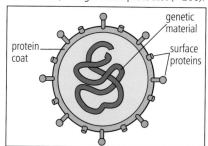

Figure 1.3.2 Structure of a human virus. Viruses are extremely small. This virus that causes influenza is about 120 nanometres in size. (One nanometre is one-thousandth of a micrometre.) Viruses are visible only under an electron microscope.

KEY POINTS

1 The cells of all living organisms have cytoplasm, a cell membrane and DNA.

2 There are five kingdoms: Plant, Animal, Prokaryote, Protoctist and Fungi.

3 Prokaryotes have cells without nuclei. They can be seen only under a light microscope.

4 Protoctists are a group of organisms with many features of eukaryoltic cells.

5 Fungi are made up of threads called hyphae. These grow over the food supply, digest it externally and then absorb it.

6 Viruses can be seen only under an electron microscope. They are not cells but a few genes inside a protein coat.

SUMMARY QUESTIONS

1 State the features present in all cells.

2 The plant kingdom and the animal kingdom contain many species that are familiar to us. Make a table to compare plants with animals. You should include similarities as well as differences.

3 Compare the structure of bacterial cells with the cells of a plant.

4 Make a table to compare the features of the five kingdoms.

5 Why are viruses not classified in one of the five kingdoms?

6 How many nanometres are there in a metre?

Vertebrates

This dace has fins, tail and a streamlined shape – adaptations for swimming.

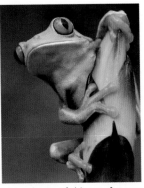

The bright red eyes of this tree frog are thought to startle predators, giving time for the frog to escape.

American crocodile, *Crocodylus acutus*.

Animals that have a vertebral column or backbone are called **vertebrates**. All vertebrates have an internal skeleton made of either bone or cartilage. They all belong to the phylum chordata which includes some invertebrates that share common features with the vertebrates.

There are five main groups of vertebrates:

fish, amphibians, reptiles, birds and mammals.

Fish

Most fish live in water permanently, but there are some species, such as the mud skipper, that can survive out of water for varying lengths of time.

Typically, fish are streamlined and have fins for swimming and for balance. They have eyes and a lateral line for detecting pressure changes in water. They breathe dissolved oxygen from the water using their gills. Their skin is covered with scales. Examples include tuna, herring, shark, catfish and cod.

Unlike fish, the other classes of vertebrates have ears for detecting sound and four limbs, although some have evolved into legless forms like snakes and some lizards.

Amphibians

These vertebrates have smooth, moist skin. Although most amphibians live on land, they return to water to breed. Fertilisation is external because sperm and eggs are released into the water (not inside a female body). Development is external: the fertilised eggs hatch into swimming tadpoles which have gills for breathing.

On land, the adult amphibians breathe using lungs, but when they are in water, they can breathe through their skin. Examples of amphibians include frogs, toads and salamanders.

Reptiles

Reptiles have dry, scaly skin to cut down water loss. They can live in dry regions as they do not have to return to water to breed.

Fertilisation takes place inside the female's body. However, development is external as they lay eggs with leathery, waterproof shells which stop them from drying out. Reptiles have lungs to breathe air. Crocodiles, lizards, snakes, turtles and tortoises are all reptiles.

Birds

Birds have feathers and their front limbs are modified as wings. Most of them are able to fly but some, like penguins and ostriches, cannot. Birds have no teeth but different species have beaks adapted to deal with different types of food.

Fertilisation is internal and development is external as the females lay eggs which are protected by hard shells.

Birds are **homeothermic** (warm-blooded). This means that they are able to regulate their body temperature. They can keep it constant even though the outside temperature changes. Examples of birds include hawks, eagles, sparrows, parrots and starlings.

Mammals

Mammals are the vertebrates that have hair or fur. Fertilisation is internal and so is development. The young develop in a womb and they are born already well developed. Female mammals suckle their young on milk from mammary glands. All mammals, even aquatic ones like whales and dolphins, use lungs for breathing.

Like birds, mammals are also homeothermic, maintaining a constant internal temperature. Leopards, bats, dolphins, bears, lemurs and wolves are all mammals.

Grey Heron, *Ardea cinerea* – a bird with a long beak for catching fish.

STUDY TIP

Make a table to help you learn the features of these five groups. Making the table will help you learn this topic.

A chimpanzee has hands and feet adapted to living in trees.

DID YOU KNOW?

Many amphibian species are threatened with extinction. Some rare Panamanian golden frogs were filmed for Sir David Attenborough's BBC TV series *Life in Cold Blood*. After the filming the frogs were moved to a zoo to protect them from disease, habitat loss and pollution.

SUMMARY QUESTIONS

1 Copy and complete the sentences using these words:

development five wings scaly moist land water classes breed fly fish fur feathers

There are _____ different groups of vertebrate. Sharks belong to the group called _____ and spend all their time in _____ . Amphibians have _____ skin and return to water to _____ . Reptiles have _____ skin and lay their eggs on _____ . Birds have front limbs that are modified to form _____ . They also have _____ and most are able to _____ . Mammals have _____ or hair and give birth to their young in an advanced state of _____ .

2 a State the features that are exhibited by all vertebrates.

  b Name the five groups of vertebrates.

  c Give an example of each group.

3 To which group of vertebrates does each of the following belong?

  a turtle b bat c whale d salamander e snake

  f hawk g shark

KEY POINTS

1 Vertebrates are animals that have a vertebral column or backbone.

2 The vertebrates are divided into five groups:

   fish, amphibians, reptiles, birds and mammals.

Invertebrates

LEARNING OUTCOMES

- Describe the external features of arthropods
- Describe the features of insects, crustaceans, arachnids and myriapods that are used to classify them into groups
- Describe how the animals in these groups are adapted to their environment

Woodlouse or slater. Notice its antennae for detecting stimuli.

Sally lightfoot crab, *Grapsus grapsus*. Notice its large claws for feeding and defence.

Invertebrates are animals that do not have a **vertebral column** or backbone.

Arthropods is the largest of the groups in the animal kingdom as it contains the largest number of species. Each arthropod species has a segmented body, an external skeleton (**exoskeleton**) and jointed legs.

The hard exoskeleton allows arthropods to live on dry land. When they grow too big for their exoskeleton, they moult and grow a new one. Some moult all through their lives, others moult only during the early stages of their life.

Four of the different types of arthropod are:

crustaceans, myriapods, insects and arachnids

Crustaceans

Crustaceans have a body divided into a **cephalothorax** (head-thorax) and abdomen. Many have a chalky exoskeleton that provides a very hard and effective protection against predators. Crustaceans have two pairs of **antennae** and **compound eyes**. They have between five and twenty pairs of legs. They breathe using **gills**.

Nearly all crustaceans live in water. Some crabs live on land but return to water to breed. Woodlice (also known as slaters) and some land crabs are exceptions as they do not use water for breeding. Examples of crustaceans include crabs, shrimps, crayfish and lobsters.

Myriapods

These are the **centipedes** and the **millipedes**. They have long bodies made up of many segments.

Their bodies are not divided into separate regions such as the thorax and abdomen. Centipedes have one pair of legs on each segment so that the total number of legs depends upon how many segments there are. Centipedes are fast-moving carnivores. They have powerful jaws and can paralyse their prey.

Millipedes have two pairs of legs on each body segment. They are slow-moving herbivores. You can often find them feeding in leaf litter.

Millipede – with its many legs for moving efficiently.

Centipede – a fast-moving carnivore.

Insects

Insects have bodies that are divided into three parts: head, thorax and abdomen.

On the thorax there are three pairs of legs and many species have two pairs of wings. They have one pair of antennae on the head and compound eyes that are made of many tiny individual components. They breathe through holes in the sides of the thorax and abdomen called **spiracles**.

Insects have colonised most habitats in the world, although there are very few species that live in the sea. Two reasons why they are so successful on land is that they are covered by a waterproof **cuticle** that stops them losing too much water, and they can fly!

This is the largest group within the arthropods. Examples include beetles, flies, locusts, cockroaches, dragonflies, butterflies, moths, bees and wasps.

Monarch butterfly, *Danaus plexippus*. Notice its wings for its migration over long distances.

Arachnids

These arthropods have bodies divided into two parts, the cephalothorax and abdomen.

Arachnids have four pairs of legs and no wings. They have no antennae but do have several pairs of simple, not compound, eyes. They paralyse their prey with poison fangs. Spiders are able to weave silken webs with their **spinnerets**. Scorpions, ticks and mites are also classified as arachnids.

Tarantula – a large spider with venom to paralyse its prey.

SUMMARY QUESTIONS

1 a Name four groups of arthropod.

  b List the key features of insects.

  c State three ways in which arachnids look different from insects.

2 Distinguish between the following on the basis of visible features:

  a centipedes and millipedes

  b crustaceans and arachnids

3 Describe the features of insects that adapt them for living successfully on land.

4 Copy and complete the table:

feature	myriapods	crustaceans	insects	arachnids
number of pairs of legs				
body regions				
number of pairs of antennae				
type of eyes				
wings				

KEY POINTS

1 Arthropods are segmented animals with jointed legs and an exoskeleton.

2 Arthropods are classified into different groups including:

crustaceans, myriapods, insects and arachnids

1.6

Ferns and flowering plants

Like all plants, ferns and flowering plants are multicellular. They are green in colour because many of their cells contain **chloroplasts**. These chloroplasts contain the green pigment **chlorophyll**, which absorbs light for photosynthesis. Each cell is surrounded by a cell wall made of cellulose.

Both ferns and flowering plants have transport systems consisting of tiny tubes. These are called **xylem vessels**, which carry water and mineral salts, and **phloem tubes**, which transport dissolved substances such as sugars. Both ferns and flowering plants are well adapted to live on land.

Ferns

Ferns are a group of plants that have become well adapted for life on land. Ferns have strong stems, roots and leaves. Their leaves have a waxy layer (the cuticle) which helps to reduce water loss. Most of them have leaves resembling those of the species in the photograph to the left. Ferns live in many different habitats including some that are quite dry. The largest are the tree ferns that have a thick stem that supports a crown of leaves at heights of several metres. Many ferns grow from a thick underground stem called a rhizome.

Unlike conifers and flowering plants, ferns do not produce seeds. If you look under the leaves of ferns at certain times of the year you will see structures that make and release microscopic spores. These spores are carried by the wind to form new plants.

The fern *Dryopteris filix-mas*.

Features of flowering plants

Flowering plants have true stems, roots and leaves. They reproduce by means of flowers which make seeds. The seeds are produced inside the ovary within the flower.

Shoots and roots

The **shoot** is the part of the plant above ground. The shoot is made up of a **stem** bearing leaves, buds and flowers. The **apical bud** is

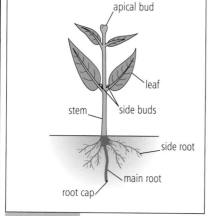

apical bud

leaf

stem — side buds

side root

main root

root cap

Figure 1.6.1 Structure of a typical dicotyledonous plant.

Magnolia is a dicotyledon.

Maize, *Zea mays*, is a monocotyledon.

the part where the stem grows new leaves. The stem supports the structures of the shoot, and spaces out the leaves so that they can receive adequate light and air (see Topic 6.6) It holds the flowers in a position which enables **pollination** to take place.

The stem allows transport of water from the soil up to the leaves and food from the leaves to other parts of the plant, such as the roots.

The **roots** are the parts of flowering plants found below ground. Roots are usually white since they do not contain chlorophyll. Roots anchor the plant firmly in the ground and prevent it from being blown over by the wind. Roots also absorb water and mineral ions from the soil (see Topic 8.2).

Dicotyledons and monocotyledons

Flowering plants can be divided into two main groups:

dicotyledons and **monocotyledons**

Dicotyledons look like the plant at the bottom of the opposite page.

Their leaves are often broad with a network of branching veins. The parts of the flower, for example the male parts known as stamens, are in multiples of four or five in each flower. Dicotyledons have *two* **cotyledons** (seed leaves) in a seed.

Grasses and cereals, like the maize plant at the bottom of page 12 are monocotyledons.

The leaves of most monocotyledons have parallel veins. Grasses and cereals have long, narrow leaves. Other monocotyledons have leaves with a variety of shapes, such as those of palm trees. The parts of the flower are in multiples of three. Monocotyledons have *one* cotyledon inside each seed.

Figure 1.6.2   Structure of a typical monocotyledon plant.

> **STUDY TIP**
>
> Be aware of the differences between flowering plants and animals. Note and learn the differences in cell structure (Topic 2.1) and reproduction (from Unit 16).

SUMMARY QUESTIONS

1 Copy and complete the sentences using these words:

> parallel   two   narrow   flowering
> broad   network   one

Monocotyledons are _____ plants that often have _____ leaves with _____ veins. They have _____ cotyledon inside the seed. Dicotyledons often have _____ leaves with a _____ of veins. They have _____ cotyledons inside each seed.

2 Find the names of five dicotyledonous plants and five monocotyledonous plants that grow where you live. Construct a dichotomous key for the ten plants you have identified.

3 Make a table of differences between ferns and flowering plants.

KEY POINTS

1 Flowering plants are multicellular – each cell is surrounded by a cellulose cell wall; those in leaves and some stems contain chloroplasts.

2 Monocotyledons have one cotyledon in their seeds and leaves with parallel veins.

3 Dicotyledons have two cotyledons and broad leaves with a network of branching veins.

4 Ferns are plants well adapted to life on land. They have xylem and reproduce by means of spores.

Sorting things out

To identify the name of a plant or animal you could look through the pictures in a book until you found the right one. However, that would take a lot of time and effort.

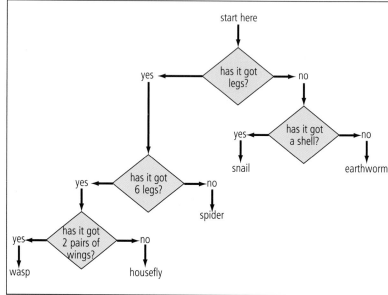

Figure 1.7.1 A branching dichotomous key.

Scientists use **dichotomous keys** to identify living things. Dichotomous means dividing into two.

A key has a number of steps – you can see them in the branching key and the numbered key on this page. At each step in the branching key you find a question or statement. Start at the beginning and answer 'yes' or 'no' to the first question or statement. This takes you to another question or to an identification. Use the branching key above to identify the animals to the left.

Use the numbered key below to identify the same animals. It is set out differently from the first key, but it works in the same way. Start at the beginning and answer the questions at each stage.

1	Has legs	Go to 2
	Has no legs	Go to 4
2	Has 6 legs	Go to 3
	Has 8 legs	Spider
3	Has 1 pair of wings	Housefly
	Has 2 pairs of wings	Wasp
4	Has a shell	Snail
	Has no shell	Earthworm

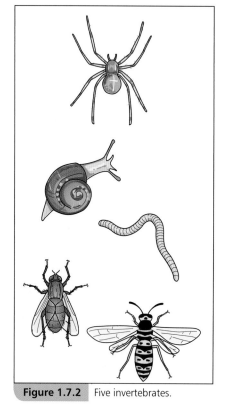

Figure 1.7.2 Five invertebrates.

Making a leaf key

Now try making a key of your own.

1 Take six different leaves and label them A to F.

2 Put the six leaves out in front of you.

3 Think of a question that will divide them into two groups.

4 Write down the question.

5 Now think up questions to divide each group into two.

6 Write these down.

7 Carry on until you come to the last pair of leaves.

8 Write out your key as a branching key or as a numbered key.

Pond animals key

Now try making a key of these pond animals below.

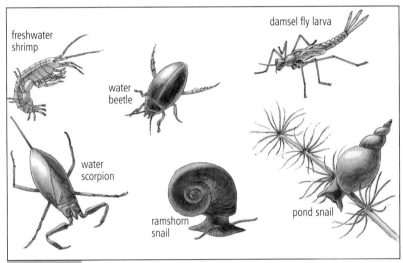

Figure 1.7.3 Pond animals.

SUMMARY QUESTIONS

1 When a scientist visited an island she discovered some insects.

She made some drawings and brought them back to the laboratory. Her drawings are shown on the right.

a Give each insect a suitable name.

b Make a dichotomous key to identify them. Present your key either as a branching key or as a numbered key.

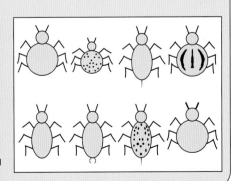

2 Write down the features that would enable you to classify a fish, an amphibian, a reptile, a bird and a mammal. (You can use the photos on pages 8 and 9 to help you.) Now make a key that you could use to identify each of these vertebrate groups.

1 Which process is carried out by all the organisms that live in a forest?

A moulting

B photosynthesis

C pollination

D respiration

(Paper 1) [1]

2 What are characteristics of all organisms?

A excretion and respiration

B ingestion and growth

C photosynthesis and egestion

D respiration and photosynthesis

(Paper 1) [1]

3 Which group of animals includes those with a segmented body, an exoskeleton and jointed limbs?

A amphibians

B arthropods

C reptiles

D vertebrates

(Paper 1) [1]

4 Which group of vertebrate includes those with dry scaly skin and four legs?

A amphibians

B fish

C mammals

D reptiles

(Paper 1) [1]

5 Each cell is surrounded by a cell membrane and contains cytoplasm.

The table shows five other features of cells. Which row shows the features shown by cells from organisms in all five kingdoms?

	chloroplasts	enzymes	nucleus	DNA	ribosomes
A	✗	✓	✓	✓	✗
B	✗	✓	✗	✓	✓
C	✓	✗	✓	✗	✓
D	✓	✓	✗	✓	✗

(Paper 2) [1]

6 Which is *not* associated with excretion from a mammal?

A exhaling carbon dioxide from the lungs

B release of heat from the skin

C removal of toxic waste products of metabolism

D removal of water that the body does not need

(Paper 2) [1]

7 The evolutionary relationships between five species are shown in the diagram.

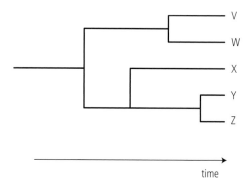

time

An analysis of base sequences from the DNA of these five species was carried out. Which two species will have the most similar sequences?

A **V** and **W**

B **W** and **X**

C **X** and **Z**

D **Y** and **Z**

(Paper 2) [1]

8 Which is an example of a morphological feature used in classifying vertebrates?

A egg laying

B type of body covering

C type of diet

D pattern of bones in the front limb

(Paper 2) [1]

9 The drawings show five arthropods (not drawn to the same scale).

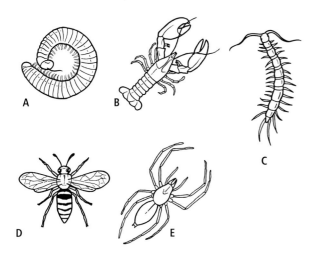

A   B   C   D   E

(a) Make a table to compare the features of the five arthropods that you see in the drawings. *[6]*

(b) Use the features that you have used in your table to make a key to the five arthropods. (The key may be either a branching key or a numbered key.) *[6]*

(Paper 3)

10 Copy and complete the table using ticks and crosses to indicate whether the four groups show the features or not.

feature	fish	amphibian	bird	mammal
backbone				
fins				
wings				
fur				
homeothermic				

(Paper 3) *[5]*

11 (a) Five of the features of living things and their definitions are listed below.

Match each of the features with its definition by pairing a letter and a number.

A movement **B** sensitivity
C respiration **D** nutrition **E** growth

Definitions

1 A permanent increase in size

2 The chemical reactions that occur in cells to break down nutrient molecules with the release of energy

3 An action by an organism to cause a change of position or place

4 The ability to detect changes in the environment and respond to them

5 The taking in of materials from the environment for energy, growth and development *[3]*

(b) Explain what is meant by the term *species*. *[2]*

(Paper 3)

12 The drawings show leaves from three species of flowering plant. They are not drawn to scale.

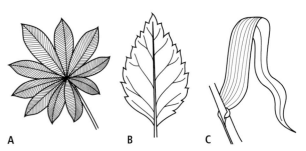

A   B   C

(a) (i) Identify the group of flowering plants to which each species belongs.

(ii) State the features that you used to make your identifications. *[5]*

(b) List five features shown by the leaves that would be useful in devising a dichotomous key to identify species of flowering plants. *[5]*

(c) (i) State three structural features that ferns have in common with flowering plants. *[3]*

(ii) Explain why fungi, which have cell walls, are not classified in the same kingdom with plants. *[3]*

(Paper 4)

Structure of cells

Cells

Cells are the small building blocks that make up all living organisms. Very small living things such as bacteria are made of only one cell.

An insect such as a fly may contain millions of cells. No one knows for certain how many cells there are in a human being – estimates vary between 10×10^{12} and 50×10^{12}.

Animal cells

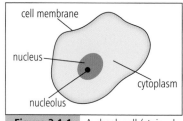

Figure 2.1.1 A cheek cell (stained with a blue dye).

Plant cells

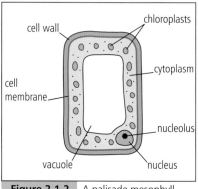

Figure 2.1.2 A palisade mesophyll cell from a leaf.

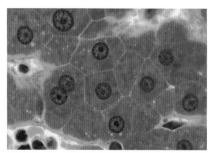

Liver cells magnified by ×300.

Differences between plant and animal cells

The differences between plant and animal cells are summarised in this table.

feature	plant cell	animal cell
cellulose cell wall	present	absent
cell membrane	present; surrounded by cell wall	present
shape	permanent shape determined by the cell wall; shapes can be nearly spherical, box-like or cylindrical	shapes vary as there is no cell wall
chloroplasts	present in some cells	absent
vacuole	large permanent vacuole in cytoplasm containing cell sap	small vacuoles in cytoplasm; do not contain cell sap
nucleus	present (often at the side of the cell close to the cell wall)	present (found anywhere within the cell)
cytoplasm	present	present

Functions of cell structures

cell structure	functions
cell membrane	• forms a barrier between the cell and its surroundings • keeps contents of cell inside • allows simple substances to enter and leave the cell, e.g. oxygen, carbon dioxide and water • controls movement of other substances into and out of the cell, e.g. glucose • often described as partially permeable (see page 28)
nucleus	• controls all activities in the cell • controls how cells develop
cytoplasm	• place where many chemical reactions take place, e.g. respiration and making proteins for the cell
chloroplast (plant cells only)	• photosynthesis • stores starch
cell wall (plant cells only)	• stops cells from bursting when they fill with water • gives shape to cells • allows water and dissolved substances to pass through freely (often described as freely or fully permeable)
sap vacuole (plant cells only)	• full of water to maintain shape and 'firmness' of cell • stores salts and sugars

(see page 28)

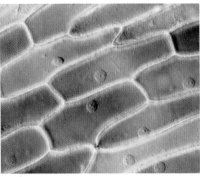

Cells of onion epidermis magnified by ×250.

1 Copy and complete this table by using ticks and crosses to indicate if the structures are present or not.

cell structures	cheek cell (animal)	onion cell (plant)	leaf cell (plant)
nucleus			
cell wall			
chloroplasts			
large vacuole			
cytoplasm			

2 State the functions of the following parts:
   a chloroplast b cell membrane c cell wall d nucleus
   e cytoplasm f vacuoles

Cell organelles

As you know, plant and animal cells contain a number of small structures present in the cytoplasm of the cell. These are called **cell organelles** and include **mitochondria**, **vesicles** and **ribosomes** on **rough endoplasmic reticulum**.

Rough endoplasmic reticulum

The endoplasmic reticulum (ER) is a complex system of double membranes. They contain fluid-filled spaces between the membranes which allow materials to be transported throughout the cell.

Where ribosomes are present on their outer surface, the membranes are called rough endoplasmic reticulum (ER). The main function of rough ER is to package and transport proteins made by the ribosomes.

Cells that produce a lot of protein, for instance those making digestive enzymes in the alimentary canal, have large amounts of rough ER.

Small pieces of rough ER may be pinched off at the ends to form small vesicles. In this way proteins can be made and stored in the rough ER and transported around the cell in the small vesicles.

Ribosomes

Ribosomes are small organelles that are found in huge numbers in all cells. They are about 20 nanometres (nm) in diameter in eukaryotic cells but smaller in prokaryotic cells where they are found free in the cytoplasm, not attached to rough ER.

The function of ribosomes is to synthesise proteins, such as those enzymes involved in respiration.

Mitochondria

Mitochondria are relatively large organelles found in all eukaryotic cells. They are often visible under the electron microscope as sausage-shaped structures about $1\,\mu m$ wide and $5\,\mu m$ long.

Each mitochondrion has a double membrane, the outer one of which controls the entry and exit of materials. The inner membrane forms many folds on which some of the chemical reactions of aerobic respiration take place.

Mitochondria are often described as the 'power plants' of the cell since they are the sites of aerobic respiration.

Cells with high rates of respiration have many mitochondria to provide sufficient energy. For example, insect flight muscle and liver cells contain vast numbers of mitochondria.

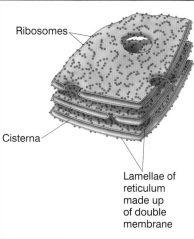

Ribosomes

Cisterna

Lamellae of reticulum made up of double membrane

Figure 2.2.1 Structure of rough endoplasmic reticulum (ER).

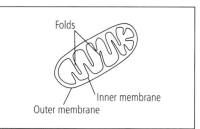

Folds

Inner membrane

Outer membrane

Figure 2.2.2 Structure of a mitochondrion.

Size of cells and specimens

Look at the photograph of some human cheek cells. The cells are 1000 times larger than in real life. This means that they have been *magnified* 1000 times.

To work out the *actual size* of a cell, measure the length of one magnified cell in millimetres. Let's say that it measures 13 mm. Now we can use this formula to work out actual size:

$$\text{actual size} = \frac{\text{image size}}{\text{magnification}}$$

So the actual size of our cell $= \dfrac{13}{1000} = 0.013\,\text{mm}$

There are 1000 micrometres (μm) in a millimetre. We can convert the answer to micrometres by multiplying by 1000 to give 13 μm.

We can use a similar technique when the image size has been reduced, as in this photograph of a goldfish. This time there has been a *reduction* of actual size in the photograph. If you calculate the fish's actual size using the same formula you will find it is 90 mm long.

To calculate the magnification of an image we reorganise the formula:

$$\text{magnification} = \frac{\text{image size}}{\text{actual size}}$$

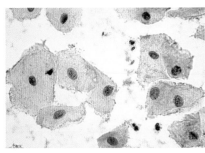

Human cheek cells (×1000).

The image of this goldfish is one third (×0.33) actual size.

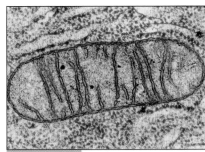

Figure 2.2.3   An electron micrograph of a mitochondrion (×80 000).

KEY POINTS

1 Ribosomes on rough ER and in the cytoplasm make proteins.

2 Rough ER packages and transports proteins.

3 Aerobic respiration occurs in mitochondria.

4 Cells with high rates of respiration have many mitochondria.

LEARNING OUTCOMES

- Identify different types of cell from diagrams and photographs
- Relate the features of these cells to their functions
- Calculate magnification and actual size of biological specimens using millimetres as units

To function efficiently, many-celled organisms have cells that are specialised to carry out certain functions. This means that the functions of the body are divided between different groups of cells. As an organism develops from a fertilised egg, new cells are produced. These cells grow and change to become specialised for certain functions. In this topic we look at some examples of specialised cells.

Specialised cells

Ciliated cells are found in the air passages in the lungs (trachea and bronchi) and in the oviducts in the female reproductive system. These cells have **cilia** on their surfaces. Cilia beat back and forth to create a current in the fluid next to the cell surfaces.

In the airways, cilia move the mucus that traps dust and pathogens up to the nose and throat. In the oviducts, cilia move the egg from the ovary to the uterus.

Root hair cells have long extensions that give them a large surface area to absorb water and ions from the soil.

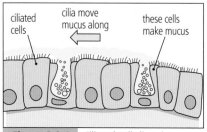

Figure 2.3.1 Ciliated cells line the airways.

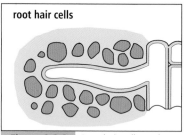

Figure 2.3.2 Root hair cells are long and thin to absorb water from the soil.

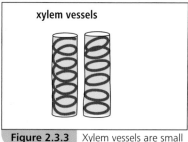

Figure 2.3.3 Xylem vessels are small tubes that carry water up the stem.

Xylem vessels are cylindrical and empty. They are arranged into columns like pipes. The cell walls are thickened with bands or spirals of cellulose and a waterproof material called **lignin**. These cells allow water and ions to move from the roots to the rest of the plant. They also help to support the stem and leaves.

Nerve cells are highly specialised cells. They have thin extensions of the cytoplasm like wires. They are able to transmit information in the form of nerve impulses around the body.

Red blood cells contain the protein **haemoglobin** that carries oxygen. They are shaped like flattened discs. This shape provides a large surface area compared with their volume which makes for efficient absorption of oxygen.

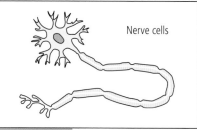

Figure 2.3.4 Nerve cells carry impulses around the body.

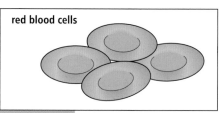

Figure 2.3.5 Red blood cells have a substance called haemoglobin that carries oxygen.

Palisade mesophyll cells

These plant cells have numerous **chloroplasts** in the cytoplasm. The chloroplasts trap light energy for **photosynthesis**. Palisade cells have a cell wall made up of tough **cellulose** which strengthens the cell. In the cytoplasm there is a large **vacuole** which is filled with cell sap. Starch grains are found in the cytoplasm. These are formed by photosynthesis and are a temporary store of energy.

Sperm cells have a tail and are adapted for swimming. The head of the sperm carries genetic information from the male parent to the female parent. Genes from the father are present in the sperm nucleus.

Egg cells are much bigger than sperm cells. They contain yolk as a store of energy. Genes of the mother are found inside the nucleus of the egg.

Look carefully at Figure 2.3.6. The cells are 500 times larger than in real life. This means that the cells have been magnified 500 times. To work out the actual size of these cells, measure the length of one of the cells in millimetres. Let's say it measures 20 mm. Now we can use a formula to work out the actual size of the cell:

$$\text{actual size} = \frac{\text{image size}}{\text{magnification}}$$

So the actual size of our cell is 20 divided by 500 = 0.04 mm

To calculate the magnification of an object we reorganise the formula:

$$\text{magnification} = \frac{\text{image size}}{\text{actual size}}$$

The actual diameter of an egg cell is 0.1 mm. To calculate its magnification in Figure 2.3.8 we divided the diameter of the drawing by 0.1 mm to give ×180.

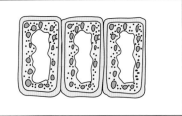

Figure 2.3.6 Leaf palisade cells contain lots of chloroplasts for photosynthesis (×500).

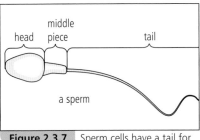

Figure 2.3.7 Sperm cells have a tail for swimming (×2000).

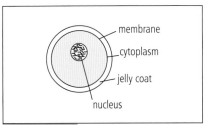

Figure 2.3.8 Egg cells contain an energy store (×180).

SUMMARY QUESTIONS

1 Draw a table matching the type of cell in the first column with its correct shape and function:

Type of cell	Shape	Function
red blood cells	hollow tube	keep the air passages free from dust
ciliated cells	like wires	transport water and ions
root hair cells	flat discs	transmit electrical impulses
nerve cells	long and thin	transport oxygen
xylem vessels	have cilia that move	absorb water from the soil

2 Here are some other specialised cells. Use this book to find out the functions of these cells and how they are adapted to carry out their functions.

    a muscle cell   b goblet cell   c white blood cell
    d pollen cell

KEY POINTS

1 During development, cells change their structure and often their shape.

2 Specialised cells have a structure that enables them to carry out a particular function, e.g. ciliated cells have cilia to move mucus along the windpipe.

Levels of organisation

- Define the terms *tissue*, *organ* and *organ system*
- Describe examples of the above that occur in plants and animals

Tissues and organs

A group of similar cells is called a **tissue**. All the cells in a tissue look the same and they work together to carry out a shared function. Muscle tissue is made up of identical **muscle** cells. These cells work together and so the muscle tissue contracts.

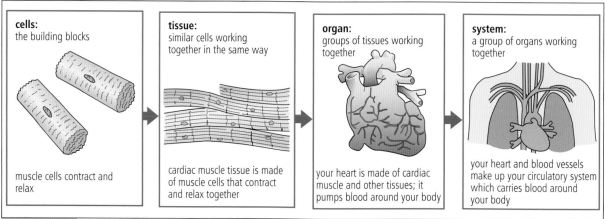

cells:
the building blocks

muscle cells contract and relax

tissue:
similar cells working together in the same way

cardiac muscle tissue is made of muscle cells that contract and relax together

organ:
groups of tissues working together

your heart is made of cardiac muscle and other tissues; it pumps blood around your body

system:
a group of organs working together

your heart and blood vessels make up your circulatory system which carries blood around your body

Figure 2.4.1 Levels of organisation in the human circulatory system.

STUDY TIP

Learn these definitions of tissue, organ and organ system. Make a list of examples of animal and plant tissues, organs and organ systems from this book.

An **organ** is made up of a group of different tissues that work together to perform specific functions.

The heart is an example of an organ. It is made up of different tissues such as cardiac muscle, nervous tissue, fibrous tissue and blood that work together to pump blood around the body.

The stomach, lungs, brain and kidneys are all organs.

Different organs work together as part of an **organ system**. Organ systems consist of a group of organs with related functions, working together to perform body functions. For example, the heart and blood vessels work together as part of the circulatory system. Here are some other organ systems:

- The digestive system is made up of the gullet, stomach, pancreas, liver and intestines.
- The excretory system is made up of the kidneys, ureters and bladder.
- The nervous system is made up of the brain, spinal cord and nerves.
- The reproductive system in females is the ovaries, oviducts, uterus and vagina; in males it is the testes, sperm ducts, prostate gland and penis.

All the different organ systems make up a living **organism**.

Plant tissues and organs

The diagrams show the tissues in a leaf. The tissue that carries out photosynthesis in leaves is called **mesophyll**. The cells making up the upper layer of the mesophyll are called palisade cells. These cells are closely packed and full of chloroplasts so that they are well adapted to absorb lots of light. The palisade cells make up the **palisade mesophyll tissue**. All the cells making up this tissue look alike and do the same function – they absorb light for photosynthesis.

A leaf is an organ. Other plant organs are roots and stems. Other structures, such as flowers and fruits, are modified leaves.

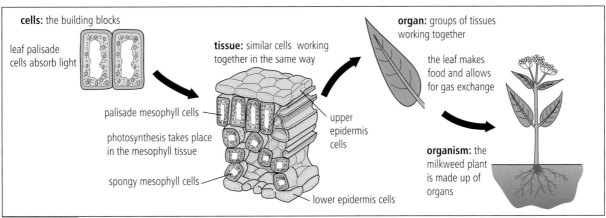

Figure 2.4.2   Levels of organisation in a flowering plant.

SUMMARY QUESTIONS

1 Copy and complete the sentences using these words:

organ system   cells   tissues   function

A tissue is made up of _____ that carry out the same _____ . An organ is formed from a group of _____ working together. An _____ is a group of organs working together to perform several body functions.

2 Copy out the organs listed on the left. Match each with the correct system from those listed on the right.

Organs	**System**
lungs and trachea	digestive
heart and blood vessels	nervous
brain and spinal cord	gas exchange
ovaries, oviducts and uterus	excretory
gullet, stomach and intestines	reproductive
kidneys and bladder	circulatory

3 Arrange the following words into the correct sequence, starting with the smallest and ending with the largest:

organ   organ system   tissue   organism   cell

1 Which structure is found *only* in plant cells?

  A cell membrane

  B chloroplast

  C cytoplasm

  D nucleus

(Paper 1)   *[1]*

2 Which structure is *not* found in animal cells?

  A cell membrane

  B cell wall

  C cytoplasm

  D nucleus

(Paper 1)   *[1]*

3 Which is the correct sequence, starting with the smallest and ending with the largest?

  A tissue, organ system, organ, cell

  B cell, tissue, organ system, organ

  C tissue, cell, organ, organ system

  D cell, tissue, organ, organ system

(Paper 1)   *[1]*

4 A student makes a drawing of a biological specimen. The length of the specimen in the drawing is 140 mm. The magnification is ×40. What is the actual size of the specimen?

  A 5600 mm   **B** 56 mm

  C 35 mm   **D** 3.5 mm

(Paper 1)   *[1]*

5 Which is *not* found in an animal cell?

  A nucleus

  B permanent vacuole

  C ribosomes

  D rough endoplasmic reticulum

(Paper 2)   *[1]*

6 The diagram shows a mitochondrion from a liver cell.

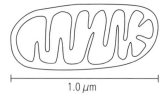

The magnification of the diagram is:

  A ×40   **C** ×4000

  B ×400   **D** ×40 000

(Paper 2)   *[1]*

7 The substances that move in and out of mitochondria are:

  A CO_2 in, O_2 out, nutrients in

  B CO_2 out, O_2 in, nutrients in

  C CO_2 in, O_2 in, nutrients out

  D CO_2 out, O_2 out, nutrients out

(Paper 2)   *[1]*

8 Cells of organisms classified in the prokaryote kingdom contain which of the following?

  A cell membrane, enzymes, nucleus, vesicles

  B cell wall, cell membrane, cytoplasm, ribosomes

  C cytoplasm, DNA, endoplasmic reticulum, ribosomes

  D cytoplasm, DNA, enzymes, mitochondria

(Paper 2)   *[1]*

9 The palisade mesophyll cell is a type of plant cell. A liver cell is a type of animal cell. Copy and complete the table to compare these two cells. Copy the table and put a tick (✔) if you think the structure is present and a cross (✗) if you think it is absent.

cell structure	palisade mesophyll cell	liver cell
cell wall		
cell membrane		
cytoplasm		
nucleus		
chloroplast		
large vacuole		

(Paper 2)   *[6]*

10 The diagram shows three animal cells.

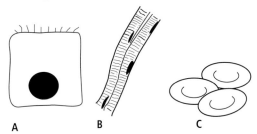

A   B   C

(a) Identify the cells **A**, **B** and **C**. [3]

(b) State the functions of the three cells. [3]

(c) State where in the body these cells are found. [3]

(d) The magnification of cell **A** is ×1000. Calculate its actual size in millimetres. [2]

(Paper 3)

11 The diagram shows three plant cells.

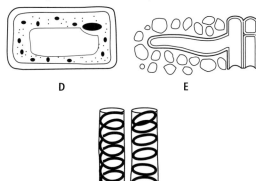

D   E

F

(a) Identify the cells **D**, **E** and **F**. [3]

(b) State the functions of the three cells. [3]

(c) For each cell, **D**, **E** and **F**, state a plant organ where it is found. In each case give a different organ. [3]

(d) The actual length of cell **D** is 0.05 mm. Calculate the magnification of the drawing. [2]

(Paper 3)

12 (a) Name:

    (i)   an animal cell which does not have a nucleus; [1]

    (ii) a plant cell that does not have a nucleus; [1]

    (iii) the organ system for transport in mammals; [1]

    (iv) the plant tissue that transports sugars. [1]

(b) Distinguish between the following pairs of terms.

    (i)   Organs and tissues [2]

    (ii) Cytoplasm and nucleus [2]

    (iii) Cell membrane and cell wall [2]

    (iv) Organ system and organism [2]

(Paper 3)

13 (a) Five cell structures and their functions are listed below.

Cell structures

  A   nucleus

  B   cell membrane

  C   cell wall

  D   large vacuole

  E   ribosome

  F   rough endoplasmic reticulum

  G   mitochondrion

Functions

  1   controls the movement of substances into and out of the cell

  2   stores water and ions

  3   makes proteins

  4   withstands pressure of water inside the cell

  5   carries out aerobic respiration

  6   transports proteins within the cell

  7   stores DNA and controls the activities of the cell

Match each of the cell structures with its function by writing a letter and a number. [5]

(b) A student looked at a cell under a light microscope. She made a drawing of the cell and showed the diameter of the nucleus in her drawing as 70 mm. She calculated the magnification of her drawing as ×10 000. What is the actual size of the nucleus in micrometres? Show your working. [2]

(Paper 4)

3.1   Diffusion

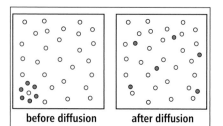

Figure 3.1.1   Diffusion in a gas. Molecules of a coloured gas spread out through the container.

Molecules in gases move about in a random way. They bump into one another and spread out to fill up all the space available. Molecules in a liquid do this as well, although it takes longer for them to fill the space. Movement in a gas is faster as the molecules are more spread out. The difference in speed between movement in gases and liquids is important for organisms. This movement of molecules is called **diffusion**.

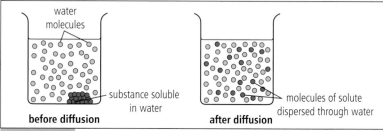

Figure 3.1.2   Diffusion in a liquid.

When molecules or ions diffuse they spread out from where there are lots of them in a given volume (a high concentration) to where there are not as many of them (a low concentration). The difference between the concentration of molecules in two places is a **concentration gradient**. Molecules carry on diffusing until they are spread out evenly. When this happens the molecules keep moving, but there is no longer a difference in concentrations so diffusion has stopped.

Diffusion is the net movement of molecules or ions from a region of high concentration to a region of lower concentration down a concentration gradient.

Cells gain some of the substances they need by diffusion from their surroundings. They also lose some of their waste substances to their surroundings by diffusion. These substances have to cross cell membranes that are **partially permeable** as they allow the movement of small molecules such as oxygen, carbon dioxide and water to pass through easily, but not larger molecules. The movement of molecules by diffusion across cell membranes is **passive movement** as cells do not need to use energy to move the molecules.

The energy for diffusion comes from the kinetic energy of random movement of molecules and ions. The more molecules present, the greater their kinetic energy and therefore their rate of movement.

Factors that affect diffusion

These factors influence the efficiency of diffusion:

- The distance molecules have to travel – note that cell membranes are very thin.
- The concentration gradient – cells use the substances that diffuse as quickly as possible, so they keep a low concentration inside the

Supplement

cytoplasm. This means that molecules keep diffusing into the cell because the cell is maintaining a steep concentration gradient.

- The surface area – some cells have cell membranes that are folded to give a large surface to allow many molecules to cross by diffusion.
- The temperature – molecules move faster and collide more often as the temperature increases. Diffusion is faster at warmer temperatures.

Gas versus liquid

Animals and plants exchange the gases oxygen and carbon dioxide with their surroundings at gas exchange surfaces. In mammals, the gas exchange surface is formed of the alveoli in the lungs (see page 131). Blood transports these two gases between the lungs and all the cells in the body. In the alveoli, oxygen diffuses across a very thin layer of cells into the blood. Carbon dioxide diffuses in the opposite direction. Breathing constantly refreshes the air in the alveoli and blood constantly removes oxygen and brings carbon dioxide, so the concentration gradients are always steep. There are many alveoli to give a very large surface area for gas exchange.

In plants, gas exchange occurs inside the leaves. The spongy mesophyll cells provide a large surface area for the exchange of gases. There are air spaces between the cells in a plant and each cell exchanges gases with this air (see page 64). This is efficient because diffusion through the air is 300 000 times faster than through water.

Water as a solvent

A **solution** is made up of two parts, the **solute** and the **solvent**. The solute dissolves in the solvent. If you dissolve sugar in water you make a sugar solution. The sugar is the solute and the water is the solvent. The solute is not always a solid like sugar. Liquids and gases can be solutes as they can dissolve in solvents too. Something which dissolves in a solvent is described as being **soluble**.

Water is sometimes called the universal solvent. About 75% of cytoplasm is water and it is the main component of transport fluids like blood, and xylem sap and phloem sap in plants. Everything transported in plants and animals has to dissolve in water and most of the chemical reactions that occur in cells happen in water. Also, water is needed for digestion and excretion to take place.

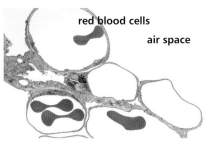

red blood cells

air space

There is a short distance between the red blood cells and the air in the alveoli. Magnification ×1500.

STUDY TIP

Always refer to concentrated solutions and dilute solutions, never to 'strong' and 'weak' solutions.

SUMMARY QUESTIONS

1 Copy and complete the sentences using these words:

liquid   random   diffusion
low   gas   high

_____ is the net movement of molecules of a _____ or a _____ from an area of _____ to an area of _____ concentration as a result of the _____ movement of molecules.

2 State how the following factors affect diffusion into cells:

a distance

b size of molecule

c surface area

d concentration gradient

e temperature

3 What do the terms *solution* and *concentration gradient* mean?

4 Explain why water is an important solvent for animals and plants.

KEY POINTS

1 Diffusion is the net movement of molecules or ions from a region of high concentration to a region of low concentration down a concentration gradient.

2 Factors that affect diffusion are: size of molecule, distance, surface area, temperature and the steepness of the concentration gradient.

Supplement

LEARNING OUTCOMES

- Define the term *osmosis*
- Describe the effect of osmosis on plant and animal tissues
- Explain the movement of water into and out of cells using the term *water potential*

STUDY TIP

If asked to define osmosis or write about it, make sure you say that it is the <u>diffusion</u> of water molecules.

Each cell is surrounded by a cell membrane. It separates the contents of the cell from the outside. The cell membrane has tiny holes in it which allows small molecules to pass through but not large ones. The cell membrane is described as being **partially permeable**.

Osmosis is a special kind of diffusion involving water molecules. It occurs when two solutions are separated by a **partially permeable membrane**.

Osmosis is the diffusion of water from a dilute solution into a more concentrated solution through a partially permeable membrane.

The tiny holes in the membrane allow small water molecules to pass through, but the large solute molecules are too big to pass through the partially permeable membrane. Water is diffusing from a place where there is a dilute solution with a **high concentration of water** to a place where there is a concentrated solution with a **low concentration of water**.

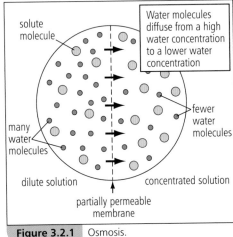

Figure 3.2.1 Osmosis.

Water potential is a way of thinking about the ability of water to move by osmosis. This is influenced by how much water is available, but also by other factors such as the pressure exerted on water in plant cells by the cell wall. It is more accurate to say that a dilute solution (containing a lot of water molecules) has a **high water potential**. A concentrated solution (containing fewer water molecules) has a **low water potential**.

In the diagram above, there is a **water potential gradient** between the two sides of the membrane. The water molecules diffuse *down* this water potential gradient, from a region of high water potential to a region of lower water potential through a partially permeable membrane.

Water molecules are free to move through the membrane in both directions by kinetic energy. However, since there are many more water molecules present on the left hand side, there will be a **net** movement of water molecules through the membrane from left to right, down the water potential gradient.

A model cell

Dialysis tubing (Visking tubing) is partially permeable. We can use dialysis tubing to represent the cell membrane and the sugar solution to represent the cytoplasm.

1 Cut two pieces of dialysis tubing, each 12 cm long. Tie one end of each with cotton.

2 Fill one model cell with a dilute sugar solution (cell A). Fill the other model cell with water (cell B).

3 Tie the other end of both model cells and weigh them on a balance. Put 'cell A' into a beaker of water and put 'cell B' into a beaker containing a concentrated solution of sugar. After 30 minutes take out the model cells and weigh them again.

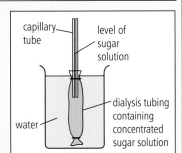

water diffuses into 'cell'

sugar solution

dialysis tubing (partially permeable membrane)

water

'cell A'

water diffuses out of 'cell'

water

dialysis tubing (partially permeable membrane)

sugar solution

'cell B'

'Cell A' increases in mass because water has diffused *into* the 'cell' by osmosis. 'Cell B' decreases in mass because water has diffused *out of* the cell by osmosis.

An osmometer

You can see the effects of osmosis if you set up this apparatus.

capillary tube

level of sugar solution

water

dialysis tubing containing concentrated sugar solution

Fill the partially permeable membrane with a very concentrated solution of sugar.

Tie it to a capillary tube and stand it in water.

Very quickly you will see the liquid moving up the tube. You can measure how fast it is moving using a ruler and a stopwatch. Use your ideas about osmosis to explain why the liquid rises in the tube.

STUDY TIP

You may be given the results of a practical demonstration of osmosis and be expected to explain the results. Make sure you remember the definition of osmosis and apply it to the results you can see. Always explain that the results are due to water molecules moving by osmosis.

SUMMARY QUESTIONS

1 Define the following terms:

   diffusion   osmosis   partially permeable membrane

2 Describe how you can find out how fast water diffuses by osmosis into a sugar solution. Remember to include all practical details.

3 Explain in terms of water potential, how water passes into plant cells placed in distilled water.

KEY POINTS

1 Osmosis is the diffusion of water molecules from a region of their higher concentration (dilute solution) to a region of their lower concentration (concentrated solution) through a partially permeable membrane.

2 A partially permeable membrane allows small molecules such as water to pass through but not large solute molecules.

3 Water molecules diffuse down a water potential gradient from a region of higher water potential to a region of lower water potential.

Supplement

Osmosis in plant and animal cells

- Describe how water can enter and leave plant cells by osmosis
- Describe how osmosis can affect animal cells

STUDY TIP

When describing the cell membrane always say that it is <u>partially</u> permeable, not semi-permeable.

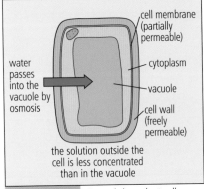

Figure 3.3.1   Osmosis in a plant cell.

STUDY TIP

In Paper 4 you may be expected to use the term 'water potential' in writing about osmosis in plant or animal cells and tissues. You can also refer to 'water potential gradients' to explain water movement into and out of cells and tissues.

Supplement

PRACTICAL

Osmosis in potato cells

1. Cut nine cores from a potato so that they are exactly the same length. Record this length.
2. Feel the cores to see how firm and 'bendy' they are.
3. Set up the following test-tubes:
   test-tube **A** – distilled water
   test-tube **B** – dilute sugar solution
   test-tube **C** – concentrated sugar solution.
4. Place three cores of potato into each test-tube and leave them for 60 minutes.
5. Remove the cores and measure their lengths. Calculate the average length of the cores in each test-tube.
   Compared to the start, cores will be:
   A – longer and firmer
   B – about the same length and firmness
   C – shorter and softer and 'bendier'.

These results are explained in the text.

A   distilled water   B   dilute sugar solution

tissue swells up   no change

C   concentrated sugar solution

tissue shrunken and flaccid

Turgidity

The cell membrane of the plant cell is partially permeable and the cell sap inside the vacuole is a solution of salts and sugars. When plant cells are placed in water, the water enters the cells. This is because there is a water potential gradient so that water molecules diffuse into the cells by osmosis.

As water enters it makes the cell swell up. The water pushes against the cell wall developing a **turgor pressure**. Eventually the cell contains as much water as it can hold. It's like a blown-up balloon. The strong cell wall stops the cell bursting. We say that the cell is **turgid**. This is what has happened to the cells in the potato cores in tube **A**. The cells have absorbed water, swollen and caused the core to get slightly longer.

Turgid cells give the plant support. They keep the stems of many plants upright. This is because the cells within a stem are supported by turgor pressure. The water pressure within the cell acts against the inelastic cell walls, keeping the cells turgid and firm. However, when these cells lose water, they are no longer firm and turgid. Plant stems and leaves that have lost water **wilt**.

Plasmolysis

When plant cells are placed into a concentrated sugar or salt solution water passes *out* of the cells by osmosis. As water passes out, the sap vacuole starts to shrink. These cells are no longer firm, and become limp. We say that they are **flaccid**. As more water leaves the cells the cytoplasm starts to move away from the cell wall. These cells are now **plasmolysed**. This is what has happened to the cells in the potato cores in test-tube **C**. The cells have decreased in volume so the whole core is shorter than at the start.

The cores in test-tube **B** did not change in length very much because the water potential of the sugar solution was about the same as the water potential of the cell sap in the potato cells. There has been no overall diffusion of water into or out of the cells so they have stayed about the same length.

Osmosis in animal cells

Figure 3.3.3 shows what happens to red blood cells when they are placed in different concentrations of a salt solution. Remember that the cell membrane is partially permeable.

The red blood cells in the picture have been placed into different liquids. Their cytoplasm is a concentrated solution of proteins, salts and sugars.

The cells in **D** were in distilled water. Water passes *into* the cells by osmosis. However, animal cells have no cell wall to stop them swelling so they burst. When red blood cells are put into a concentrated salt solution (**F**), they shrink as water passes *out* of the cells by osmosis. The cells in **E** have not changed in size as they are in a solution which has the same water potential as the cells. This is how they are in the blood when surrounded by blood plasma.

Figure 3.3.3 This shows what happens when red blood cells are put into distilled water (**D**) and a concentrated solution of salt (**F**). **E** shows red blood cells as they appear when suspended in blood plasma.

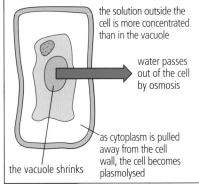

the solution outside the cell is more concentrated than in the vacuole

water passes out of the cell by osmosis

as cytoplasm is pulled away from the cell wall, the cell becomes plasmolysed

the vacuole shrinks

Figure 3.3.2 Water passes out of a plant cell by osmosis.

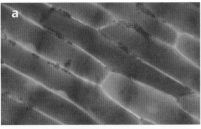

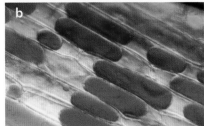

These epidermal cells from a red onion are (a) turgid and (b) plasmolysed.

SUMMARY QUESTIONS

1 Explain what is meant by each of these terms:

   turgid   turgor pressure   flaccid   plasmolysis

2 Some potato cores were weighed and then placed into a dilute sugar solution. After 2 hours, they were taken out of the solution, dried on a paper towel and weighed again.

   The mass of the potato cores remained unchanged.

   a   What does this tell you about the concentration of the sugar solution?

   b   Explain your answer to part a in terms of osmosis.

3 Explain, using the term *water potential*, what happens to cores of potato when they are placed into distilled water for 60 minutes.

KEY POINTS

1 Water passes into plant cells by osmosis. A plant cell that is full of water is turgid. Turgid cells provide support for leaves and young stems.

2 If plant cells are placed into a concentrated sugar solution, water passes out by osmosis. These cells are no longer firm, they are flaccid. As the vacuole shrinks, the cell membrane moves away from the cell wall – the cell is now plasmolysed.

3 Animal cells burst if they are placed into water as they have no cell wall to resist the increase in size.

Active transport

Active transport

Cells take up molecules and ions and keep them in high concentrations.

Look at the concentration of magnesium ions in the root hair cell and the concentration of magnesium ions in the soil solution:

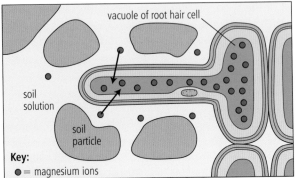

Figure 3.4.1   Root hair cells absorb ions by active transport.

The concentration of magnesium ions is far greater inside the vacuole of the root hair cell than it is in the water in the soil. We might expect magnesium ions to diffuse out of the root hair cell into the soil water down a diffusion gradient. The magnesium ions are maintained at a high concentration inside the root hair cell by a process called **active transport**.

Active transport is the movement of ions or molecules in or out of a cell through the cell membrane against a concentration gradient, using energy released during respiration.

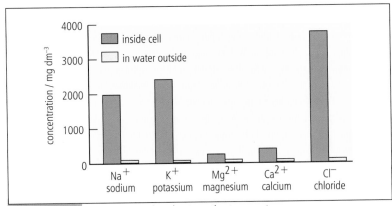

Figure 3.4.2   Ion concentration due to active transport.

The bar chart shows the concentrations of some ions inside the cells of a freshwater plant and in the water in which it lives. These ions cannot have been taken into the plant by diffusion. They are taken in against a concentration gradient by active transport.

Active transport needs energy

The cell membrane contains **carrier proteins**. These carrier proteins span the cell membrane and provide means by which ions and molecules can enter or leave a cell by active transport. First the molecule or ion combines with a carrier protein. Energy from respiration enables the carrier protein to change its shape to carry the ion or molecule to the inside of the membrane. The molecule or ion is released to the inside of the membrane and the carrier protein reverts to its original shape.

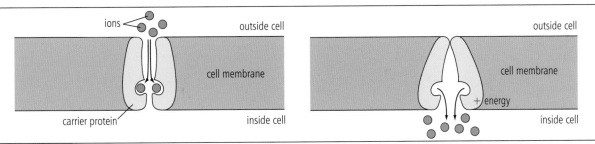

Figure 3.4.3   Carrier proteins in membranes carry out active transport.

Epithelial cells lining the **villi** in the small intestine and the kidney tubules absorb glucose by active transport (see Topic 7.8). These cells have high rates of respiration to provide energy for this active transport.

Active transport relies upon respiration to take up ions or molecules against a concentration gradient. Any factor that affects the rate of respiration will also affect the rate of active transport. So a lack of oxygen would reduce respiration rate and active transport.

An increase in temperature would increase the rate of respiration, up to a point, so would also have the same effect on active transport. The presence of poisons such as cyanide can stop respiration, so active transport would stop altogether.

SUMMARY QUESTIONS

1 Explain what is meant by active transport.

2 a Describe the role of carrier proteins and respiration in active transport.

  b Give two examples of active transport taking place.

3 Make a table to compare diffusion with active transport.

  Make sure that you have three columns headed 'features', 'diffusion' and 'active transport'. For features you can include 'needs energy from the cell', 'concentration gradient'. You may be able to think of some other features to use in your table.

4 State what effect each of the following would have on active transport, in each case give reasons for your answer:

  a a lack of oxygen

  b an increase in temperature

  c the presence of a poison, such as cyanide

KEY POINTS

1 Active transport is the movement of ions or molecules across the cell membrane, against a concentration gradient, using energy from respiration.

2 Active transport enables root hair cells to take up ions, and epithelial cells of the villi to take up glucose.

1 Which is *not* partially permeable?

  A cell membrane of palisade mesophyll cell

  B cell membrane of red blood cell

  C cell wall

  D dialysis tubing

(Paper 1) [1]

2 Which is the *best* definition of diffusion?

  A movement of solvent molecules through a partially permeable membrane down a concentration gradient

  B net movement of molecules down a concentration gradient

  C net movement of molecules against a concentration gradient

  D random movement of molecules in a gas or a liquid

(Paper 1) [1]

3 Root hair cells are surrounded by soil water, which is a very dilute solution of mineral ions, such as nitrate ions and magnesium ions. Plant cells have higher concentrations of these ions. How do root hair cells absorb ions from soil water?

  A active transport

  B diffusion

  C osmosis

  D random movement

(Paper 1) [1]

4 A student cut up a potato into pieces that looked like chips. The student described these pieces as 'fairly firm'. The pieces were put into three different liquids. Which row shows the results that the student obtained?

	very concentrated salt solution	very dilute salt solution	water
A	fairly firm	swollen and very firm	soft
B	soft	fairly firm	swollen and very firm
C	soft	soft	fairly firm
D	swollen and very firm	soft	fairly firm

(Paper 1) [1]

5 Which is the *best* definition of osmosis?

  A the diffusion of water molecules through a partially permeable membrane

  B the net diffusion of water molecules down a water potential gradient

  C the net diffusion of water molecules down a water potential gradient through a partially permeable membrane

  D the net diffusion of water molecules through a partially permeable membrane from a solution with a low water potential to a solution with a high water potential

(Paper 2) [1]

6 Some fresh plant tissue was put into a concentrated salt solution for 60 minutes. Which *best* explains why the tissue became softer?

  A water diffused down a water potential gradient from the cells to their surroundings

  B water diffused into the cells so that they became turgid

  C water diffused out of the cells so that they became plasmolysed

  D the cells lost turgor pressure and became flaccid

(Paper 2) [1]

7 (a) Explain the importance to humans of the following:

  (i) diffusion of oxygen in the alveoli [2]

  (ii) diffusion of carbon dioxide in the alveoli [2]

  (iii) absorption of glucose by diffusion in the small intestine. [2]

(b) Explain the importance to plants of the following:

  (i) diffusion of carbon dioxide into leaves [2]

  (ii) absorption of magnesium ions by root hair cells. [2]

(c) Osmosis is a type of diffusion. Explain how osmosis differs from diffusion. [3]

(Paper 3)

8 A group of students investigated osmosis. They began by peeling 50 small onions. They divided the peeled onions into five batches of 10 onions and weighed them. Each batch was placed into a solution of different concentrations of salt (sodium chloride). After immersion for two hours each batch was surface dried and reweighed. The students calculated the percentage change in mass. The table shows their results.

conc. of salt / $g\,dm^{-3}$	mean mass of onions / g		percentage change in mass
	before immersion	after 2 hours immersion	
0	147	173	+ 18.0
25	153	165	+ 8.0
50	176	172	−2.0
100	154	149	
150	149	142	−4.5
200	183	175	−4.5

Information and data used to compile the table from Practical osmosis in vegetable pickling, Ray W James. Journal of Biological Education (1993) 27 (2), pages 90–91

(a) Calculate the percentage change in mass for the onions kept in the $100\,g\,dm^{-3}$ salt solution. Show your working. *[2]*

(b) State why the students calculated the percentage change in mass. *[1]*

(c) Plot a graph of the results. *[5]*

(d) Use your graph to find the salt solution in which there is no change in mass. *[1]*

(Paper 4)

9 An experiment was set up to investigate the factors influencing the uptake of ions by plant roots. Some roots were cut from a plant, washed and placed in three solutions, **A**, **B** and **C**, containing potassium ions. A mixture of gases was bubbled through each of the solutions. A gas mixture rich in oxygen was bubbled through solution **A**; solution **B** received a gas mixture with a very low concentration of oxygen; solution **C** received no oxygen in the gas mixture. The roots were left for 24 hours and then the rate of uptake of potassium ions was determined.

The rate of uptake of potassium ions was highest in solution **A** and lowest in solution **C**.

(a) Explain how the supply of oxygen to roots influences their uptake of potassium ions. *[3]*

The experiment was repeated with solutions containing roots kept at different temperatures. They were all provided with the gas mixture that had been given to solution **A**. The results are shown in the table.

temperature / °C	rate of uptake of potassium ions / arbitrary units
5	3
10	5
20	10
30	20
40	15

(b) Draw a graph of the results. *[6]*

(c) Describe the results shown in your graph. *[4]*

(d) Explain the effect of temperature on the uptake of potassium ions by the roots. *[3]*

(Paper 4)

10 (a) Explain the term *concentration gradient*. *[2]*

(b) Explain why diffusion is described as the *net* movement of molecules or ions. *[2]*

Some resources required by cells, such as molecules and ions, are often in very low concentrations in their surroundings.

(c) Describe how epithelial cells of the villi absorb molecules of glucose from the gut contents. *[5]*

(Paper 4)

4 Biological molecules

4.1   Biological molecules

Biological molecules are complex chemicals like carbohydrates, proteins and fats. They are useful chemicals that are needed by living organisms for **metabolism**. By metabolism we mean all the chemical reactions taking place in the cells of the body. These reactions include the release of energy in respiration, protein synthesis, and the growth and repair of cells.

Green plants make the complex chemical compounds that they need from simple raw materials. Carbon dioxide and water are the raw materials for **photosynthesis**. The simple sugars produced in photosynthesis are used to make a wide range of other compounds. Plants need minerals to make some of these complex compounds.

Animals eat plants and/or other animals that feed on plants. Biological molecules that animals require are present in their diet. The different biological molecules needed for a balanced diet in humans are carbohydrates, proteins, fats, vitamins, minerals, fibre and water.

Carbohydrates

These contain the elements carbon (C), hydrogen (H) and oxygen (O). Carbohydrates include sugars and starches.

Glucose is a simple sugar which is made in photosynthesis, used in respiration and transported in the blood. It consists of six carbon atoms arranged into a ring.

Sucrose is a double sugar molecule made up of two molecules of simple sugars joined by chemical bonds. Complex carbohydrates are made by joining many simple sugar molecules together by chemical bonds. Plants store **starch** as an energy store. They have enzymes to catalyse the reactions that join glucose together into long chains. **Glycogen**, (sometimes called animal starch) is another complex carbohydrate made from glucose by animals as a store of energy. It is stored in the liver and muscles.

Starch and glycogen, unlike sugars, are insoluble and do not taste sweet. **Cellulose** is a complex, structural carbohydrate made up of thousands of glucose units. The glucose units are held together by bonds forming long and unbranched chains. These cellulose molecules are linked together to form fibres, which give plant cells walls their strength and rigidity.

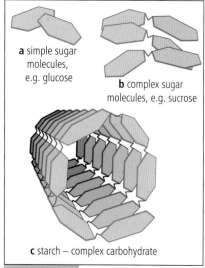

a simple sugar molecules, e.g. glucose

b complex sugar molecules, e.g. sucrose

c starch – complex carbohydrate

Figure 4.1.1   Different types of carbohydrate.

STUDY TIP

Biological molecules are made by organisms. They only become food molecules when eaten by an animal. The term nutrient (or nutrient molecule) is often used to refer to the biological molecules described in Topics 4.1 and 4.2. Beware – it is also used to refer to ions, such as nitrate ions, sodium ions and magnesium ions, which organisms need from their environment.

Proteins

Proteins are complex molecules made up of carbon, hydrogen and oxygen, but they also contain nitrogen (N) and many have sulfur (S). Proteins are long-chain molecules made up of smaller molecules called **amino acids**. After formation they are either folded into different shapes (see Topic 5.1) or become arranged into long fibres.

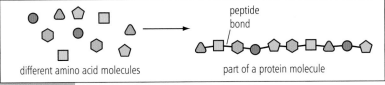

different amino acid molecules   part of a protein molecule

peptide bond

Figure 4.1.2 How a protein is made from amino acids.

There are about 20 different types of amino acid. Molecules of amino acids are made into chains as you can see in the diagram. It is the sequence of the different amino acids in the chain which determines the type of protein that is formed. Each individual amino acid joins the chain by means of a chemical bond called a peptide bond.

Supplement

The different sequences of amino acids give different shapes to protein molecules. These different shapes of protein molecules can be related to their function. Enzymes are proteins that provide a surface for reactions to take place called the 'active site' (see page 47). **Antibodies** are proteins with a structure that has binding sites on its surface. This enables them to bind with chemicals called **antigens** on the surface of pathogens and make the pathogens stick together (see page 120).

Fats

Fats and oils are made up of the elements carbon, hydrogen and oxygen. Each fat molecule is made up of one molecule of **glycerol** and attached to this are three **fatty acids**. There are different types of fatty acid and these can form different fats with different properties. Fats are used for energy storage and thermal insulation in the body.

STUDY TIP

You may be asked to compare these three groups of biological molecules. Make a table to compare them using headings such as 'Name of biological molecules group' and 'Elements'. You will be able to add to this table in Topic 7.2. When naming the elements always use their full names, not their symbols.

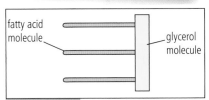

fatty acid molecule   glycerol molecule

Figure 4.1.3 A molecule of fat.

KEY POINTS

1 Carbohydrates and fats are made up of the elements carbon, hydrogen and oxygen. Proteins are made up of the same elements plus nitrogen and sulfur.

2 Carbohydrates are simple sugars (e.g. glucose), and complex carbohydrates such as starch, glycogen and cellulose.

3 Small molecules are joined together to make long chain molecules. Sugars are made into starch, glycogen and cellulose. Amino acids are made into proteins.

4 A molecule of a fat is made by combining three fatty acid molecules with a molecule of glycerol.

SUMMARY QUESTIONS

1 a List the components of a balanced diet for a human.
   b Give one example of each of the following:
      i a simple sugar ii a complex carbohydrate found in plants
      iii a complex carbohydrate found in animals

2 a Which elements are present in proteins, but are not found in carbohydrates and fats?
   b Name the small molecules that are joined together to make protein molecules.
   c Give three examples of proteins that are made in the body.
   d State the function of each protein that you named in part c.

3 a Name the molecules that are reacted together to make a molecule of fat.
   b Give two uses of fats in the body.

Chemical tests for biological molecules

State how the following chemical tests are carried out:

- iodine test for starch
- Benedict's test for reducing sugars
- biuret test for protein
- ethanol emulsion test for fat
- DCPIP test for vitamin C

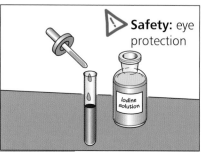

⚠ **Safety:** eye protection

Figure 4.2.1 Test for starch.

STUDY TIP

These tests are often known as 'food tests', but they can be carried out on any plant and animal material or on solutions made up in the laboratory.

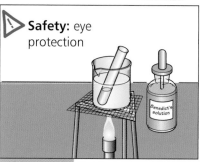

⚠ **Safety:** eye protection

Figure 4.2.2 Test for reducing sugars.

Simple chemical tests can be used to identify starch, sugars, proteins, fat and vitamin C.

First, it is important to carry out these tests on pure forms of these compounds. For example, if you are testing a food for reducing sugars, you should first carry out the chemical test on a glucose solution. Then keep the results to compare with your results from testing other materials, such as foods or animal and plant tissues. You should also do a test with water so that you can see the negative result as well.

You will need to make an **extract** from the material you are testing. This involves grinding up a small amount of the material with some water with a pestle and mortar or putting it into a blender. The chemicals will be in solution in the extract.

Safety: Some of the chemicals used in these tests are corrosive, so always wear eye protection.

Testing for starch

- Half fill a test-tube with the food extract you wish to test for starch.
- Add two or three drops of **iodine solution**.
  Iodine solution usually looks yellow or light brown.
- A positive result for starch is if the iodine solution turns **blue–black**. If the extract remains a yellow or light brown colour it does *not* contain starch.

Testing for reducing sugars

- Put a known volume of the extract you wish to test for reducing sugars in a test-tube.
- Place a beaker on a heat-proof mat.
- Carefully half fill the beaker with boiling water from a kettle (or place the beaker on a tripod and gauze and boil the water with a Bunsen burner).
- Add the same volume of **Benedict's solution** to the test-tube containing the food extract and put it into the hot water.
- Benedict's solution is bright blue.
- A positive test for simple sugars is when Benedict's solution turns red or orange (if you look carefully you can see it turn green and then yellow before turning orange). If you leave the test-tube to cool you will also see a precipitate.
- You can use Benedict's test to tell how much simple sugar is present. If the colour changes to green, the extract contains only a little of the reducing sugars. If it turns a deep orange colour then it contains a lot of reducing sugars.
- If the colour remains blue then the extract does *not* contain any reducing sugars.

Testing for protein

- Half fill a test-tube with the extract you wish to test for protein.
- Add five to six drops of **biuret solution** (this solution contains copper sulfate solution and sodium hydroxide solution).

Safety: Take care as sodium hydroxide solution is corrosive.

- Biuret solution usually looks blue in colour.
- A positive test for protein is if the biuret solution turns purple, violet or lilac.
- If the colour remains blue, then the extract does *not* contain protein.

Testing for fats

Fats will not dissolve in water but they will dissolve in **ethanol**. If a solution of fat in ethanol is added to water a cloudy white emulsion is formed.

- Chop up or grind a small amount of material you wish to test for fats. (Do not add water to make the extract this time.)
- Put the extract into a clean test-tube and add enough ethanol to cover it.
- Put a stopper over the test-tube and shake up the contents.
- Add some distilled water to make the test-tube half full.
- Shake the contents of the test-tube once more.
- A white emulsion that looks cloudy white or a milky colour is a positive test for fats.
- If this does not happen, the extract does *not* contain fat.

Testing for vitamin C

Vitamin C is in juices, such as freshly-squeezed lemon juice. A solution of vitamin C can be made from vitamin C tablets. DCPIP is a blue liquid that loses its colour when it comes into contact with vitamin C.

1 Put a known volume of DCPIP solution in a test-tube.

2 Fill a syringe or dropping pipette with a solution of vitamin C or with a juice, e.g. orange juice.

3 Add the liquid one drop at a time.

If the colour of DCPIP disappears the test is positive for vitamin C. If the blue colour persists then the test is negative.

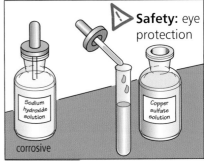

Safety: eye protection

corrosive

Figure 4.2.3   Test for protein.

SUMMARY QUESTIONS

1 Describe what you would see in each of the following:
 a a positive test for starch
 b a positive test for protein
 c a positive test for a reducing sugar
 d a negative test for starch.
 e a positive test for vitamin C

2 Describe how you would carry out each of the following procedures:
 a making an extract of a food to test for reducing sugar
 b demonstrating how to carry out the reducing sugar test using a glucose solution

3 You are given four test-tubes each containing a different concentration of glucose. They are labelled **A**, **B**, **C** and **D** and you do not know the concentration of glucose in each tube.

 You have been told to find out which solution is the most concentrated and which is the least concentrated. Describe the practical procedures that you would follow and the observations you would expect. Remember to give safety precautions.

4 Describe how you would use the DCPIP test to compare the vitamin C content of different fruit juices.

KEY POINTS

1 Iodine solution gives a blue–black colour when added to starch.

2 Benedict's solution turns orange when boiled with simple sugars.

3 Biuret solution gives a purple or violet colour when added to protein.

4 A white emulsion forms when fat dissolved in ethanol is added to water.

5 Vitamin C decolourises DCPIP solutions.

4.3

DNA

- Describe the structure of DNA as two strands coiled together to form a double helix
- State that DNA strands contain chemicals called bases
- Describe how the bases always pair up in the same way

STUDY TIP

A nucleotide is a 'building block' of DNA. The diagrams show how the bases, A, C, T and G, are arranged in DNA. You do not need to remember the structure of a nucleotide shown in Figure 4.3.2.

The structure of DNA

Each chromosome is made up of thousands of **genes** arranged like beads in a necklace. It is the genes that carry the genetic information that affects how we grow and what we look like. For instance, there are genes for eye colour, hair colour and height. Some genes code for the production of enzymes that control all the chemical reaction that take place in cells.

If we could unravel a chromosome, it would form an extremely long thread. The thread would be made up of a chemical called **DNA (deoxyribonucleic acid)**. A gene is made up of a short length of DNA so the long thread that makes up a chromosome contains hundreds of genes.

DNA belongs to a complex group of biological molecules known as **nucleic acids**. Each DNA molecule is made up of thousands of units each called a **nucleotide**. A single nucleotide is made up of three molecules:

- a phosphate   - a sugar   - a base

The sugar and phosphate molecules join up and form the backbone of the DNA strand. The bases are attached to the sugar molecules. If you look at Figure 4.3.1 you will see that DNA is made up of *two* strands of nucleotides. It is rather like a ladder. The whole molecule is twisted into a **double helix** – a bit like a spiral staircase with the bases as the steps.

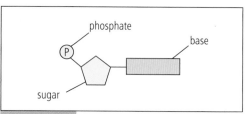

| **Figure 4.3.2** | A single nucleotide. The sugars and phosphates make up the 'uprights' of the ladder and the bases make up the 'rungs'. |

Base pairing

So how is the DNA molecule held together?

If you look at Figure 4.3.3 you can see that the bases join together. Each pair of bases is held together by weak bonds. There are four different bases in DNA:

- **thymine (T)** - **adenine (A)** - **cytosine (C)** - **guanine (G)**

(You don't have to remember each of the names, just their letters.)

The bases always pair up in the same way:

- **adenine (A) pairs with thymine (T)**
- **cytosine (C) pairs with guanine (G)**

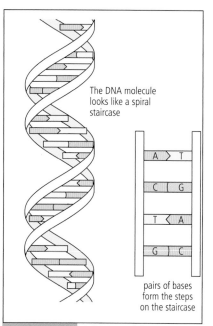

The DNA molecule looks like a spiral staircase

pairs of bases form the steps on the staircase

| **Figure 4.3.1** | A DNA model. |

Although the bonds holding the two chains of nucleotides together are weak, there are many of them. So altogether they keep the DNA double helix in shape.

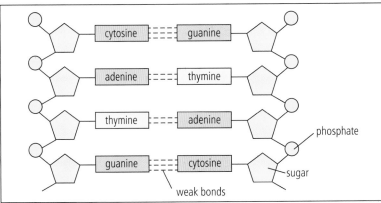

Figure 4.3.3 How the bases pair up in DNA.

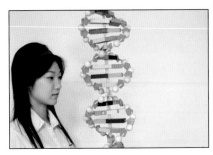

A student examines a model of DNA that shows the pairs of bases. In this model G (guanine) is yellow and C (cytosine) is red.

SUMMARY QUESTIONS

1 Figure 4.3.4 shows a partially completed section of the DNA molecule. Copy and complete the diagram by writing in the letters of the missing bases.

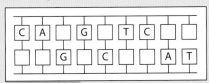

Figure 4.3.4

2 The table below shows the percentages of the four bases in DNA from four different organisms.

organism	percentage of each base			
	A	C	G	T
human	31	19	19	31
locust	29	21	21	29
yeast	32	18	18	32
bacterium	15	35	35	15

a Explain the pattern shown in the table.

b In an organism 26% of the bases in DNA are found to be A. What percentage would be C? Show your working.

c Explain how it is possible to have two very different organisms such as humans and yeast with very similar percentages of DNA bases.

James Watson and Francis Crick worked out the structure of DNA by building models. They are posing with their final model that they made in Cambridge in the early 1950s.

KEY POINTS

1 The DNA molecule consists of two strands coiled together to form a double helix.

2 Each strand contains chemicals called bases A, C, G and T, which bond together.

3 A always bonds with T; C always bonds with G.

4 A chromosome is made up of a long super-coiled strand of DNA.

5 A gene is a length of DNA that codes for the production of a particular protein.

1 What are the chemical elements in proteins?

  A carbon, hydrogen and oxygen

  B carbon, hydrogen, nitrogen and oxygen

  C carbon, hydrogen, nitrogen, oxygen and sulfur

  D carbon, nitrogen, oxygen and sulfur

(Paper 1)   *[1]*

2 The small molecules that are built up into proteins are:

  A amino acids   **C** glucose

  B fatty acids   **D** glycerol

(Paper 1)   *[1]*

3 Which row gives the positive results for the four chemical tests?

	Benedict's test	biuret test	iodine test	emulsion test
A	green	violet	yellow	no suspension
B	blue	blue	yellow	no suspension
C	yellow	red	blue	cloudy suspension
D	orange	violet	blue-black	cloudy suspension

(Paper 1)   *[1]*

4 Water is an important molecule in organisms because it is a:

  A solute   **C** solvent

  B solution   **D** suspension

(Paper 1)   *[1]*

5 Enzymes and antibodies are proteins. Special regions of these proteins have specific shapes that bind to other molecules. Proteins can form specific shapes because they have:

  A molecules of different lengths

  B molecules with different numbers of amino acids

  C molecules with different sequences of amino acids

  D molecules with other molecules or ions attached to them

(Paper 2)   *[1]*

Use the diagram of DNA for questions 6–8.

DNA molecules are composed of sub-unit molecules known as nucleotides. The diagram shows a small part of a DNA molecule.

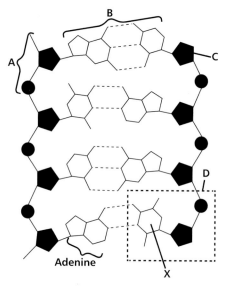

6 Which of the labelled parts of DNA, A, B, C or D, indicates a base pair?

(Paper 2)   *[1]*

7 Which of the labelled parts of DNA, A, B, C or D, indicates a nucleotide?

(Paper 2)   *[1]*

8 What is the structure labelled X?

  A adenine (A)   **C** guanine (G)

  B cytosine (C)   **D** thymine (T)

(Paper 2)   *[1]*

9 The diagram below shows how a starch molecule increases in length.

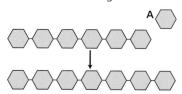

(a) Name molecule **A**.   *[1]*

(b) Name two places where starch is stored in plants.   *[2]*

(c) Describe how you would find out whether a plant tissue contained starch. Include the practical details of the test in your answer.   *[4]*

(Paper 3)

10 The diagram shows a protein molecule:

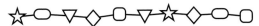

(a) Name the sub-unit molecules that are assembled into the protein. *[1]*

(b) Name two proteins found in the human body. *[2]*

(c) State two places in the human body where many proteins are made. *[2]*

(d) Some seeds are rich in proteins. Describe the chemical test that you would use to determine whether different types of seed are rich in proteins. Include the practical details of the test in your answer. *[4]*

(Paper 3)

11 The diagram shows a molecule of a fat.

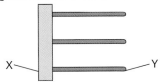

(a) Name the components of the molecule labelled **X** and **Y**. *[2]*

(b) List two places in the human body where fats are stored. *[2]*

(c) State three roles of fats in the human body. *[3]*

(d) Oils have the same molecular structure as fats. Many seeds are rich in oils. Describe a chemical test to find out whether seeds from different flowering plant species contain oil or not. Include the practical details of the test in your answer. *[4]*

12 A student used the DCPIP test to find out how much vitamin C is present in different fruit juices. The concentration of the DCPIP solution was 10 g dm⁻³. The student followed these instructions.

1 Put 2 cm³ of DCPIP solution into a test tube.

2 Use a graduated pipette or burette to add a 10 g dm⁻³ vitamin C solution drop by drop to the DCPIP solution. Shake the tube gently after adding each drop. Continue to add the vitamin C solution until the colour of the DCPIP solution disappears.

3 Record the volume of vitamin C solution that was added.

4 Repeat the procedure two more times and calculate a mean volume.

5 Repeat steps 1 to 4 with the juices to be tested.

The results are shown in the table.

test substance	volume added to DCPIP solution / cm³			
	1	2	3	mean
10 g dm⁻³ vitamin C solution	1.8	1.9	2.1	1.9
orange juice	2.4	2.6	2.3	2.4
grapefruit juice	3.2	3.1	2.9	3.1
apple juice	9.6	9.1	9.4	9.4

(a) Explain why the student took three readings for each test substance. *[3]*

(b) Use the results to calculate the concentration of vitamin C in the three fruit juices. Show your working. *[3]*

(Paper 4)

13 (a) DNA molecules have a double helix structure. Explain what this means. *[2]*

(b) State the precise site of DNA in an animal cell. *[2]*

The sequence of bases in a small sample of DNA is:

ATAGATCCCGAA

(c) Write out the sequence of bases in the opposite strand in this DNA molecule. *[1]*

(d) The DNA sequence determines a small part of the structure of a protein. Use the sequence above to explain how base sequences determine the sequences of amino acids in a protein. *[1]*

(e) In 2007, scientists compared the degree of similarity between some of the DNA taken from the nuclei of white blood cells of a domesticated cat and several other mammalian species. The results are below.

percentage similarity between domesticated cat and:					
wild cat	human	dog	cow	rat	mouse
99	90	82	80	69	67

What do you conclude from these results? *[3]*

(Paper 4)

5 Enzymes

5.1 Structure and action of enzymes

LEARNING OUTCOMES

- Define the term *catalyst* as a substance that speeds up a chemical reaction and is not changed by the reaction
- Define *enzymes* as proteins that act as biological catalysts to speed up the rate of chemical reactions
- Explain how an enzyme works using the 'lock and key' model

A **catalyst** speeds up a chemical reaction and remains unchanged at the end of the reaction. Enzymes are proteins, produced by organisms, that speed up chemical reactions. They are known as **biological catalysts**.

How enzymes work

Many chemical reactions take place in organisms. These reactions happen too slowly to keep organisms alive unless they are speeded up by enzymes. There are many different types of enzyme as each one catalyses a different reaction. Most enzymes work inside cells, but many of those that we will discuss here work outside cells, for example in the gut (see Topics 7.5 and 7.7).

The reactions that enzymes catalyse can be divided into three types.

1 Breaking large molecules into small ones

This is important in nutrition when large food molecules are broken down into small ones so that they can be absorbed and then used. Bacteria and fungi release enzymes to break down their food and we release enzymes into the gut for the same reason.

2 Building up large molecules from small ones

Small molecules, such as glucose, are joined together to make large molecules. These enzymes work inside cells to speed up the formation of storage molecules, such as starch, and structural molecules such as cellulose for cell walls of plants.

3 Converting one small molecule into another

Many of the chemical reactions that occur inside cells involve small changes to molecules, such as adding or removing atoms or groups of atoms. For example, there are enzymes that remove hydrogen from compounds during respiration.

Figure 5.1.1 shows the way in which an enzyme catalyses the breakdown of a molecule.

This computer-generated image of an enzyme (on the right) shows its 3D shape.

STUDY TIP

When writing about enzymes always say that enzymes catalyse reactions.

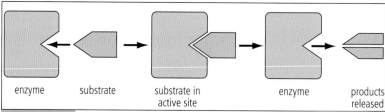

| enzyme | substrate | substrate in active site | enzyme | products released |

Figure 5.1.1

Figure 5.1.2 shows how an enzyme is involved in building a molecule from two smaller molecules.

Enzymes are made of protein

All enzymes have five important properties.

1 They are all proteins.
2 Each enzyme catalyses one reaction.
3 They can be used again and again.
4 They are influenced by temperature.
5 They are influenced by pH.

Enzymes are made of protein molecules. These molecules can be folded into many different shapes. Each type of enzyme molecule has a shape that makes it suitable for catalysing one type of reaction. This explains why there are many different enzymes – one enzyme for each reaction.

Enzymes catalyse reactions in which **substrates** are converted into **products**. Look at Figure 5.1.1. Notice that the shape of part of the enzyme matches the shape of the substrate molecule. The enzyme and the substrate have shapes that are **complementary** so they fit together. Once they fit together the reaction can take place. Other substrates have the wrong shape to fit into the enzyme so will not be involved in the reaction catalysed by this enzyme. When the reaction is over the product or products leave the enzyme and another substrate molecule enters.

Supplement
As you can see in Figure 5.1.2 the part of the enzyme where the substrate(s) fit and where the reaction takes place is the **active site**. This way of describing how enzymes work is known as the 'lock and key' model. The enzyme and substrate combine like a key entering a lock to form an **enzyme-substrate complex**.

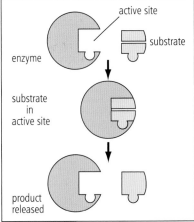

Figure 5.1.2 This shows how an enzyme can join two molecules together.

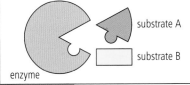

Figure 5.1.3 Only substrate A will fit the enzyme's active site.

SUMMARY QUESTIONS

1 Copy and complete the sentences using these words:

speed   catalysts   proteins   reactions

Enzymes are biological _____ that _____ up the rate of chemical _____ . Enzymes are all _____ and each enzyme acts on one type of _____ .

Supplement
2 Use the terms *complementary* and *active site* to explain what is meant by the 'lock and key' model of enzyme action.

3 Use the 'lock and key' model to explain:
a   why each enzyme will act only on one substrate
b   why an enzyme can be used to catalyse a large quantity of substrate
c   why destroying the active site by heating the enzyme to a high temperature stops the enzyme from working

KEY POINTS

1 Enzymes are biological catalysts that increase the rate of chemical reactions.

2 Enzymes catalyse reactions in which substrate molecules are converted to product molecules, either by building up or breaking down.

3 All enzymes are protein molecules. Part of the enzyme where the reaction occurs has a complementary shape to the substrate molecule(s).

4 The active site is the part of the enzyme where the substrate fits and where the reaction occurs.

Supplement

Factors affecting enzyme action: temperature

LEARNING OUTCOMES

- Carry out an investigation into the effects of temperature on enzyme activity
- Describe the effects of temperature on enzyme activity
- Explain the effects of temperature on enzyme activity

STUDY TIP

When you are describing a graph like this one, always use some figures taken from the graph in your answer. Here you can say the highest rate is 5.6 mg product per minute at 40°C.

STUDY TIP

You may be asked to describe the effect of temperature and pH on enzymes, but only in Paper 4 will you be asked to explain how these factors influence enzymes. When you explain, remember that enzymes are proteins and are denatured by high temperatures and extremes of pH.

The activity of an enzyme is determined by measuring the rate of the reaction that the enzyme catalyses. This may be done either by measuring how much product is formed or by measuring how much substrate is used over a period of time. The rate is like the speed of the reaction measured in quantity of product or substrate per unit of time, e.g. per minute.

Effect of temperature on enzymes

The activity of enzymes is influenced by temperature. This graph shows the effect of increasing temperature on the rate of an enzyme-catalysed reaction.

Look at the graph and observe that the rate of reaction:

- is slow at low temperatures, e.g. at 10°C
- increases as the temperature increases to 40°C
- reaches a maximum at 40°C
- decreases at temperatures greater than 40°C
- is zero at 60°C.

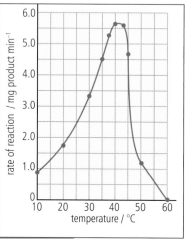

Figure 5.2.1

The temperature at which the maximum rate of reaction occurs is called the **optimum temperature**. This is the best temperature for the enzyme. Here are some examples of optimum temperatures:

- fungal and plant enzymes: approximately 20°C (see page 250)
- human enzymes: 37°C (body temperature – see Topics 7.5 and 7.7)
- some of the enzymes produced by bacteria for use in industry: 90°C.

As we saw in Topic 5.1, an enzyme molecule is folded into a shape that accepts the substrate molecules. The shape of an enzyme's active site is maintained by bonds between different parts of the molecule. Remember that the shape of the active site determines whether the substrate will fit into the enzyme for the reaction to occur.

At first, increasing the temperature of an enzyme-controlled reaction will increase the rate. This is because enzyme and substrate molecules have greater **kinetic energy**. They move around more quickly and there are more chances of them colliding, the substrate fitting into the active site and a reaction taking place.

At higher temperatures the bonds holding the enzyme molecule together start to break down. This changes the shape of the active site, so the substrate no longer fits.

We say that the enzyme has been **denatured** and it can no longer catalyse the reaction.

Investigating the effect of temperature on enzyme action

Amylase is an enzyme that breaks down starch to maltose. Starch changes iodine solution blue-black. When starch is broken down, iodine solution will not change colour. Draw a table for your results like this.

temperature at which amylase and starch was kept / °C	time taken for starch to be fully broken down / min

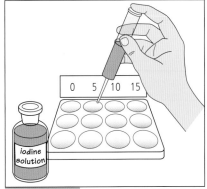

Figure 5.2.2 Testing the solution for starch in iodine.

1 Put 5cm^3 of starch solution into each of four test tubes labelled A, B, C, and D.

2 Put test tube A into the refrigerator and note the temperature.

3 Put test tube B into a test tube rack and note the temperature.

4 Put test tube C into a water bath at about 35°C and note the temperature.

5 Put test tube D into a water bath at about 80°C and note the temperature.

6 Collect 10cm^3 of amylase solution in a clean boiling tube.

7 Put a drop of iodine solution into each of the wells on a spotting tile.

8 Add 2cm^3 of amylase solution to each of the four tubes and note the time.

9 At two-minute intervals use a clean pipette to remove some of the test solution from each test tube and add it to a drop of iodine solution on the spotting tile.

10 When the amylase solution has broken down all the starch the iodine solution will no longer turn blue-black. They will stay light brown. What does this tell you about the effects of temperature on the amylase?

SUMMARY QUESTIONS

1 a Sketch a graph to show the effect of increasing temperature on the rate of a reaction catalysed by a human enzyme, such as salivary amylase.

b Describe, in words, what is shown by your graph.

2 Suggest what each of the following would do to the rate of reaction catalysed by a human enzyme:

a a temperature below 10 °C

b a temperature of 37 °C

c a temperature of 50 °C

3 Use the 'lock and key' model to explain what happens when enzymes are denatured at high temperatures.

KEY POINTS

1 Increasing the temperature of an enzyme-controlled reaction increases the rate of reaction up to a maximum, which occurs at the optimum temperature.

This is because greater kinetic energy is causing a greater number of collisions between enzyme and substrate molecules.

2 At higher temperatures the rate of reaction decreases until it stops acting as a catalyst.

This is due to a change in the shape of the active site which means the substrate can no longer fit. The enzyme is now denatured.

Enzymes and pH

LEARNING OUTCOMES

- Describe the effects of pH on enzyme action

- Carry out an investigation into the effects of pH on enzyme activity

- Explain the effects of pH on enzyme action

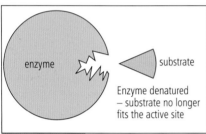

Figure 5.3.1

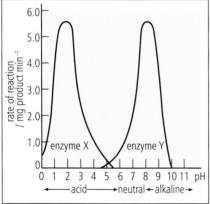

Figure 5.3.2

STUDY TIP

Never say 'the enzyme is killed'. Enzymes are not organisms. They are protein molecules. The correct word to use is denatured.

The rate of enzyme-controlled reaction can be affected by the pH of its surroundings. Hydrochloric acid is present in the stomach, so the enzymes that are active in the stomach work best in acidic conditions, i.e. at a low pH.

Alkaline bile is present in the first part of the small intestine, so the enzymes that are active here work best in alkaline conditions, i.e. at a high pH.

Most enzymes work best inside cells where the conditions are neutral. The optimum pH of these enzymes is therefore about pH 7.0.

Enzymes and pH

Enzymes are influenced by the pH of their surroundings.

Many enzymes work best in neutral conditions, but some work best in acidic conditions and some in alkaline conditions.

Look at the graph in Figure 5.3.2 showing the action of enzymes X and Y.

- Enzyme X works best at pH 2.0 – that is its optimum pH.
- The optimum pH for enzyme Y is pH 8.0.
- Up to pH 2.0 the rate of reaction increases for enzyme X and then between pH 2.0 and pH 5.5 it decreases. There is no reaction above pH 5.5.
- Between pH 4.5 and pH 8.0 the rate of reaction increases for enzyme Y and then between pH 8.0 and pH 10.0 it decreases. There is no reaction below pH 4.5 and above pH 10.0.

As we have seen, the three-dimensional shape of an enzyme is vital if it is to function. Many of the chemical bonds holding the structure of the enzyme are weak bonds. If these bonds that hold the enzyme molecule in shape are broken by changes in pH, then the shape of the active site can be altered. When the rate of reaction is zero, the shape of the active site has changed so much that the substrate molecules will no longer fit.

At these values of pH enzymes are denatured. Small changes in pH can affect the rate of reaction without denaturing the enzyme, but at the extremes of its pH range an enzyme becomes unstable and denatures.

The effect of pH on the action of a protease

Egg white contains a lot of protein. In this experiment you will investigate the effect of protease from the stomach on the protein in egg white. You will also see how pH affects the way the enzyme works.

1 Set up a water bath at 40 °C and label three test tubes 1–3.

2 Using a syringe put 5 cm³ of egg white into each test tube.

3 Add 2 cm³ of sodium carbonate (an alkali) to tube 1.

4 Add 2 cm³ of water to tube 2.

5 Add 2 cm³ of dilute hydrochloric acid to tube 3.

6 Compare the pH of each test tube by using a clean glass rod for each tube to dip into the solution and then touch the pH test paper.

7 Place all three test tubes in the water bath at 40 °C for five minutes.

8 Add 1 cm³ of protease to each test tube.

9 Compare the appearance of the tubes every minute until there is no further change.

Complete the table below.

tube	egg white + protease plus:	pH	appearance after 5 minutes
1	2 cm³ of sodium carbonate solution		
2	2 cm³ of water		
3	2 cm³ of hydrochloric acid		

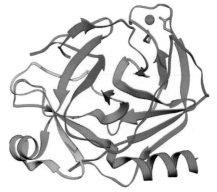

This is a model of protease that breaks down proteins to amino acids in the small intestine. It works best at a pH of about 8.5 and at 37 °C.

KEY POINTS

1 Most enzymes work best at one value of pH.

2 At either side of their optimum pH value, enzyme activity decreases.

3 At some values of pH the shape of the active site changes so that substrate molecules no longer fit.

Supplement

SUMMARY QUESTIONS

1 Sketch a graph to show the effect of increasing pH on the rate of an enzyme-catalysed reaction, where the optimum pH is 7.0 and there is no activity below pH 4.0 and none above pH 10.0.

2 Use the 'lock and key' model to explain what happens when enzymes are denatured at extremes of pH.

3 An investigation was carried out into the effects of pH on the action of the enzyme amylase on starch. Eight test tubes were set up at a different pH. They were incubated in a water bath at 30 °C for 1 hour. The concentration of reducing sugar (product) was then estimated. The results are shown in the table.

pH	4.0	5.0	6.0	6.5	7.0	8.0	9.0	10.0
concentration of reducing sugar produced	1	12	26	32	33	27	13	5

a Plot a graph to show these results.

b Explain the effects of pH on the action of amylase in this investigation.

1 Which is the correct statement about enzymes?

A Enzymes are biological catalysts that function inside and outside cells.

B Enzymes are catalysts made of protein that only function inside cells.

C Enzymes are continually made by cells as they are used up during the reactions that they catalyse.

D Enzymes only catalyse reactions in which substances are broken down into smaller molecules.

(Paper 1) [1]

2 The equation shows a reaction catalysed by an enzyme.

lipase
fat + water ⟶ fatty acids + glycerol

The progress of the reaction can be followed by detecting the decrease in pH. This is because:

A fat is an acidic substance

B fatty acids are one of the products and they lower the pH

C the reaction stops when the pH decreases no further

D water is a reactant

(Paper 1) [1]

3 Starch and human salivary amylase were mixed together at different temperatures. Which temperature will give the fastest rate of starch digestion?

A 20 °C

B 60 °C

C 37 °C

D 10 °C

(Paper 1) [1]

4 The pH of enzyme-controlled reactions may be changed from pH 7 to pH 8. How does this affect the rate of the reactions?

A always increases the rate

B always decreases the rate

C has no effect on the rate

D may change the rate or have no effect on the rate.

(Paper 1) [1]

5 In an enzyme-catalysed reaction, the substrate binds to the part of the enzyme molecule known as the:

A active site

B complementary site

C reaction site

D substrate site

(Paper 2) [1]

6 Albumen is a protein found in egg white and in the blood. A solution of protease was kept at 75 °C and then added to a solution of albumen. No reaction occurred.

Which is the most likely explanation?

A 75 °C is below the optimum temperature of the protease

B albumen cannot be broken down by proteases

C the protease is denatured

D there was not enough kinetic energy for the reaction

(Paper 2) [1]

7 The diagram shows an enzyme and four possible substrates.

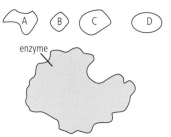

Which of the substances shown is the substrate for this enzyme?

(Paper 2) [1]

8 The optimum temperature for an enzyme-catalysed reaction is the temperature at which:

A least substrate is broken down

B most product is formed

C the rate of reaction is at its fastest

D the reaction is completed in the longest time

(Paper 2) [1]

9 (a) Explain the following statements.

    (i) Only small quantities of enzymes are required inside cells. *[2]*

    (ii) Cells make many different types of enzymes, not just one. *[2]*

The table shows the effect of temperature on the relative activity of an enzyme kept in a solution at pH 7.

temp / °C	relative activity
5	4
15	8
25	16
35	32
45	30
55	7

(b) Explain why all the reaction mixtures were kept at the same pH? *[2]*

(c) Draw a graph of the results in the table. *[6]*

(d) Describe the results shown in the graph. *[4]*

(Paper 3)

10 (a) Explain the following statements.

    (i) Amylase digests starch, but proteases do not. *[1]*

    (ii) Human enzymes catalyse reactions at 37 °C, but not at 73 °C. *[2]*

The table shows the relative activity of a human enzyme in solutions of different pH kept at 35 °C.

pH	relative activity
3	5
5	14
7	32
9	10
11	4

(b) Explain why all the reaction mixtures were kept at 35 °C? *[3]*

(c) Draw a graph of the results in the table. *[6]*

(d) Describe the results shown in the graph. *[4]*

(Paper 3)

11 The diagram shows an enzyme-catalysed reaction.

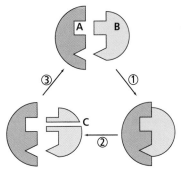

(a) State the names of **A**, **B** and **C**. *[3]*

(b) Use the diagram to describe what happens when an enzyme catalyses a reaction. *[4]*

(c) Explain why the model is known as 'lock and key'. *[2]*

(d) Use the model to explain what happens when the enzyme molecule changes shape in high temperatures or extremes of pH. *[4]*

(Paper 4)

12 Hydrogen peroxide is a very toxic substance which is produced by cells as part of their metabolism. The enzyme catalase breaks down hydrogen peroxide and is found in organisms in all kingdoms.

$$\text{Hydrogen peroxide} \xrightarrow{\text{catalase}} \text{oxygen} + \text{water}$$

A student investigated the activity of the enzyme catalase in yeast by measuring the volume of oxygen collected over time. The graph shows the results.

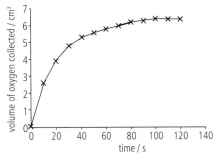

(a) Use the results shown to describe the effect of catalase on hydrogen peroxide. *[4]*

(b) Explain why catalase is the only enzyme to catalyse the breakdown of hydrogen peroxide. *[2]*

(c) Describe what happens when an enzyme catalyses a reaction. *[4]*

(Paper 4)

6 Plant nutrition

6.1 Photosynthesis

LEARNING OUTCOMES

- Define the term *photosynthesis*
- State the word equation for photosynthesis
- Describe how to test a green leaf for starch
- State the balanced chemical equation for photosynthesis
- Describe the role of chlorophyll

These plant cells are full of chloroplasts.

STUDY TIP

Make sure the equation is balanced when you write it out. Make sure there are the same number of atoms of carbon, hydrogen and oxygen on both sides of the arrow.

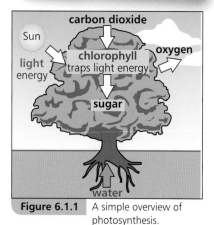

Figure 6.1.1 A simple overview of photosynthesis.

The process of photosynthesis

When you look at some leaf cells under the microscope you see many round, green structures called **chloroplasts**. Green plants use light energy to convert carbon dioxide and water from their surroundings into simple sugars. This process is called **photosynthesis**.

Photosynthesis is the process by which plants make carbohydrates from raw materials using energy from light.

Photosynthesis can be summarised in this word equation:

$$\text{carbon dioxide} + \text{water} \xrightarrow{\text{light and chlorophyll}} \text{glucose} + \text{oxygen}$$

The balanced chemical equation for photosynthesis is:

$$6CO_2 + 6H_2O \xrightarrow{\text{light and chlorophyll}} C_6H_{12}O_6 + 6O_2$$

This equation suggests that photosynthesis is one simple reaction. This is not the case. There are many reactions in photosynthesis, each catalysed by a different enzyme.

Light energy is absorbed by chlorophyll. The energy from light is transferred by chlorophyll as chemical energy to drive the reactions that form carbohydrates from water and carbon dioxide. During this process the energy is used to split water into hydrogen ions and oxygen. The hydrogen ions are used to reduce carbon dioxide to carbohydrate. As a result, the light energy absorbed by chlorophyll becomes the chemical bond energy in the simple sugars that are produced. Glucose is one simple sugar that is produced; it is converted to starch as a store of energy in leaves and other parts of plants.

Photosynthesis provides energy for plants and for all other organisms that feed on plants directly or indirectly (see page 244). The equations show that oxygen is produced as a by-product. The plant may use oxygen in its own respiration or it may diffuse out into the atmosphere where it is used by other organisms.

Investigating photosynthesis

Green plants make simple sugars from carbon dioxide and water. However, if too much sugar is dissolved in the cell sap of plant cells it would make a very concentrated solution. This would make water move in from other cells by osmosis and cause the cells to swell so much that they would need to make very thick cell walls. To prevent this, glucose molecules are linked together to form larger starch molecules which are insoluble and have no effect on osmosis. You can use iodine solution to test for the presence of starch in leaves. As you will see in Topic 6.2 we can use this test to find out whether photosynthesis has been happening or not.

Testing a leaf for starch

1 Submerge a leaf in boiling water for one minute. This kills the leaf as it destroys membranes, making it easier to extract the chlorophyll.

2 At this point, **turn off** any Bunsen burners used to boil the water.

3 Put the leaf into a test-tube of ethanol. The chlorophyll is extracted by dissolving into the ethanol.

4 Stand the test-tube in a beaker of hot water for about 10 minutes.

5 Wash the leaf in cold water. This removes the ethanol and rehydrates the leaf which softens it and makes it easy to spread out.

6 Spread the leaf out flat on a white surface and put some drops of iodine solution on it.

7 If the leaf goes blue-black, starch is present. If it stays red/brown there is no starch.

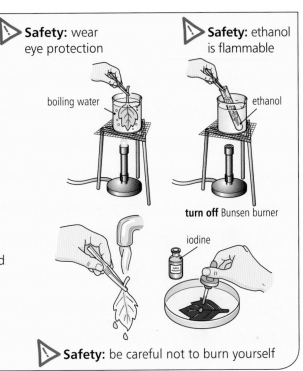

▷ **Safety:** wear eye protection

boiling water

▷ **Safety:** ethanol is flammable

ethanol

turn off Bunsen burner

iodine

▷ **Safety:** be careful not to burn yourself

SUMMARY QUESTIONS

1 Copy and complete the sentences using these words:

**oxygen   starch   chlorophyll   carbon dioxide
energy   water   photosynthesis**

Green plants make their own food by _____ . Green _____ in the chloroplasts of the leaves traps the Sun's _____ . Raw materials for this process are carbon _____ and _____ . Simple sugars are made in the leaves and some are changed to _____ . The waste product of this process is the gas _____ .

2 To find out if photosynthesis has taken place in a leaf we can carry out a test.

Copy out the stages listed on the left. Match each one with the correct reason on the right.

Stage in test	Reason
wash the leaf in cold water	to test for starch
boil the leaf in ethanol	to soften it
add drops of iodine solution to the leaf	to remove the ethanol and rehydrate the leaf
dip the leaf in boiling water	to extract the chlorophyll

3 Write out the word equation for photosynthesis and identify the raw materials and products.

4 Using only information in this topic, explain why plants are vital to our survival.

KEY POINTS

1 Chlorophyll absorbs light energy that is used in photosynthesis.

2 Photosynthesis is the process by which plants make glucose from raw materials using energy from light.

3 Carbon dioxide and water are the raw materials for photosynthesis.

4 The products of photosynthesis are glucose and oxygen.

5 A leaf that has been exposed to light will give a positive test for starch.

What is needed for photosynthesis?

It is important to understand the factors that affect photosynthesis in order to increase crop yields.

STUDY TIP

Note that carbon dioxide and water are the raw materials; light is the source of energy, and is not a raw material.

The requirements for photosynthesis

Green plants need the following in order to carry out photosynthesis:

- **light**, which provides energy for the process
- **chlorophyll**, a green pigment that absorbs the energy from light. Chlorophyll is in the chloroplasts of the leaf
- **carbon dioxide**, which diffuses into the leaves from the air
- **water**, which is absorbed by the plant's roots from the soil.

Water and carbon dioxide are the **raw materials** for photosynthesis.

Investigations demonstrate that chlorophyll, carbon dioxide and light are needed for photosynthesis.

If a plant carries out photosynthesis it will make simple sugars and store them as starch. If it cannot photosynthesise then it will not make starch.

To be certain that the investigation is valid, we must make sure that the leaves have no starch at the start. To do this, a plant is left in the dark for at least 48 hours. This is called **de-starching**. The plant is then given all the things that it needs *except* for the substance that we are testing. (It is difficult to show that water is needed for photosynthesis, as water is required by a plant for so many other things; most of the plant is water so you can hardly 'take it away'.)

PRACTICAL

Showing that chlorophyll is needed for photosynthesis

Take a de-starched, variegated plant such as a geranium.

('Variegated' means some parts of the leaves are white because there is no chlorophyll there.)

Place the plant in sunlight for about 6 hours.

Draw one leaf to show the white and green parts.

Now test this variegated leaf for starch using the starch test described in Topic 6.1.

Variegated geranium leaves.

You should find that only the green parts of the leaf go blue-black.

The green parts contain chlorophyll which is needed for photosynthesis to make starch.

The white parts contain no chlorophyll, so no photosynthesis occurs here.

Therefore the white parts of the leaf give a negative result with the starch test.

▷ **Safety:** wear eye protection

Showing that carbon dioxide and light are needed for photosynthesis

You could prove that a plant needs **carbon dioxide** to make its own food by providing it with everything it needs for photosynthesis *except* carbon dioxide. Then test to see if it has made starch.

PRACTICAL

Showing that carbon dioxide is needed for photosynthesis

1 Take a de-starched plant. Enclose it in a plastic bag with a chemical that absorbs carbon dioxide. (**Soda lime** absorbs carbon dioxide.)

2 Leave the plant in the light for a few hours. Test a leaf for starch as described in Topic 6.1. The leaf should show a negative result for the starch test. Deprived of carbon dioxide the leaf is unable to photosynthesise and make starch.

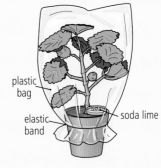

A **control** experiment should be set up in exactly the same way but without the soda lime. This means the plant in the control experiment does not have carbon dioxide removed. Then we can be sure it was the absence of carbon dioxide that caused the lack of starch, and not keeping the plant inside the plastic bag.

You can also show that a plant needs **light** to carry out photosynthesis.

PRACTICAL

Showing that light is needed for photosynthesis

1 Take a de-starched plant. Cover part of the leaf with some aluminium foil to prevent light getting through.

2 Leave the plant in the light for a few hours.

3 Test the leaf for starch as described in Topic 6.1. Only the parts of the test leaf that were left uncovered go blue-black. The parts of the leaf that were covered did not receive light and could not carry out photosynthesis and so could not make starch.

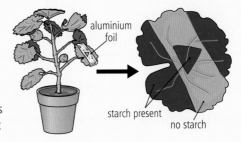

SUMMARY QUESTIONS

1 a Describe how you would de-starch a plant.

   b Explain why it is necessary to use a de-starched plant to show that plants need light to carry out photosynthesis.

   c What is a variegated leaf?

   d Describe how you could carry out an investigation to show that carbon dioxide is needed for photosynthesis.

2 State whether the following statements are true or false.

   a Plants get all their food from the soil.

   b Plants need chlorophyll to carry out photosynthesis.

   c Water for photosynthesis is absorbed through the leaves.

   d Leaves that are de-starched give a blue-black colour when tested with iodine.

3 Explain what is meant by a *control experiment*.

KEY POINTS

1 Plants need light, chlorophyll, carbon dioxide and water in order to carry out photosynthesis.

2 We can test to see that light, chlorophyll and carbon dioxide are needed for photosynthesis by not giving plants each of these and testing leaves for starch.

Products of photosynthesis

Photosynthesis makes food in these wheat plants.

How much of this food depends upon plants?

The products of photosynthesis are simple sugars, such as glucose, and oxygen.

Most of the sugars which a plant needs for food are made in the leaves.

- Some of the **glucose** is used for respiration in the leaf.
- Some of the glucose is changed into **starch** and stored in the leaves for use in the future, e.g. at night.
- Some of the glucose is used to make **cellulose**, which is needed to make cell walls.
- Glucose is converted to **sucrose** and transported to other parts of the plant in the phloem (see Topic 8.4).

Glucose can also be converted to other substances:

- Plants get nitrogen by absorbing nitrate ions from the soil.
- Glucose and nitrate are used to form **amino acids**, which are built up into **proteins**.
- Plants need proteins for growth and cell repair and for making enzymes and hormones.
- **Sugars** are converted to **oils**, which are an efficient way to store energy in seeds.

The importance of photosynthesis

Think about the food you have eaten in the last 24 hours. A lot of what you eat comes from plants like rice, maize, wheat, potatoes, soya beans and nuts.

If you eat meat or fish then that comes from animals that have eaten plants.

Products like cooking oil and margarine are made from plants.

Oxygen is a by-product of photosynthesis

Oxygen is a product of photosynthesis. It is not the main product. Some is used by the plant's respiration, but usually there is more than is needed, so most diffuses out of the leaves into the atmosphere. That is why it is called a **by-product**.

Plant products

Plants provide us with our food and also food for our livestock. There is more about this on page 248.

Plants also provide a range of raw materials for industry, such as timber and cotton. Many medicines have been discovered in plants including digitalis, which is a heart drug, and two anti-cancer drugs extracted from the rosy periwinkle from Madagascar.

Plants are the dominant organisms in almost all environments. As such they provide habitats for animals and microorganisms. The rainforests make up only 6% of the Earth's land surface, but they support more than half the world's species of animals and plants.

The concentrations of gases in the atmosphere are kept constant by photosynthesis. Without green plants, the concentration of carbon dioxide in the air would increase and the concentration of oxygen would decrease.

PRACTICAL

Oxygen produced in photosynthesis

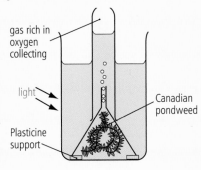

1 Set up the apparatus to collect bubbles of gas given off by the Canadian pondweed (*Elodea canadensis*).

   gas rich in oxygen collecting

   light

   Canadian pondweed

   Plasticine support

   A lamp provides the light energy that the plant needs. Carbon dioxide is dissolved in the water. Add some sodium hydrogencarbonate powder to the water and stir to dissolve.

   This is to make sure that there is always enough carbon dioxide in the water.

   If there is not enough carbon dioxide then photosynthesis slows down.

2 Place the apparatus in the light and allow a few hours for the gas to collect.

3 Test the gas for oxygen with a glowing splint.

   Standing the funnel on plasticine gives a gap to allow carbon dioxide dissolved in the water to reach the pondweed so photosynthesis does not slow down.

   A water plant is used because it is easy to collect the gas by downward displacement of water.

STUDY TIP

Remember that some of the oxygen produced in photosynthesis is used by the plant for its respiration. This means that some oxygen produced is not released into the atmosphere or water.

SUMMARY QUESTIONS

1 What are the following products made in plants used for?
   a glucose b starch c cellulose d protein e oils

2 a Describe how you would show that plants produce oxygen as a by-product of photosynthesis. Remember to include all practical details.

   b How would you show that light is necessary for oxygen production?

3 Plants are important because they provide us with:
   a food b fuels c building materials d medicines

   Give three examples of plants that provide us with each of these.

KEY POINTS

1 Products of photosynthesis include glucose, which is used to make starch, cellulose, amino acids (and then proteins).

2 Photosynthesis is important as it provides us with food and medicines; it helps to keep constant concentrations of carbon dioxide and oxygen in the atmosphere.

When plants carry out photosynthesis they produce oxygen. We can determine how fast or slow photosynthesis occurs by measuring the products that are produced. This can be done by measuring how much starch or oxygen is made. Starch production can be measured by finding out the change in dry mass, but it is not an easy method to use. Instead, measuring how much oxygen is produced is much easier as we can use an aquatic plant as described in Topic 6.3. This is done by counting bubbles or by measuring the volume of oxygen produced.

Environmental conditions such as light intensity, temperature and carbon dioxide concentration influence the rate of photosynthesis.

PRACTICAL

Photosynthesis and light intensity

You can use this apparatus to measure the effects of light intensity on the rate of photosynthesis:

1 Cut a piece of pondweed about 5 cm in length.

2 Put a paperclip on the pondweed to stop it floating to the surface.

3 Put a lamp close to the plant and measure the distance between the plant and the lamp.

4 Count the number of bubbles released over 5 minutes. Repeat several times and calculate the average.

5 Repeat this procedure with the lamp at different distances from the plant.

The number of bubbles should decrease as the distance between the lamp and the plant increases.

When the lamp is close to the plant the light intensity is high. This gives the plant lots of energy so the rate of photosynthesis is high. As the lamp is moved further away the light intensity decreases; there is less energy for the plant and the rate of photosynthesis decreases.

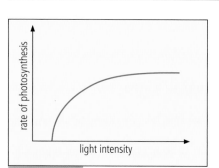

Figure 6.4.1 The effect of increasing light intensity on the rate of photosynthesis.

The graph shows some results of an experiment to measure the effect of increasing light intensity on the rate of photosynthesis.

In the dark there is no photosynthesis. At low light intensities there are very few bubbles of oxygen produced. The rate of photosynthesis increases as the light intensity increases until at a certain light intensity the rate does not increase any further. Increasing the light intensity increases the energy available to the plant for photosynthesis.

When investigating the effect of different light intensities it is important to keep the other environmental conditions the same. Temperature can be kept constant by putting a thermometer into the beaker of water and adding hot or cold water if the temperature changes. The lamp may give off heat, so the temperature may increase. The concentration of carbon dioxide in the water can be kept constant by adding sodium hydrogencarbonate to the water and letting it dissolve. This ensures that the plant will not run out of carbon dioxide.

Photosynthesis and temperature

To find the effect of temperature similar apparatus can be used. Pieces of pondweed are placed into test-tubes which are put inside beakers of water at different temperatures. The minimum is five different temperatures across a range of 10 °C to about 40 °C. The lamp is kept at one distance all the time so that the light intensity is constant. Sodium hydrogencarbonate is added to the water in the test-tubes to make sure carbon dioxide concentration is constant. The number of bubbles of oxygen produced at different temperatures can be counted in the same way.

Figure 6.4.2 shows the results you might expect. The maximum rate of photosynthesis occurs at the **optimum temperature**. The actual temperature depends on the plant that is used. Plants that grow in temperate latitudes have optimum temperatures that are lower than those that grow in the tropics.

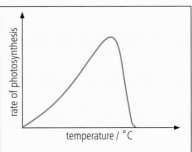

Figure 6.4.2   The effect of increasing temperature on the rate of photosynthesis. The rate decreases at high temperatures because enzymes in the chloroplasts are denatured.

The effect of carbon dioxide on the rate of photosynthesis

The same apparatus can be modified for this investigation.

Adding different quantities of sodium hydrogencarbonate to the water increases the concentration of carbon dioxide. At least five different test-tubes can be used each with a different mass of sodium hydrogencarbonate added.

The light intensity and the temperature must be kept constant.

You can see that the effect of increasing carbon dioxide on the rate of photosynthesis is the same as the effect of increasing the light intensity. If there is no carbon dioxide there can be no photosynthesis. As the carbon dioxide concentration increases there is an increase in the rate of photosynthesis. Above a certain concentration the rate remains constant.

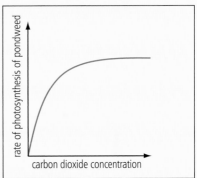

Figure 6.4.3   The effect of carbon dioxide on the rate of photosynthesis.

SUMMARY QUESTIONS

1 List the main environmental factors that influence the rate of photosynthesis.

2 A student used the apparatus shown opposite. He wrote down the distances and the numbers of bubbles per minute: 20 cm and 25, 30 cm and 24, 40 cm and 18, 60 cm and 10, 100 cm and 6.

   a   Explain how the student kept other conditions constant.

   b   Make a table of the results and plot them as a line graph.

   c   Explain why your graph does not look like Figure 6.4.1.

3 Describe the effect on the rate of photosynthesis of a increasing carbon dioxide concentration, and b increasing the temperature.

KEY POINTS

1 The release of oxygen from aquatic plants is used to measure the rate of photosynthesis.

2 Light intensity, temperature and the concentration of carbon dioxide are three environmental factors that influence the rate of photosynthesis.

6.5

Glasshouse production

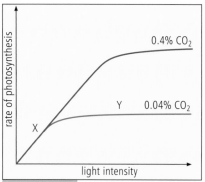

Figure 6.5.1 The effect of increasing light intensity on the rate of photosynthesis at two concentrations of carbon dioxide.

STUDY TIP

When you use a graph to answer a question, always move a ruler along the graph from left to right and follow the trends or patterns. Here you can rule lines on the graph to find the conditions when the rate starts to remain constant.

We have seen that light intensity, temperature and carbon dioxide concentration influence the rate of photosynthesis in plants.

- Light intensity determines the energy available to photosynthesis.
- Temperature influences the activity of enzymes in the chloroplasts.
- Carbon dioxide is a raw material for photosynthesis, so if the concentration increases there is more of it available for chloroplast enzymes to use to make carbohydrates.

These three factors are **limiting factors**.

Limiting factors

A limiting factor is something present in the environment in such short supply that it restricts life processes.

Light intensity influences the rate of photosynthesis. If the light intensity is low, it does not matter if the plant has lots of carbon dioxide and water and a warm temperature because there is a shortage of energy for photosynthesis and the rate cannot be very high. In this case light intensity is the limiting factor for the rate of photosynthesis. In Figure 6.5.1 you can see that as the light intensity increases the rate of photosynthesis increases. The rate increases because light intensity is the limiting factor. You can also see that the rate becomes constant however much the light intensity is increased. Light intensity is now no longer the limiting factor, it must be something else such as temperature or carbon dioxide concentration.

Chemical reactions catalysed by enzymes increase with **temperature**. You can see from the graph in Figure 6.4.2 on page 61 that the rate of photosynthesis increases until a certain temperature when the rate reaches a maximum and then decreases. The rate increases because the kinetic energy of the substrates and enzymes in the chloroplast increase and they collide more often (see Topic 5.2). The rate of photosynthesis decreases at higher temperatures because the enzymes in chloroplasts are denatured (see Topic 5.3).

Look at the graph in Figure 6.5.1. Light intensity is limiting the rate of photosynthesis at point X. Another factor is limiting the rate of photosynthesis at point Y. This could be carbon dioxide concentration or temperature.

However, you can see that the rate of photosynthesis increases if carbon dioxide concentration is increased from 0.04% to 0.4% as there is more of this raw material available. This means that the limiting factor at Y is carbon dioxide concentration.

Glasshouse production

Growers try to improve the yield of their crops by giving them the best possible conditions for photosynthesis to take place.

Conditions inside a glasshouse allow plants to:
- grow earlier in the year
- grow in places where they would not normally grow well.

The following conditions inside glasshouses are controlled.

- **Temperature**
  Sunlight heats up the inside of the glasshouse.
  The glass stops a lot of this heat from escaping.
  Electric heaters are used in cold weather.
  Ventilator flaps are opened to cool the glasshouse on hot days.

- **Light**
  The glass lets in sunlight. Artificial lighting can be used to grow plants when light intensity gets too low. Blinds keep out very strong light and shading lowers the temperature in tropical countries.

- **Carbon dioxide**
  Growers can pump carbon dioxide into glasshouses to increase carbon dioxide concentration. They can also burn butane or natural gas which provides carbon dioxide and also heat to raise the temperature of glasshouses in cold weather.

- **Water**
  Many glasshouses have automatic watering systems using sprinklers and humidifiers which ensure plants always get enough water.

Glasshouses can control limiting factors, such as temperature, light intensity and humidity.

Water misting of a glasshouse crop.

All these factors are monitored and controlled by computers so few staff are needed. Sensors for carbon dioxide concentration, humidity, light intensity and temperature detect changes in these limiting factors. Computers process data from the sensors and control all heating, ventilation, lighting and shading in the glasshouse.

SUMMARY QUESTIONS

1 Explain what is meant by the term *limiting factor*. Use the process of photosynthesis to help explain your answer.

2 State which factors would be limiting photosynthesis in the following. Give your reasons in each case.
   a Plants growing on a high mountainside early in the morning.
   b Plants growing in a dense tropical forest at midday.
   c A field of maize early on a very bright morning.
   d The same field late in the afternoon.
   e A house plant in a shaded position in a warm room.

3 Explain how each of the following can be controlled in glasshouses to give optimum conditions for growth:
   a carbon dioxide concentration
   b light intensity   c temperature

KEY POINTS

1 A limiting factor is a factor in the environment that is in such short supply that it restricts a life process, such as photosynthesis.

2 Light intensity, temperature and carbon dioxide concentration are limiting factors of photosynthesis.

3 Carbon dioxide concentration, light intensity, humidity and temperature are controlled in modern glasshouse systems to give optimum conditions. This ensures that the rate of photosynthesis is kept high so plants produce maximum yield.

6.6 Leaves

LEARNING OUTCOMES

- Identify and label the parts of a leaf
- Explain how the structure of a leaf is adapted for photosynthesis

A leaf has an ideal shape for photosynthesis.

The shape of a leaf makes it ideally adapted to carry out its functions of absorbing light for photosynthesis and allowing gases to diffuse into and out of the leaf tissues.

Leaves have:

- **a large surface area** – to absorb light rays
- **a thin shape** – so gases can diffuse in and out easily
- **many chloroplasts** – to absorb light for the reactions that take place in photosynthesis
- **veins** – to support the leaf surface and to carry water and ions to the leaf cells, and to take sucrose and amino acids away from the leaf to all other parts of the plant.

Inside a leaf

To find out how a leaf works we can look at the structure of a typical leaf.

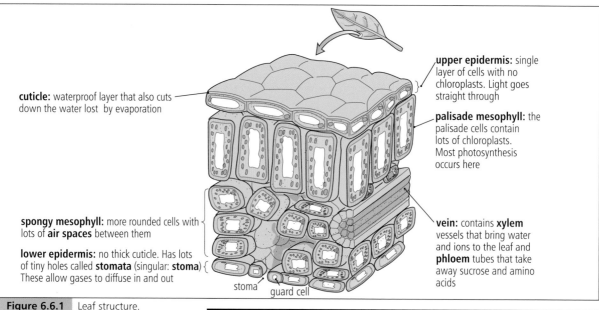

cuticle: waterproof layer that also cuts down the water lost by evaporation

spongy mesophyll: more rounded cells with lots of **air spaces** between them

lower epidermis: no thick cuticle. Has lots of tiny holes called **stomata** (singular: **stoma**). These allow gases to diffuse in and out

upper epidermis: single layer of cells with no chloroplasts. Light goes straight through

palisade mesophyll: the palisade cells contain lots of chloroplasts. Most photosynthesis occurs here

vein: contains **xylem** vessels that bring water and ions to the leaf and **phloem** tubes that take away sucrose and amino acids

stoma   guard cell

Figure 6.6.1 Leaf structure.

PRACTICAL

Looking at a leaf section

1 Observe a transverse section of a leaf on a slide under the microscope.

Look for the tissues in Figure 6.6.1 under low power.

Now look at each type of tissue under high power.

Some of the internal features are adaptations for photosynthesis.

- Palisade mesophyll cells are packed tightly together near the upper surface of the leaf to maximise absorption of light where its intensity is highest.
- There are many chloroplasts in the palisade mesophyll cells to absorb as much light as possible.
- Stomata (usually in the lower epidermis) open to allow carbon dioxide to diffuse into the leaf. Carbon dioxide is a raw material for photosynthesis.
- Leaves are thin so that carbon dioxide does not have to diffuse far from the atmosphere to the cells of the palisade and spongy mesophyll.

- There are large intercellular air spaces within the spongy mesophyll layer. This makes it easy for carbon dioxide to diffuse to all the mesophyll cells. Diffusion through air is much faster than diffusion from cell to cell.
- Xylem in veins bring water and ions to the mesophyll cells. Water is a raw material for photosynthesis and magnesium ions are needed by the cells to make chlorophyll.
- The sugar produced in photosynthesis is converted to sucrose and transported away from the leaf in the phloem in veins.

Stomata

Stomata are small pores (holes) in the epidermis that allow gases to diffuse into and out of the leaf. Stomata are usually in the lower epidermis, but some plants like water lilies have them in the upper epidermis.

In sunlight:

- carbon dioxide diffuses in for photosynthesis
- oxygen made in photosynthesis diffuses out
- water vapour diffuses out.

Opening and closing

Stomata are opened and closed by **guard cells**.

Stomata usually open during the day. Water passes into the guard cells by **osmosis**. This makes them bend so the stoma opens.

Carbon dioxide diffuses into the leaf for photosynthesis and oxygen diffuses out. Water vapour also diffuses out.

At night the stomata close. Water passes out of the guard cells by osmosis and they straighten and move closer together so closing the stomata pores. The stomata also close in hot, dry weather to help prevent the plant wilting.

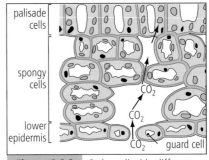

Figure 6.6.2 Carbon dioxide diffuses into the leaf for photosynthesis.

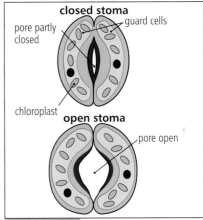

Figure 6.6.3 Closed stoma and open stoma.

STUDY TIP

See Topic 3.2 to remind yourself about osmosis, which is how water moves into and out of guard cells.

SUMMARY QUESTIONS

1 Match each of the leaf parts on the left with their correct function on the right:

stomata	carry water up from the stem
palisade mesophyll	allow gases to pass into and out of leaf
spongy mesophyll	contain chloroplasts for photosynthesis
cuticle	contain chloroplasts for photosynthesis
veins	prevent too much water being lost

2 a Which gas would pass into a leaf during daylight?

b Which gas passes out of a leaf during daylight?

3 Leaves have a large surface area to absorb light.

Lay a leaf onto graph paper, draw around it and work out its area by counting the squares.

Find out if leaves from the top and bottom of a plant have the same leaf area. Record your results and explain them.

KEY POINTS

1 The diagram of the internal structure of a leaf shows different tissues which each have important functions to carry out.

2 Stomata control the diffusion of gases into and out of a leaf.

LEARNING OUTCOMES

- Describe the importance of nitrate ions and magnesium ions in plant nutrition
- Explain the effects of nitrate and magnesium deficiency on plants
- Use hydrogencarbonate solution to investigate gas exchange of an aquatic plant kept in the light and in the dark

Plants need more than light, carbon dioxide and water for healthy growth. They also need **mineral salts** (also called **plant nutrients**). Nutrients are absorbed from the soil in small quantities as ions by active transport in the roots (see Topic 3.4).

Plant nutrients

Nutrients are needed for healthy growth of plants. They are used for a variety of purposes in plants. If these nutrients are lacking in the soil then plants do not grow well and show certain symptoms known as deficiency symptoms. Plants need **nitrate** ions to make amino acids which are used to make proteins. As proteins are required for growth, plants deficient in nitrate show poor growth.

Magnesium ions are absorbed by plants and used to make chlorophyll. Leaves of plants deficient in chlorophyll look yellow, a condition known as chlorosis. Plants need phosphate for making compounds such as DNA and for respiration.

If the soil does not contain enough nutrients, farmers and growers may add more as **fertiliser** to replace the missing nutrients.

This plant has magnesium deficiency.

PRACTICAL

The effects of nutrients on growth

Some plants are grown in water culture to show that plants need different nutrients.

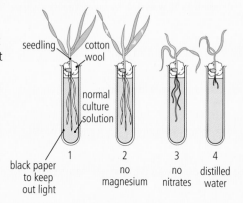

1 Set up some test-tubes as shown here.

Cover the tubes with black paper to prevent algae growing in the water.

Each solution lacks a certain nutrient except test-tube 1 which has all of them.

2

3 Leave the apparatus in good light for 6 weeks.

4 After 6 weeks you could examine the colour and size of leaves.

5 You could measure length of stems and roots to compare growth.

Gas exchange in plants

All living organisms, including plants, carry out respiration all the time. Plants are no different from other organisms. To carry out respiration they need a supply of oxygen and they produce carbon dioxide as a waste. At night, plants exchange these gases with their surroundings. During the day, plants carry out photosynthesis, as well as respiration. In bright light, when there is a high rate of photosynthesis, some of the oxygen produced in their chloroplasts is used by mitochondria for aerobic respiration. The rest of the oxygen is not required so diffuses out of the plant. The carbon dioxide produced by mitochondria in respiration is used by chloroplasts for photosynthesis. However, this is not enough so carbon dioxide diffuses in from the surroundings.

Gas exchange in aquatic plants is investigated with hydrogencarbonate indicator solution. This solution contains hydrogencarbonate to provide carbon dioxide and two pH indicators. Carbon dioxide is an acid gas. When it dissolves in water it forms carbonic acid, which is a weak acid. Hydrogencarbonate indicator solution is prepared by bubbling atmospheric air through it. This gives the indicator a red colour.

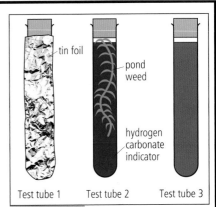

Figure 6.7.1

SUMMARY QUESTIONS

1 Explain why plants need mineral ions.

2 Describe the likely appearance of a plant growth without nitrate ions.

3 Describe and explain the gas exchange that occurs between a plant and its surroundings **i** at night, **ii** in bright sunshine.

PRACTICAL

Photosynthesis and respiration in plants

1 Set up three test tubes as shown in Figure 6.7.1.

2 Add hydrogencarbonate indicator solution to each test tube.

3 Cut two pieces of pondweed to 10 cm in length and place one in each of test tubes 1 and 2.

4 Put test tube 1 in darkness by covering it in foil.

5 Leave the test tubes near a light for 2–3 hours.

If carbon dioxide is added to the water by the plant, the hydrogencarbonate indicator solution will turn from red to yellow. If carbon dioxide is taken up from the water by the plant, the hydrogen carbonate indicator solution will turn from red to purple.

- In test tube 1, the indicator turned yellow because carbon dioxide was released by the pondweed during respiration.
- In test tube 2, the indicator turned purple because carbon dioxide was taken up by the pondweed for photosynthesis.
- The purpose of test tube 3 was to act as a control.

The indicator solution only responds to changes in carbon dioxide. The colour changes are nothing to do with the oxygen released by the plant in the light or absorbed by the plant in the dark. When oxygen dissolves in water it does not change the pH. If you wanted to monitor the changes in oxygen concentration you would have to use an oxygen probe and a data logger.

If pond weed is kept in a test tube of hydrogencarbonate indicator solution at low light intensity, the colour of the solution usually remains red as the rates of photosynthesis and respiration are the same.

KEY POINTS

1 Plant roots absorb nutrients including nitrate ions needed for healthy growth.

2 Nitrate ions are needed to make amino acids; magnesium ions are used by the plants to make chlorophyll.

3 If mineral ions are lacking, a plant develops deficiency symptoms.

4 In light plants use up carbon dioxide in photosynthesis and release oxygen. At night plants produce only carbon dioxide by respiration.

5 Hydrogencarbonate indicator solution is used to detect changes in carbon dioxide concentrations as a result of gas exchange by aquatic plants.

1 Photosynthesis occurs in palisade mesophyll cells. Which structure inside these cells carries out photosynthesis?

   A   chloroplast

   B   cytoplasm

   C   nucleus

   D   vacuole

   (Paper 1) [1]

2 The raw materials for photosynthesis are:

   A   carbon dioxide and nitrate ions

   B   light energy and water

   C   light energy, carbon dioxide and water

   D   water and carbon dioxide

   (Paper 1) [1]

3 Magnesium ions are used by plants for making:

   A   cellulose

   B   chlorophyll

   C   fats and oils

   D   proteins

   (Paper 1) [1]

4 Which represents the movement of gases through the open stomata of a leaf during the day?

	gas into leaf	gas out of leaf
A	oxygen	carbon dioxide
B	water vapour	carbon dioxide
C	carbon dioxide	oxygen
D	oxygen	water vapour

   (Paper 1) [1]

5 Which does *not* occur during photosynthesis?

   A   absorption of light by chlorophyll

   B   conversion of glucose to sucrose

   C   reduction of carbon dioxide to a carbohydrate

   D   splitting of water molecules using light energy

   (Paper 2) [1]

6 Which row of the table (top of next column) would provide the conditions for the highest rate of photosynthesis of a tropical plant?

	light intensity	carbon dioxide concentration	temperature / °C
A	high	low	30
B	low	high	20
C	low	low	30
D	high	high	35

   (Paper 2) [1]

7 Which feature is an adaptation to ensure a good supply of carbon dioxide to the cells inside a leaf?

   A   large internal air spaces

   B   many chloroplasts in palisade mesophyll cells

   C   thin cuticle on the lower epidermis

   D   xylem and phloem in veins

   (Paper 2) [1]

8 Hydrogencarbonate indicator solution is red when atmospheric air is bubbled through it. A student placed a piece of pondweed inside test tube Y that contained this indicator solution. Tube Y was covered completely in foil. This was repeated for test tube Z, but the tube was left uncovered. Both tubes were left in bright sunshine. After 60 minutes the colour of the indicator in tube Y was yellow and in tube Z was purple. What is the correct reason for this?

   A   pondweed in Y had absorbed carbon dioxide and the pondweed in Z had produced carbon dioxide

   B   pondweed in Y had produced carbon dioxide and the pondweed in Z had produced oxygen

   C   pondweed in Y had produced carbon dioxide, but the pondweed in Z had absorbed carbon dioxide

   D   pondweed in Y had produced oxygen, but the pondweed in Z had absorbed oxygen

   (Paper 2) [1]

9 The diagram on the next page shows a transverse section through a leaf as you would see it in a microscope.

   (a) Name the tissues labelled **E**, **F**, **G** and **H**. [4]

   (b) Describe the functions of the parts of the leaf labelled **I** and **J**. [3]

   (c) Explain how cells in the tissues labelled **F** and **G** are adapted for photosynthesis. [3]

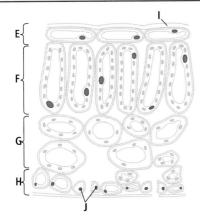

(Paper 3)

10 Two plants, **K** and **L**, were placed in the dark for several days. They were then placed into the apparatus shown below in the light to find out if carbon dioxide is required for photosynthesis. **L** was provided with carbon dioxide.

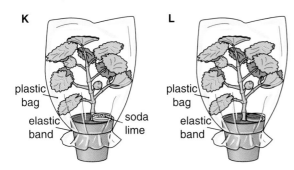

(a) Explain why the plants were put in the dark for several days. [2]

(b) Explain why the plants were well watered and kept in bright light. [2]

(c) Describe how you would show that plant **K** did not carry out any photosynthesis while inside the bag, but plant **L** did. [4]

(d) Explain why plant **L** was included. [2]

(Paper 3)

11 A student investigated the effect of varying the light intensity on the rate of photosynthesis using the apparatus shown below.

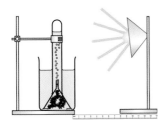

(a) State how the student would:

(i) vary the light intensity; [2]

(ii) determine the rate of photosynthesis; [2]

(iii) maintain a constant temperature and constant carbon dioxide concentration throughout the investigation. [4]

(b) The graph below shows the student's results.

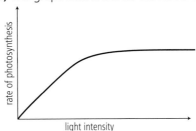

Using the term *limiting factors*, explain the results shown in the graph. [4]

(Paper 4)

12 Two groups of seedlings were grown in solutions that were deficient in two ions that plants need. Group **M** was grown without nitrate ions; group **N** was grown without magnesium ions. Group **O** received all the ions that plants require.

The results are shown in the table.

group of seedlings	colour of leaves	growth
M	yellow (especially the lower leaves)	reduced growth but not as much as seedlings in group **N**
N	lower leaves – yellow upper leaves – pale green	severely reduced growth
O	green	normal growth

(a) Explain the effects of the deficiencies of nitrate ions and magnesium ions as shown in the table. [4]

(b) Why was it necessary to include the seedlings in group **O**? [2]

(c) Many crops throughout the world are grown in glasshouses. Explain how conditions are controlled in glasshouses to achieve high yields. [6]

(Paper 4)

7 Animal nutrition

A balanced diet

A **balanced diet** provides an adequate intake of the biological molecules and energy needed to sustain the body and ensure good health and growth.

Figure 7.1.1   Which is the better-balanced meal?

A balanced diet has these components:

carbohydrates, proteins, fats, vitamins, minerals, water and fibre

Each of these carries out one or more of three basic functions.

- **To provide energy** – this is mainly the role of carbohydrates and fats. Proteins are used for energy only if they are in excess of requirements for growth, development, repair and replacement.
- **To allow growth and repair** of body cells and tissues. Proteins in the diet provide a source of amino acids for cells to make their own proteins (see Topic 4.1).
- **To regulate the body's metabolism** – vitamins and minerals are needed in very small quantities in the diet to help regulate our metabolism.

Even if you are lying in bed and completely inactive you are still using energy to keep your heart beating, your lungs working and your body temperature constant. The chemical reactions in your body, such as those involved in growth and repair, are occurring and require energy.

The energy required for these body functions is the **basal metabolic rate (BMR)**. The BMR varies from person to person, but an adult requires about 7000 kJ per day. So even if you lie down and do very little you still need this energy.

Diet depends on age, sex and activity

Children have a greater energy requirement than adults because they have a higher basal metabolic rate (BMR).

A young child may weigh less than an adult but may have a higher BMR per kilogram per day because they are still growing.

	energy used in a day / kJ	
	male	female
8-year-old	8 500	8 500
teenager, aged 14	12 500	9 700
adult office worker	11 000	9 800
adult manual worker	15 000	12 500
pregnant woman		10 000
breast-feeding mother		11 500

Young children also need more protein per unit of body mass because they are constantly developing and need to make new cells. Elderly people generally have lower energy and protein needs and have a lower BMR. However, they need to eat a balanced diet if they are to stay healthy. Women have a relatively higher fat content in their bodies than men. Fat is stored in fat tissue, for example under the skin. Fat tissue has a lower metabolic rate than muscle, so women generally have a lower energy requirement than men.

Compare the daily energy requirements of the people shown in Figure 7.1.2 and try to explain the differences.

The energy needs of occupations that involve physical activity are greater than less active jobs that involve sitting at a desk. People who take part in vigorous physical sports also require high energy and high protein diets. The extra protein is required for muscle development. Women need extra nutrients during pregnancy and when they are breast-feeding. For example, a woman in the last three months of pregnancy may require about 0.8 MJ more energy than the 9 MJ recommended for a non-pregnant woman.

Why can fat be bad for us?

Fat provides us with more energy per gram than carbohydrates or proteins. However, we should be aware that there are two main types of fat: **saturated fat** that comes from animals and **unsaturated fat** that comes from fish and plants. Saturated fats in the diet may increase the concentration of **cholesterol** in our blood. Cholesterol is a chemical made in the liver and found in the blood. The quantity of cholesterol that we produce depends upon our diet as well as the genes that determine our metabolism which we inherit. High fat diets tend to increase the risk of producing more cholesterol and having a high concentration in the blood.

High concentrations of cholesterol in the blood are linked to a narrowing of the arteries (see page 108), an increased risk of developing high blood pressure and heart disease. There are two types of unsaturated fat. Monounsaturated fats have little effect on blood cholesterol. Polyunsaturated fats may help to reduce cholesterol concentrations which will help to reduce heart disease.

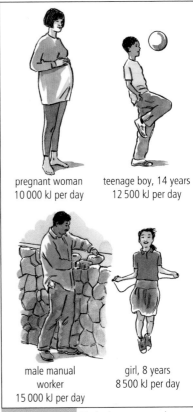

pregnant woman
10 000 kJ per day

teenage boy, 14 years
12 500 kJ per day

male manual worker
15 000 kJ per day

girl, 8 years
8 500 kJ per day

Figure 7.1.2 Typical energy needs.

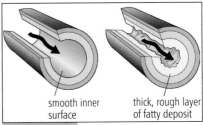

smooth inner surface

thick, rough layer of fatty deposit

Figure 7.1.3 A healthy artery (left) and an artery with fat deposited in the wall.

SUMMARY QUESTIONS

1 a What is meant by the term *balanced diet*?

b State how the protein and energy requirements of the following will differ from those of a 16-year-old female who does not take much exercise:

   i   an elderly person   ii   a 2-year-old child

   iii   a pregnant mother   iv   a male athlete

2 a What are the health risks associated with a diet high in fat?

b Explain how a person may reduce the fat content of their diet. (You may find it useful to look at pages 73 and 74 before answering this question.)

KEY POINTS

1 A balanced diet provides energy and proteins, carbohydrates, fats, vitamins, minerals, water and fibre in the correct quantities.

2 Requirements of the diet depend on a person's age, gender and how active they are.

3 Saturated fats increase blood cholesterol which is linked to narrowing of the arteries and heart disease.

Sources of nutrients

Eating healthily

Diets in different parts of the world vary. However, we should all eat from each of the food groups shown in the food guide pyramid.

Carbohydrates

Carbohydrates include sugars and starches, which provide us with a ready source of energy that is easily respired. Simple sugars are absorbed almost immediately by the stomach into the blood to give an immediate source of energy. This energy is released as a result of respiration. Good sources of carbohydrates are rice, potatoes, bread, yams, sugar and honey.

Proteins

In order for your body to grow and develop, it needs protein to make new cells. You digest protein into amino acids and then use these to make your own proteins. There are 20 different types of amino acid and your body must have all of these to make its own protein (see Topic 4.1).

Cell membranes and cytoplasm contain a great deal of protein. Your body may need to replace old or damaged cells so you need enough protein in your food for this as well. If protein is not used for growth and repair it may be respired to provide energy. Good sources are meat, fish, milk and nuts.

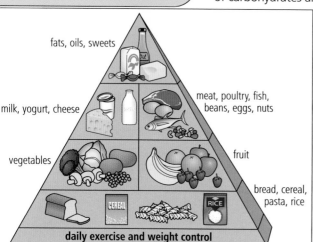

fats, oils, sweets

milk, yogurt, cheese

meat, poultry, fish, beans, eggs, nuts

vegetables

fruit

bread, cereal, pasta, rice

CEREAL   RICE

daily exercise and weight control

Figure 7.2.1   The food guide pyramid.

Fats

Fats are important as a long-term energy store. The fat is stored under the skin and around the heart and kidneys. One gram of fat releases more than twice as much energy than can be released from one gram of carbohydrate or protein. When we are short of energy our body uses the fat.

Fats are also good thermal insulators, since they cut down heat loss. Fats also give buoyancy to marine animals, for example whales have a thick layer of blubber. Good sources are butter, cheese, fat in meats and fish, nuts.

Water

Water makes up two-thirds of your body mass. You take in water when you drink or eat.

You could go without food for a number of weeks, but you would die in a few days without water. Water is needed for chemical reactions to take place in solution. The blood transports substances dissolved in water. Waste chemicals are passed out of our bodies in solution in the urine, and water in our sweat cools us down. It is important that your intake of water each day equals your loss of water in urine, faeces, sweat and breath.

Vitamins and minerals

We need small, regular amounts of vitamins and minerals if we are to be healthy. If these are lacking in the diet we can develop symptoms of deficiency diseases:

name	rich food source	use in body	deficiency symptoms (deficiency disease)
vitamin C	oranges, lemons, other citrus fruits	tissue repair, resistance to disease	bleeding gums (scurvy)
vitamin D	fish oil, milk, butter (also made by skin in the Sun)	strengthens bones and teeth	soft bones, legs bow outwards (rickets)
iron	liver, meat, cocoa, eggs	used in formation of haemoglobin in red blood cells for transport of oxygen	tiredness, lack of energy (anaemia)
calcium	milk, fish, green vegetables	strengthens bones and teeth	weak, brittle bones and teeth (rickets), muscle weakness and cramps

Tangerines are a good source of vitamin C.

Fibre

Dietary fibre or roughage comes from plants. It is made up mainly of cellulose from plant cell walls. Although it cannot be digested, it is an important part of the diet. High-fibre foods include bran cereals, cabbage, sweetcorn and celery.

Fibre adds bulk to our food. Since it is not digested, it passes down the entire gut from mouth to anus and does not provide any energy.

The muscles of the gut wall need something to push against. Fibre helps the movement of food in the alimentary canal by peristalsis so preventing constipation (see Topic 7.3). Fibre absorbs poisonous wastes from bacteria in our gut. Many doctors believe that a high-fibre diet lowers the concentration of **cholesterol** in the blood. Fibre reduces the risk of heart disease and bowel cancer.

DID YOU KNOW?

In 1747, James Lind, a Scottish ship's surgeon, was able to show that citrus fruits, such as oranges and lemons, successfully alleviated the symptoms of scurvy. The disease affected sailors on long voyages for years; vitamin C in the citrus fruits provided the remedy.

KEY POINTS

1 Carbohydrates are a quick source of energy, while fats are an energy store.

2 Protein is needed for growth and repair of tissues.

3 Lack of calcium and vitamin D can cause rickets (soft bones); lack of iron causes anaemia and lack of vitamin C causes scurvy.

4 Roughage helps the passage of food through the gut.

SUMMARY QUESTIONS

1 List the components of a balanced diet and give a good source of each one.

2 Explain why the diet must contain each component.

3 Explain the term *deficiency disease*.

Balancing energy needs

Many children do not eat this sort of meal at lunchtime!

Exercise helps fight heart disease and obesity.

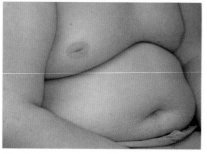

Obesity is a growing problem.

Getting the balance right

The food you eat in a day should provide you with enough energy to get through that day.

What happens if we eat too much?

If you eat more food than you need, your body stores the extra as fat.

Your **energy intake** is the energy you get in your food in a day in proteins, carbohydrates and fats.

Your **energy output** is the energy your body uses in a day. If energy intake is greater than energy output, then fat is stored in the body and body mass increases. We may run the risk of becoming overweight or even **obese**. Obesity is defined in the box on the next page. People with a low basal metabolic rate (BMR) are more likely to become overweight or obese than those with higher BMRs. It is likely that their diet contains fattening foods, such as high fat foods and refined foods with a lot of added sugar.

Obesity

In rich, industrialised countries, like Britain and the USA, more people suffer from overeating than from eating too little. As we have seen, if a person takes in more energy in food than they use, then the excess is stored as fat. Taken to extremes, people can put on so much weight that they become obese.

Major causes of obesity include:

- high intake of fatty foods and refined foods containing a lot of added sugar
- too little exercise
- social and emotional stress, leading to 'comfort eating'.

People who are overweight and obese are much more likely than thinner people to have health problems such as heart disease (see Topic 9.5), high blood pressure, diabetes caused by high blood sugar (see Topic 10.3) and arthritis (worn joints).

How could someone lose weight?

They could:

- eat less high-energy foods (lower their energy intake)
- take more exercise (increase their energy output).

A sensible approach to slimming should combine:

- a balanced diet with a lower intake of energy
- a gradual increase in exercise
- having an ideal, but achievable, target for weight reduction.

Obesity

There are two ways in which people can identify being obese:

- being 20% above the recommended weight for his or her height
- having a **body mass index (BMI)** greater than 30.

$$\text{body mass index} = \frac{\text{body mass / kg}}{\text{height / metres}^2}$$

A person with a BMI of less than 20 is underweight; between 25 and 30 is overweight; more than 30 is obese; more than 40 is severely obese. If your BMI is in the range of 20 to 24, then your body mass is in the acceptable range.

These calculations are for adults. The BMI is calculated in the same way for children but the interpretation is different.

STUDY TIP

You do not have to remember how to calculate the body mass index, but you should be aware that someone is classified as obese if their BMI is greater than 30.

Constipation

Roughage or **fibre** is important for a healthy, balanced diet. It is indigestible but adds bulk to our food. This is important for keeping food moving down the alimentary canal. The muscles of the gut wall contract to squeeze the food along. Soft foods do not stimulate the muscles to contract as effectively as the harder, indigestible foods that form roughage. If the movement of food is slow it will result in **constipation** making it difficult to defecate.

A diet containing fruit and vegetables contains lots of roughage; these are the best foods to relieve constipation.

STUDY TIP

Malnutrition refers to diets that are not balanced because there is too little energy or too much energy, a shortage of certain components or far too much of those components. Constipation is a form of malnutrition as the diet is lacking in roughage.

SUMMARY QUESTIONS

1 Copy and complete the sentences using these words:

**exercise   energy   sugar   obesity
high-energy   fat   fatty**

If you take in more _____ than you use, the excess is stored as _____ . Eating _____ foods and foods with added _____ over long periods of time can lead to _____ . A person can lose weight by eating less _____ foods and taking more _____ .

2 List four health problems that are linked to obesity.

3 Calculate the body mass index for the following people and name the BMI category in each case:

a   a man 1.70 m in height with a body mass of 110 kg

b   a woman 1.75 m in height with a body mass of 50 kg

c   a man 1.85 m in height with a body mass of 75 kg

4 a   State two good sources of roughage in the diet.

b   Explain the effects of a lack of roughage in the diet.

KEY POINTS

1 Some people put on weight if their energy intake (in their food) is greater than their energy output (in exercise).

2 Obesity can be caused by a high intake of fatty foods and refined foods containing a lot of added sugar plus the effects of too little exercise.

3 A diet lacking in roughage (fibre) can lead to constipation.

Starvation and nutrient deficiency

LEARNING OUTCOMES

- Describe the effects of starvation
- Describe the effects of scurvy
- Discuss the causes and effects of vitamin D and iron deficiencies

People can survive for many weeks without food, provided they have water. This is because during periods of starvation, the body draws upon its stores of carbohydrate, fat and protein for energy, but there are no stores of water. During starvation the basal metabolic rate (BMR) is reduced. The body uses stored fat and then breaks down protein in the muscles to use as a source of energy. This leads to muscle wasting and an emaciated appearance.

Worldwide, the most common form of undernutrition is **protein energy malnutrition (PEM)**. As the term suggests, this is caused by a lack of dietary energy and protein. In its worst form it can lead to **kwashiorkor** or **marasmus**, which are terms to describe the appearance of children with PEM. These are two different ways in which the body responds to lack of food. In both conditions, children are underweight, though more so in marasmus. However, it is thought that children with marasmus have adapted better to lack of food and respond better when given food. The symptoms of these conditions are shown on the left.

In both cases, children are in danger and feeding should involve frequent servings of small quantities of food. This is because PEM causes cells of the pancreas and intestine to die, so fewer digestive enzymes are secreted and the surface area for absorption of digested food is reduced. During starvation, there are deficiencies of the protective vitamins; people are at risk of developing infectious diseases.

Dietary surveys of poor communities in developing nations have shown that protein, which should be used for the growth and repair of cells, has to be used as an energy source. This shows that children with these conditions are suffering from a lack of energy in the diet.

Children who develop the symptoms may be fed on tropical crops such as cassava and sweet potato; while having a high energy content, these are low in protein, so they need to be supplemented with other protein foods. Staple foods such as wheat and rice will satisfy energy intake and also provide enough protein.

Nutrient needs

Vitamin C is found in oranges, lemons and other citrus fruits. As we have seen, a deficiency of vitamin C leads to the disease called **scurvy**. This can cause bleeding in parts of the body, particularly the gums. Vitamin C helps bond cells together and helps in the use of calcium by bones and teeth. Vitamin C is used to form an important protein used to make skin, tendons, ligaments and blood vessels. It also helps in the healing of wounds and the formation of scar tissue.

Symptoms of kwashiorkor:

oedema (swelling of the abdomen and legs); sparse, dry hair; flaky skin; fat accumulation in liver.

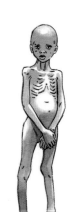

Symptoms of marasmus:

very low body mass; thin arms and legs; little muscle or fat; wizened, old-looking face.

Figure 7.4.1 Typical symptoms of two forms of protein energy malnutrition.

Vitamin D is needed for the small intestine to absorb calcium. It is also needed to regulate the deposition of calcium in bone cells. Deficiency symptoms include the lack of calcium in bones causing **rickets** in children. Bones fail to grow properly and become soft, so when children start walking the bones bend with the weight of the body. Rickets can be prevented by eating foods rich in vitamin D like fish-liver oil, butter, eggs and milk. In adults, deficiency gives rise to the condition known as **osteomalacia**, which leads to a softening of the bones and an increased chance of fractures.

Iron is needed to make the blood protein **haemoglobin** which is found in red blood cells. Haemoglobin is a protein combined with iron. It is needed to carry oxygen around the body.

Foods rich in iron include liver, meat, cocoa. If there is a lack of iron in the diet, a disease called **anaemia** can occur. An adult needs about 16 mg of iron each day. At puberty, girls need to make sure they have enough iron in their diet as they start to **menstruate** (have periods).

The lack of iron leads to a reduction in the number of red blood cells. The main symptoms of anaemia are tiredness and lack of energy, shortness of breath, heart palpitations and a pale complexion.

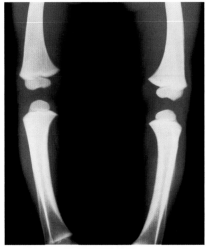

An X-ray showing the weakened bones and bowed legs of a child with rickets.

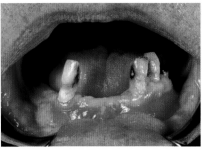

The mouth of a person who has scurvy. The gums are inflamed and the teeth have fallen out as a result of not enough vitamin C in the diet.

SUMMARY QUESTIONS

1 a Explain the main causes of starvation in terms of an inadequate diet.

  b What are the symptoms of:

    i kwashiorkor   ii marasmus?

  c Explain why treatment of these two conditions involves frequent servings of very small amounts of food.

2 The information below is from a box containing breakfast cereal.

	Amount per 100 g
Energy / kJ	1800
Protein / g	6
Carbohydrate / g	83
Fat / g	3

  a How much energy would there be in a 150 g serving of the cereal?

  b How much fat would there be in a 75 g serving of the cereal?

  c The mass of protein a young adult female needs per day is 60 g. How much more cereal would a female have to eat to get this mass of protein?

3 Make a table to compare iron, vitamin C and vitamin D in the diet. Include in your table information about their roles in the body, good dietary sources and the deficiency diseases associated with each nutrient.

KEY POINTS

1 Starvation results in a very low body mass with lack of fat and muscle wasting, plus a lack of resistance to infections.

2 Scurvy is caused by a deficiency of vitamin C in the diet and can cause bleeding in parts of the body, particularly the gums.

3 Lack of iron in the diet can cause anaemia which affects the ability of the blood to transport of oxygen.

4 Vitamin D deficiency can cause a softening of the bones causing rickets and osteomalacia.

- Define *mechanical* and *chemical digestion*
- Define *ingestion, absorption, assimilation* and *egestion*
- State the function of amylase, protease and lipase, listing the substrates and the end-products

During the passage of food along the alimentary canal, a number of different processes can be identified.

- **Ingestion** is the taking of substances (food and drink) into the body through the mouth.
- **Digestion** is the breaking down of large insoluble molecules into small soluble molecules so that they can pass through the gut wall into the blood.
- **Absorption** is the movement of small food molecules and ions through the wall of the intestine into the blood.
- **Assimilation** is the movement of digestive food molecules into the cells of the body where they are used, becoming part of the cells.
- **Egestion** is the passing out of food that has not been digested or absorbed, as faeces through the anus.

The large food molecules are starch, proteins and fats. Digestion occurs mechanically and chemically. This happens in the alimentary canal, or gut, which together with the liver and pancreas form your digestive system (see Topic 2.4). When the food has been digested, it is **absorbed** through the wall of the small intestine into the blood. Before the food molecules can go through the gut wall they must be dissolved. Large food molecules are insoluble; they will not dissolve, therefore cannot get through the wall of the small intestine. On the other hand, small food molecules are soluble. They will dissolve so they can get through the wall of the small intestine and enter the blood and the lymph, which is another body fluid that absorbs and transports fat (see page 114).

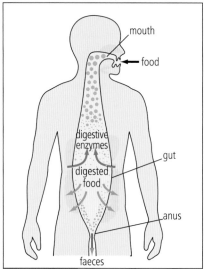

Figure 7.5.1 Digestion in the alimentary canal.

Mechanical digestion

Mechanical digestion is the breakdown of large pieces of food into smaller pieces of food without changing the food molecules. This starts in the mouth where chewing breaks down food into smaller pieces that can be swallowed. Muscular contractions of the stomach continue this process. In the small intestine, large globules of fat are broken into smaller globules by emulsification by bile. There is more about mechanical digestion in the next few topics.

Chemical digestion

Before the body can use food that has been eaten, the food must be broken down into small molecules.

Chemical digestion is the breakdown of the large insoluble food molecules into smaller soluble molecules by the action of **enzymes**. This occurs in the mouth, stomach and small intestine. Mechanical digestion gives a larger surface area for the enzymes to work on.

The three types of enzymes in the alimentary canal are: proteases, carbohydrases and lipases. Each type of food molecule needs a specific enzyme because they have different shapes. See Topic 5.1 to remind yourself about the reason for having different enzymes to breakdown proteins, carbohydrates and fats.

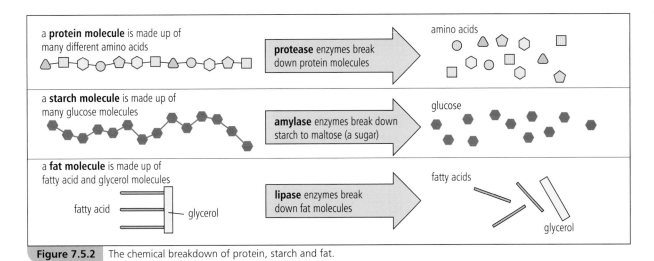

Figure 7.5.2 The chemical breakdown of protein, starch and fat.

The carbohydrase enzyme that breaks down starch is called **amylase**.

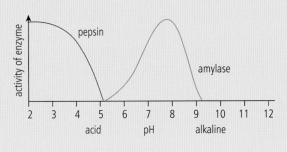

Teeth

The teeth carry out mechanical digestion when chewing food. This breaks down food into smaller pieces so making it easier to swallow. The surface area of food is increased so making it easier for enzymes to catalyse the reactions needed for chemical digestion.

Adult humans have 32 teeth, but not all of the teeth are the same. They have different shapes for performing different functions:

- incisors are chisel-shaped for biting and cutting
- canines are pointed for piercing and tearing
- premolars have uneven 'cusps' for grinding and chewing
- molars are like premolars and are for chewing up food.

Counting the number of each type of tooth, you should find that there are:

8 incisors, 4 canines, 8 premolars, and 12 molars, which makes 32 teeth in all.

During our lives we have two sets of teeth. The first set, or **milk teeth**, are small teeth and there are only 8 molars as our jaws are small. Between the ages of 6 and 12 these teeth gradually fall out, to be replaced by our **permanent teeth**. The last of our permanent teeth will come through when we are at least 18, if at all! These are our back molars or **wisdom teeth**.

A child usually has 20 milk teeth.

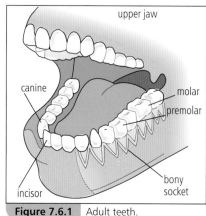

Figure 7.6.1 Adult teeth.

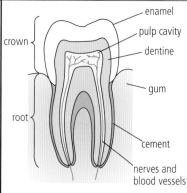

Figure 7.6.2 Vertical section through a molar.

Enamel forms the hard, outer layer of the crown of the tooth, which is the part above the gum. Inside this is softer **dentine**, which is more like bone in structure. A layer of **cement** fixes the root of the tooth into a bony socket in the jaw. The root is the part that is below the gum.

The **pulp cavity** is a space in the tooth containing nerves and blood vessels.

Healthy teeth and gums

Tooth decay is caused by bacteria in your mouth. They mix with saliva to form **plaque**, which is a layer that sticks to your teeth and gums.

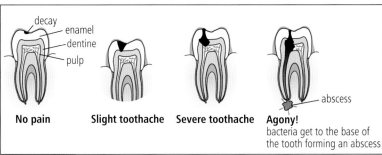

No pain   **Slight toothache**   **Severe toothache**   **Agony!**
bacteria get to the base of
the tooth forming an abscess

Figure 7.6.3   Stages in tooth decay.

After a meal, particles of sugary food may be left between your teeth. Bacteria in the plaque changes the sugar into acid because they are respiring anaerobically. The acid can attack the enamel on the surface of a tooth and this starts off tooth decay. When the enamel is worn away, acid will attack the dentine. If the cavity reaches the pulp cavity this will be very painful and cause severe toothache. The acid can also make your gums red and swollen. Figure 7.6.4 shows how our teeth should be brushed.

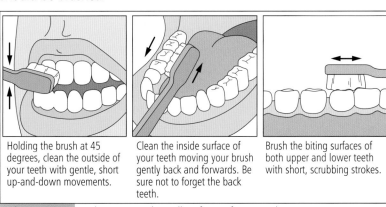

Holding the brush at 45 degrees, clean the outside of your teeth with gentle, short up-and-down movements.

Clean the inside surface of your teeth moving your brush gently back and forwards. Be sure not to forget the back teeth.

Brush the biting surfaces of both upper and lower teeth with short, scrubbing strokes.

Figure 7.6.4   Make sure you clean all surfaces of your teeth.

SUMMARY QUESTIONS

1 Copy and complete the sentences using these words:

dentine   crown   mechanical   socket
root   bite   enamel   chew   bone

The teeth are used to _____ and _____ your food. This is a type of _____ digestion. Each tooth consists of a _____ above the gum and a _____ which is fitted into a bony _____ . The _____ is the hard, outer layer of the tooth; inside this is the softer _____ which is similar to _____ .

2 a   Name the four main types of teeth.

   b   For each type in part **a**, describe its shape and its function.

   c   What is the difference between the milk teeth and the permanent teeth?

   d   Describe the stages of tooth decay.

   e   State three ways in which tooth decay can be prevented.

STUDY TIP

Note that fluoride strengthens the enamel of the teeth, giving them a hard exterior surface.

KEY POINTS

KEY POINTS

1 Incisors are for biting and cutting, canines for piercing and tearing, and premolars and molars are for grinding and chewing.

2 Tooth decay is caused by bacteria in plaque, which changes sugar to acid that attacks the enamel and dentine of the tooth.

3 Tooth decay can be avoided by regular brushing with a good toothbrush, avoiding sugary foods and visiting the dentist regularly.

Mouth, oesophagus and stomach

STUDY TIP

The first enzyme to act on our food is salivary amylase. Make notes on what happens to starch, protein and fat as they pass from the mouth to the end of the small intestine. List the enzymes that act on them, where the enzymes are produced and what they do.

Chewing and swallowing

Food and drink are ingested into the body through the mouth. When food enters the mouth, the incisor teeth and canine teeth bite it into chunks. The premolar and molar teeth grind these chunks of food into much smaller pieces. It is important to chew your food well as this makes digestion easier. The tongue mixes the food with **saliva** and the moistened food is chewed by the teeth. The food is then rolled into a ball or **bolus**.

The **salivary glands** make saliva, which contains:

- **mucus**, a slimy substance that lubricates the passage of the food bolus down the throat
- **amylase**, the enzyme that catalyses the breakdown of starch to maltose.

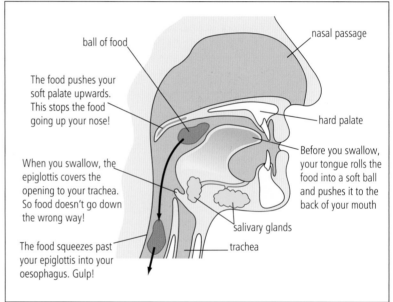

ball of food

nasal passage

The food pushes your soft palate upwards. This stops the food going up your nose!

hard palate

Before you swallow, your tongue rolls the food into a soft ball and pushes it to the back of your mouth

When you swallow, the epiglottis covers the opening to your trachea. So food doesn't go down the wrong way!

salivary glands

The food squeezes past your epiglottis into your oesophagus. Gulp!

trachea

Figure 7.7.1 Events during swallowing.

The oesophagus

Food passes down the **oesophagus** from the mouth to the stomach. The oesophagus (also known as the gullet) has circular and longitudinal muscles in its wall.

The movement of food down the oesophagus occurs by a wave of muscular contraction called **peristalsis**. The circular muscles contract and the longitudinal muscles relax behind the food bolus to push it along. In front of the bolus, the circular muscles relax and the longitudinal muscles contract to widen the oesophagus to allow the bolus of food to move along. Movement of food in the small and large intestines also occurs by peristalsis, though to a lesser extent.

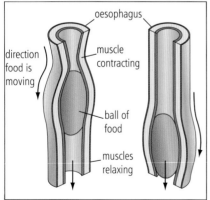

oesophagus

direction food is moving

muscle contracting

ball of food

muscles relaxing

Figure 7.7.2 Peristalsis in the oesophagus is a wave of contraction to pass food to the stomach.

The stomach

The stomach is a muscular bag that usually holds about 1 litre of food but can stretch to accommodate more. When food reaches the stomach a number of things happen to it.

The walls of the stomach make a digestive juice known as **gastric juice**. This contains the protease **pepsin** which starts the digestion of proteins to smaller molecules called polypeptides.

Gastric juices also contain **hydrochloric acid** which carries out two important functions due to its low pH.

- It kills any bacteria in the food. The low pH denatures the enzymes in any harmful microorganisms in the food.
- The acid pH of about 1.5 to 2.0 gives the optimum conditions for the action of the protease enzyme pepsin.

The mixture of food, gastric juice and hydrochloric acid is called **chyme**. The muscular walls of the stomach churn up the food making sure that it is mixed well with the juices. This muscular action is a type of mechanical digestion. After 2–3 hours of churning, the contents of the stomach are like a runny liquid.

A ring of muscle called the **pyloric sphincter**, opens to let the food pass a little at a time into the **duodenum** which is the first part of the small intestine.

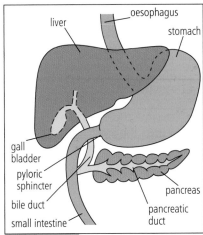

Figure 7.7.3 Upper part of the digestive system.

KEY POINTS

1 Food is chewed up by the teeth. It is then swallowed by the actions of the tongue, saliva, soft pallette and epiglottis.

2 Peristalsis involves a wave of muscular contractions that squeeze the food bolus, carrying it down the oesophagus to the stomach.

3 In the stomach, food is churned up and mixed with gastric juice and hydrochloric acid. This starts the digestion of proteins.

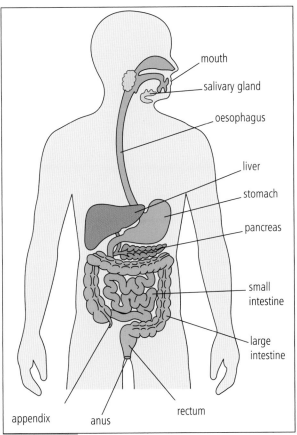

Figure 7.7.4 The digestive system.

SUMMARY QUESTIONS

1 With reference to the following, describe the events that take place when food is chewed and swallowed:

  a saliva   c the tongue

  b mucus   d the epiglottis

2 a Describe one function of the muscle layers in the stomach wall.

  b i Name the type of protease produced by the walls of the stomach.

    ii Describe the function of this type of enzyme.

  c Explain briefly why a different enzyme is required for the digestion of each type of food substance.

Small intestine and absorption

LEARNING OUTCOMES

- Describe the digestion of food in the small intestine, including the role of bile in emulsifying fats
- Define the term *absorption*
- Describe how the small intestine is adapted for efficient absorption of food

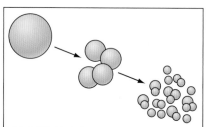

Figure 7.8.1 Large globules of fat are emulsified to smaller globules of fat to increase their surface area.

STUDY TIP

Emulsification increases the surface area of fats for lipase enzymes to act upon. Emulsification is not a type of chemical digestion as the fat is not changed chemically.

The small intestine

The small intestine is not really so small as it is about 6 metres long; however its diameter is less than that of the large intestine. The first part, after the stomach, is the **duodenum** that leads into the **ileum**.

The **pancreas** is connected to the duodenum by the **pancreatic duct**. **Pancreatic juice** flows down the duct to meet food arriving from the stomach. Pancreatic juice contains:

- **amylase** – breaks down starch to maltose
- **trypsin** – a protease that breaks down proteins and polypeptides to peptides
- **lipase** – breaks down fats to fatty acids and glycerol.

These enzymes do not work well in an acidic environment, so pancreatic juice contains the alkali, sodium hydrogencarbonate. This neutralises the acidic food which enters the duodenum from the stomach.

Bile also enters the duodenum along a tube called the **bile duct**. Bile is a yellow-green fluid made in the **liver** and stored in the **gall bladder**.

Bile is alkaline and also neutralises the acid which was added to the food in the stomach. This gives the best pH for enzymes in the small intestine to work.

Bile **emulsifies** fats by breaking down large globules of fats into smaller globules. These minute droplets have a much larger surface area over which the enzyme lipase can act to break down molecules of fat into fatty acids and glycerol.

Cells lining the ileum make enzymes that complete the digestion of food.

- Proteases which break down peptides to amino acids.
- Sucrase which breaks down sucrose to glucose and fructose.
- Maltose is broken down by maltase to glucose on the membranes of the epithelium lining the small intestine.

Absorption

Absorption is the movement of digested food molecules through the wall of the intestine into the blood or the lymph. Digested food includes simple sugars, amino acids, fatty acids and glycerol. These molecules pass through the wall of the small intestine either by diffusion or by active transport.

The small intestine is adapted for efficient absorption of food as it has:

- a very large surface area of about 9 square metres
- a thin lining (only one cell thick), so digested food can easily cross the wall into the blood and lymph.

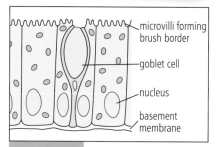

microvilli forming brush border

goblet cell

nucleus

basement membrane

Figure 7.8.2 Detailed structure of the epithelial cells lining the small intestine.

It is possible to fit such a large surface area into such a small space because the small intestine is very long (at least 6 metres in an adult). It has a folded inner lining with millions of tiny, finger-like projections called **villi** (singular: **villus**). The epithelial cells lining the villi have microscopic projections called **microvilli**. These vastly increase the absorptive area of the cell membrane of the epithelial cells (see Figure 7.8.2).

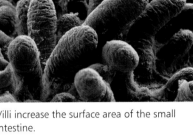

Villi increase the surface area of the small intestine.

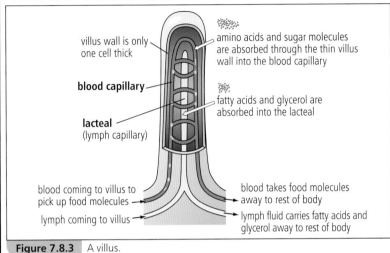

villus wall is only one cell thick

amino acids and sugar molecules are absorbed through the thin villus wall into the blood capillary

blood capillary

fatty acids and glycerol are absorbed into the lacteal

lacteal (lymph capillary)

blood coming to villus to pick up food molecules →

lymph coming to villus →

← blood takes food molecules away to rest of body

← lymph fluid carries fatty acids and glycerol away to rest of body

Figure 7.8.3 A villus.

The digested food reaches the capillaries and **lacteals** (lymph capillaries) in the villi. Absorbed food molecules are transported quickly to the liver by the **hepatic portal vein**. Fatty acids and glycerol are transported in the lymph. Movement of the gut empties the lacteal and the lymph moves slowly through lymphatic vessels, which are thin-walled like veins, eventually to enter into the blood near the heart. As a result, fat does not enter the bloodstream too quickly (see page 115).

Supplement

SUMMARY QUESTIONS

1 Match each part of the digestive system in the first column with its correct function in the second column:

oesophagus	makes bile
stomach	moves food from the stomach to the ileum
liver	makes pancreatic juice
gall bladder	carries food to the stomach by peristalsis
pancreas	mixes food with gastric juice
duodenum	stores bile

2 a State: i the components of bile, ii the organ where bile is produced, iii the organ where bile is stored.

   b Describe two ways in which bile aids the digestion of fats.

3 a How are conditions in the stomach different from the conditions in the small intestine?

   b State the composition of: i gastric juice, ii pancreatic juice.

KEY POINTS

1 Bile emulsifies fats in the duodenum. The smaller fat globules have a larger surface area upon which lipase can act.

2 Enzymes produced by the pancreas and the small intestine complete chemical digestion.

3 Absorption is the movement of digested food molecules through the wall of the intestine into the blood and lymph.

4 The small intestine is adapted for absorption by being very long, having a folded inner lining with millions of tiny villi.

Large intestine and intestinal disease

LEARNING OUTCOMES

- Describe the large intestine and its function
- Describe diarrhoea as a symptom of cholera
- Explain how cholera bacteria cause diarrhoea and dehydration

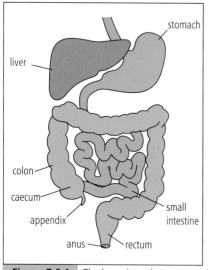

Figure 7.9.1 The large intestine.

Vibrio cholerae as seen in an electron microscope (×13 000).

The large intestine

The large intestine is about 1.5 metres long and can be divided into the caecum, the appendix, the colon and the rectum.

By the time food gets to the **colon** there is not much useful food left. It is now mainly fibre, dead cells, bacteria and some water, as most of it has been absorbed in the small intestine.

The small intestine absorbs about 5–10 dm³ of water every day. As food passes along the colon, more water is absorbed into the blood – about 0.3–0.5 dm³ per day.

The solid waste, or **faeces**, are stored in the **rectum**. Eventually the faeces are egested through the **anus**. The passing out of digestible food through the anus, in the form of faeces, is called **egestion**.

Cholera

Cholera is a water-borne disease caused by the bacterium *Vibrio cholerae*. The main symptom of cholera is diarrhoea as a result of dehydration. Diarrhoea is the loss of watery faeces from the body.

Some people infected with the cholera bacterium may not actually suffer from the disease and are called **symptomless carriers**.

Transmission

Cholera tends to occur in areas where there is a lack of proper sanitation, an unclean water supply or contaminated food. Water supplies become contaminated when infected people pass out large numbers of the bacteria in their faeces.

Cholera is transmitted by the ingestion of water or, more rarely, food that is contaminated by faecal material containing the bacteria.

Transmission can occur by drinking contaminated water, by infected people handling food or cooking utensils without washing their hands.

Effects on the body

Almost all cholera bacteria ingested by humans are killed by the acidic conditions in the stomach. However, some pathogens may survive and reach the small intestine. Here they burrow through the mucous lining the wall of the intestine and start to produce a toxin.

Molecules of this toxin enter the epithelial cells of the intestinal wall. Here they disrupt the functioning of the cell surface membrane of the epithelial cells. Chloride ions pass out of the cells into the space inside the intestine (the intestinal lumen). This accumulation of ions in the lumen creates a water potential gradient. Water flows by osmosis from the cells and from the blood into the lumen. Movement of ions and water from the blood into the lumen of the intestine cause the symptoms of cholera – dehydration and diarrhoea.

Oral rehydration therapy

To treat diarrhoeal diseases like cholera it is important to rehydrate infected people as soon as possible. Just giving them water is not effective because water is not absorbed by the intestine, and drinking water, or bottled water, does not contain the ions that are being lost. Cholera is treated by giving people a solution of salts and glucose to drink. This is **oral rehydration therapy** (ORT). An ORT solution contains:

* water to rehydrate the blood and other tissues
* sodium ions to replace the ions lost from blood and tissue fluid
* glucose to provide energy for the active uptake of sodium ions from the intestine
* other ions such as potassium and chloride to replace ions lost in diarrhoea.

During a cholera outbreak, sachets with a powder containing salts and glucose are made available. The contents of each sachet are made into a solution with boiling water. This takes little training for a person to administer and is given regularly and in large volumes until people recover.

When these ions are reabsorbed, the water potential of the epithelial cells decreases and water is absorbed to make up for the water that was lost in diarrhoea.

Cholera is most easily transmitted where there is a lack of clean water and sanitation is poor or non-existent.

A child receives oral rehydration therapy during the cholera epidemic in Haiti in 2010.

SUMMARY QUESTIONS

1 a What sort of organism causes cholera?

  b How does the organism enter the body?

  c What are the symptoms of cholera?

2 a What sort of conditions encourage the spread of cholera?

  b How can effective treatment of sewage and the chlorination of drinking water help control the spread of cholera?

  c Why are these measures not put into place in some developing countries?

3 Explain in detail how the cholera bacterium brings about dehydration, diarrhoea and loss of salts from the blood in the intestine.

KEY POINTS

1 Diarrhoea is the loss of watery faeces from the intestine.

2 Cholera is a disease caused by a bacterium.

3 Diarrhoea caused by cholera can be treated by using oral rehydration therapy.

4 The cholera bacterium produces a toxin that causes chloride ions to pass into the intestine. This causes water to pass out the cells by osmosis, resulting in diarrhoea and dehydration loss of ions from the blood.

1 The breakdown of large, insoluble food molecules into small, water-soluble molecules is:

A absorption   **B** digestion

C egestion   **D** ingestion

(Paper 1)   *[1]*

2 Which is a correct description of the action of a protease?

A breaks down fat into fatty acids and glycerol

B breaks down protein into amino acids

C breaks down starch into glucose

D converts amino acids into proteins

(Paper 1)   *[1]*

3 The drawing shows a vertical section through an incisor tooth.

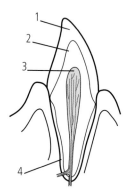

Which row is the correct identification of the parts shown in the drawing above?

	1	2	3	4
A	cement	pulp	enamel	dentine
B	dentine	enamel	pulp	cement
C	enamel	dentine	pulp	cement
D	pulp	cement	dentine	enamel

(Paper 1)   *[1]*

4 Diarrhoea is a symptom of

A anaemia   **B** cholera

C rickets   **D** scurvy

(Paper 1)   *[1]*

5 Which row of the table (top of next column) gives the deficiency symptoms for vitamin C, vitamin D, calcium and iron?

	vitamin C	vitamin D	calcium	iron
A	gums bleed	bowing of leg bones	muscle weakness and cramps	tiredness
B	bowing of leg bones	gums bleed	tiredness	muscle weakness and cramps
C	scurvy	rickets	rickets	anaemia
D	tiredness	muscle weakness	soft bones	gums bleed

(Paper 2)   *[1]*

6 The drawing shows part of the digestive system.

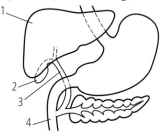

What is the function of the liquid produced by part 1, stored in part 2 and released into part 3?

A to digest proteins to amino acids

B to increase the surface area of fat droplets

C to acidify the contents of part 4

D to digest fats to fatty acids and glycerol

(Paper 2)   *[1]*

7 The following events occur during the passage of starch through the alimentary canal.

1 Amylase is secreted in saliva from salivary glands.

2 Maltase is secreted onto the surface of epithelial cells in the small intestine.

3 Maltase catalyses the breakdown of maltose to glucose.

4 Glucose enters the blood plasma.

5 Glucose is absorbed by active transport into the epithelial cell.

6 Amylase catalyses the breakdown of starch to maltose.

Which is the correct sequence for these events?

A 1,6,2,3,5,4   **B** 6,1,3,2,4,5

C 2,3,1,6,5,4   **D** 3,2,6,1,4,5

(Paper 2)   *[1]*

8 Which feature of a villus increases its surface area for absorption?

   A capillaries   **B** goblet cells

   C lacteal   **D** microvilli

(Paper 2) [1]

9 Copy the diagram from question 6.

  (a) On the diagram label the following: oesophagus, stomach, pancreas and duodenum. [4]

  (b) Draw arrows on the diagram to show the direction taken by food as it passes through the parts of the alimentary canal on the diagram. [2]

  (c) On the diagram use label lines and the following letters to show:

    E   the site of secretion of hydrochloric acid

    F   an organ that secretes molecules of a protease

    G a place with an alkaline pH

    H the organ that stores bile [4]

(Paper 3)

10 A balanced diet is essential for good health.

  (a) State what components should be provided in a balanced diet. [4]

  (b) It is important that a woman's diet should include an increased quantity of certain components when she is breast-feeding. State three components that should be increased and explain why it is important that they should be increased. [6]

(Paper 3)

11 The liver performs important functions in the digestion and assimilation of food. One of these is the production of an alkaline digestive juice that breaks down large globules of fat.

  (a) (i) Name the digestive juice secreted by the liver. [1]

    (ii) Explain how this digestive juice reaches the food in the small intestine. [3]

    (iii) Explain why this digestive juice is alkaline. [2]

    (iv) Outline how this digestive juice breaks down large globules of fat. [4]

  (b) Explain what is meant by *assimilation*. [2]

(Paper 4)

12 The diagram shows a villus from the alimentary canal.

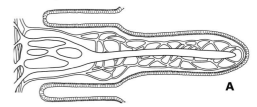

  (a) Copy the diagram and label the capillaries and the lacteal. [2]

  (b) State the part of the alimentary canal where villi are located. [1]

  (c) Name three substances that pass from **A** on the diagram into the villus. [3]

  (d) With reference to the diagram, explain the significance of villi in the alimentary canal. [6]

(Paper 4)

13 A meal contains bread and cheese. Bread contains starch, and cheese contains protein.

  (a) Describe how starch is digested in the alimentary canal. In your answer state where the changes that you describe take place. [5]

    The diagram shows a small protein molecule and the places along the molecule where proteases **1** and **2** act on the molecule to help break it down.

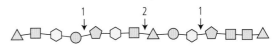

  (b) How many fragments will be produced after the molecule has been digested by the two proteases? [1]

  (c) Suggest why protease 2 cannot act on the protein in the same places as protease 1. [4]

  (d) State the organs in the alimentary canal where proteases digest protein. [2]

  (e) The fragments produced by digestion by proteases 1 and 2 are not absorbed into the blood. Explain why this is so. [2]

  (f) Describe what happens to these fragments so that the protein is completely broken down. [2]

(Paper 4)

Eucalyptus regnans is the tallest flowering plant, growing up to 90 metres.

Most living organisms need a transport system. Only very small organisms, such as bacteria, protoctists and some worms, do not have transport systems. Their bodies are so small, or so flat, that they can rely on diffusion alone to obtain oxygen and remove carbon dioxide.

Vertebrates and flowering plants are large organisms and have transport systems to carry substances, such as food and water, throughout their bodies. These transport systems move fluids through tubes so that all of the fluid moves in the same direction within each tube. This type of transport is called **mass flow**.

Vertebrates have a circulatory system consisting of a pump (the heart) and blood vessels. Flowering plants have two separate transport systems: **xylem** and **phloem**.

Xylem and phloem

Xylem and phloem are plant tissues composed of cells that are specialised for transport (see Topic 2.3). These tissues are found throughout the plant body in roots, stems and leaves.

Xylem tissue transports water and mineral ions. The roots absorb the water and mineral ions from the soil. These enter the xylem in the root and travel upwards in the stem to the leaves, flowers and fruits.

Transport in the xylem, unlike in the phloem, is in one direction only – from roots via stem to leaves.

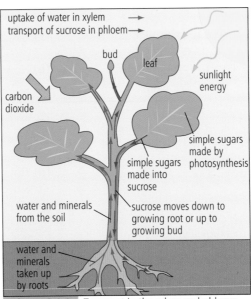

Figure 8.1.1   Transport in the xylem and phloem.

Phloem tissue transports sucrose, amino acids and hormones throughout the plant. Sucrose is a soluble, complex sugar which is made especially for transporting energy. It is made in leaves with the sugars from photosynthesis and from starch in storage organs, such as swollen roots and stems.

Substances are transported in the phloem in two directions: downwards from leaves to roots and upwards from leaves to flowers, fruits and buds; also from storage organs to new stems and leaves.

Hormones control cell division for growth of the stem, roots and leaves. They also control the growth of flowers and fruits (see page 170).

Inside the root and stem

To find the transport tissues of roots and stems, we cut across them to give transverse sections.

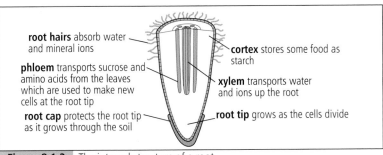

root hairs absorb water and mineral ions

cortex stores some food as starch

phloem transports sucrose and amino acids from the leaves which are used to make new cells at the root tip

xylem transports water and ions up the root

root cap protects the root tip as it grows through the soil

root tip grows as the cells divide

Figure 8.1.2   The internal structure of a root.

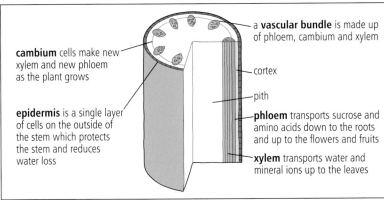

cambium cells make new xylem and new phloem as the plant grows

a **vascular bundle** is made up of phloem, cambium and xylem

cortex

pith

epidermis is a single layer of cells on the outside of the stem which protects the stem and reduces water loss

phloem transports sucrose and amino acids down to the roots and up to the flowers and fruits

xylem transports water and mineral ions up to the leaves

Figure 8.1.3   The internal structure of a stem.

See Topic 6.6 for the positions of xylem and phloem in the leaves.

This celery was left standing in a red dye for several hours. The red strands are the xylem vessels filled with dye.

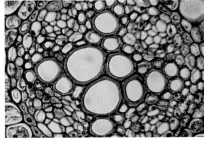

The tissue in the centre of this root is xylem, which is stained red. The green areas between the xylem are phloem tissue.

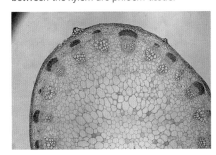

A transverse section through a young stem. Note the vascular bundles.

SUMMARY QUESTIONS

1 Copy and complete the sentences using these words:

**leaves   sucrose   water   amino   blood vessels
mineral ions   fruits   heart   storage**

Vertebrates have a transport system made up of a _____ and _____. Flowering plants have two transport systems. Xylem tissue transports _____ and _____ up the stem from the roots to the _____, flowers and _____. Phloem tissue transports _____ and _____ acids from leaves to all parts of the plant, such as _____ organs.

2 Using the diagrams above, state the functions of each of these issues:

a   phloem   b   xylem   c   cambium   d   epidermis
e   root cap   f   root cortex

KEY POINTS

1 Xylem vessels transport water and mineral ions from the roots up the stem to the leaves.

2 Phloem vessels transport sucrose, made in the leaves during photosynthesis and in storage organs, to other parts of the plant.

3 Xylem and phloem are distributed in a central core inside a root. Inside a stem they are organised into vascular bundles.

Water uptake

Water uptake by roots

The main functions of roots are to anchor the plant in the soil and to take up water and mineral ions. Just behind the root tip are small **root hairs**. Water passes into the root hairs by **osmosis**.

Soil water is a dilute solution of various solutes including mineral ions, such as nitrate and potassium. Root hairs have thin, permeable cell walls and provide a large surface area to absorb water. The cell sap in the root hair cells is a more concentrated solution than soil water. This is because root hair cells also absorb mineral ions and have other solutes such as sugars. The cell membrane is partially permeable so water diffuses from the soil into the root hair cells by osmosis.

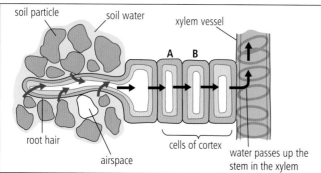

Figure 8.2.1 Absorption of water by a root.

Water then passes across the cortex of the root. Most of the water moves through the cell walls and in the spaces between the cells. Some moves from cell to cell. Water eventually reaches the xylem in the centre of the root. From here water moves up the xylem through the stem and to the leaves where it enters the spongy mesophyll cells. Much of the water enters the cell walls, evaporates to form water vapour and then diffuses through stomata to the atmosphere.

We describe the movement of water into a root hair, across the cortex and into a xylem vessel in terms of **water potential** (see Topic 3.2). Water passes *down* a **water potential gradient**, from a high water potential in the solution in the soil to a lower water potential in the root hair cell. Water in turn, passes from the root hair cell to a lower water potential in the cells of the cortex and eventually to an even lower water potential in the xylem vessel.

PRACTICAL

Looking at root hairs

Look closely at a section of a young root with a hand lens and then with a microscope.

Look at the root hairs.

Root hairs are long and thin. You can also look at root hairs on a germinating seedling such as a bean or a radish.

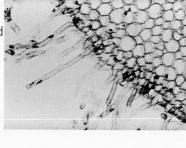

They have a large surface area through which water is absorbed by osmosis and mineral ions are absorbed by active transport.

Root hairs of a germinating bean seed.

Xylem vessels are like pipes as they are empty except for the water that fills them. Water is not *pushed* from the roots up to the leaves, instead it is *pulled* by the evaporation of water.

Transpiration pull

Transpiration is the evaporation of water from the surfaces of mesophyll cells in the leaves and the loss of water vapour to the atmosphere. Water vapour can diffuse out of the leaf when the stomata are open.

Water is 'pulled' up the xylem in the stem from the roots to the leaves by **transpiration pull**.

The water is used for photosynthesis and to stop the plant from wilting. As water is used up or lost from the leaves, more is sucked up from the xylem vessels, rather like water being sucked up a straw.

This mass flow of water up the xylem relies on two properties of water:

- **cohesion** – the water molecules tend to attract each other, sticking together
- **adhesion** – the water molecules also tend to stick to the inside of the xylem vessel.

So there is a continuous flow of water from the roots to the leaves. This movement of water up the xylem is called the **transpiration stream**.

SUMMARY QUESTIONS

1 Copy and complete the sentences using these words:

**stomata   osmosis   large   ions
transpiration   hair   area**

Water passes into a root _____ cell by _____. Mineral _____ are also taken up. A root hair cell has a _____ surface _____ for taking up water and mineral ions. Water is lost from the leaves by _____. This is controlled by the opening and closing of the _____ .

2 Look at the diagram of the root hair in the soil:

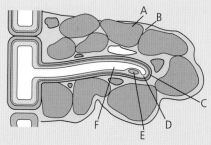

a Name the parts labelled A to F.
b Explain how water gets into a root hair cell from the soil.
c Give two other functions of root hair cells.

3 Explain, using the term *water potential*, how water passes into root hair cells, across the cells of the cortex, into the xylem vessels and up to the leaves, where it is lost to the atmosphere.

KEY POINTS

1 Water enters a root hair cell by osmosis. It then passes across the cells of the root cortex by osmosis before passing into the xylem and then up the stem to the leaves.

2 Root hairs are well adapted for the absorption of water since they have thin cell walls and have a large surface area.

3 Loss of water from the leaves 'pulls' more water up the stem from the roots in the transpiration stream.

Supplement

LEARNING OUTCOMES

- Define the term *transpiration*
- Describe how water vapour loss is related to cell surfaces, air spaces and stomata
- Describe how variations in temperature, humidity and light intensity can affect transpiration rate
- Describe how wilting occurs

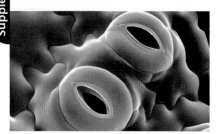

Two open stomata.

STUDY TIP

When water evaporates it forms water <u>vapour</u>. Do not refer to it as just 'water'.

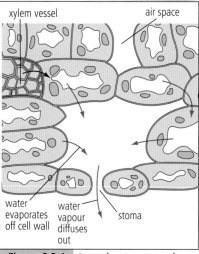

xylem vessel — air space

water evaporates off cell wall — water vapour diffuses out — stoma

Figure 8.3.1 Loss of water vapour from a leaf.

Plants absorb large quantities of water from the soil. Much of this water passes into roots, moves up through stems and enters leaves where it is lost to the atmosphere as water vapour. If you look at a diagram of a leaf you will see that there are many air spaces inside. These are lined by mesophyll cells which have damp cell walls. Water evaporates from this huge internal surface so the air spaces become saturated with water vapour.

There is usually more water vapour in the air spaces than there is in the outside air, so water vapour diffuses through the stomata into the air. The water lost from the spongy mesophyll cells is replaced by more water from the xylem.

Transpiration is the loss of water vapour from plant leaves by the evaporation of water at the surfaces of the mesophyll cells, followed by the diffusion of water vapour through the stomata.

Wilting

More transpiration takes place during the day than at night because the stomata are open during the day and closed at night. The stomata open during the day so that carbon dioxide can diffuse into the leaf. Carbon dioxide is a raw material for photosynthesis and diffuses into chloroplasts in mesophyll cells.

The stomata close at night to reduce the volume of water lost by transpiration. They may also close in hot, dry conditions during the day as water lost in transpiration is not being replaced by water from the soil.

The stomata close up to reduce transpiration. If the plant still does not get enough water it will start to **wilt** (see above on the right). Its cells have lost so much water that they are no longer turgid or full of water. Turgid cells are firm and give the plant support. If the cells become flaccid then the plant becomes soft. The stem is no longer upright and the leaves droop.

Wilting is not a bad thing. The leaves move downwards so are out of the direct rays of the Sun so do not get as hot. When the temperature decreases and they can absorb more water than is lost by transpiration the leaves will recover.

Factors affecting transpiration

Other factors affecting transpiration are environmental. You can see the effects of two of these factors in Figure 8.3.2.

As **light** intensity increases, the rate of transpiration increases up to a maximum that is determined by the other conditions, such as the humidity of the air and temperature.

Humidity is a measure of the quantity of water vapour in the air. In very dry conditions when the humidity is low, the rate of transpiration is high. As the air becomes more and more humid, the rate of transpiration decreases.

The rate of transpiration increases as the air **temperature** increases.

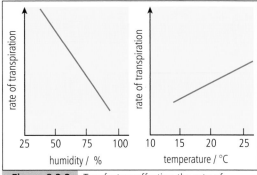

Figure 8.3.2 Two factors affecting the rate of transpiration.

Explanations

Light causes stomata to open. As the light intensity increases, stomata open wider. As light intensity decreases when it is cloudy or towards evening, stomata close so that less water is lost by transpiration. In **humid** conditions, there may be nearly as much water vapour in the atmosphere as in the air inside the leaves. This means that there is no concentration gradient for water vapour so the rate of transpiration is low. In **dry** conditions the concentration gradient is very steep so water vapour diffuses out of the leaves through the stomata. Temperature influences the rate of evaporation of water from surfaces, such as the cells inside the leaf. As temperature increases, water molecules on the cell surfaces have more kinetic energy and enter the air inside the leaf as water vapour.

Measuring transpiration

It is not easy to measure the rate of transpiration but you can use a **potometer** to measure the **rate of water uptake**. The volume of water lost is slightly less than the volume of water taken in by the roots. This is because some of the water is used up in photosynthesis and in keeping cells turgid.

The potometer is filled by submerging it in water. It is set up as shown, making sure the stem in the stopper is air-tight. An air bubble is allowed to form in the capillary tube. The distance moved by the bubble is a measure of the rate of water uptake.

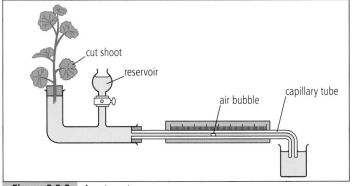

Figure 8.3.3 A potometer.

Translocation

LEARNING OUTCOMES

- Define the term *translocation*
- State that sucrose and amino acids move in phloem from a source to a sink
- Explain that some parts of a plant may act as a source and a sink at different times during the life of a plant
- Compare transpiration with translocation

Food is made in the leaves by photosynthesis. The soluble products include sucrose, amino acids and fatty acids. These are carried to all parts of the plant in solution in the **phloem**. This transport is called **translocation** which means 'from place to place'.

The transport of food takes place from regions of production (the leaves) to the regions of storage or to regions where respiration or growth take place. The simple sugars produced by photosynthesis are converted in the leaves into sucrose. This is transported to parts of the plant where growth or storage occur to provide energy. Sucrose is:

- broken down by an enzyme to give simple sugars that are used in respiration
- changed to starch for storage in the root cortex and in seeds
- used to make cellulose for new cell walls at the growing root tip and shoot tip
- stored in some fruits to make them sweet and attract animals.

Leaves use the simple sugars from photosynthesis and nitrate ions to make amino acids. These are needed all over the plant to make proteins, especially where new cells are made in shoot and root tips. Leaves also make fatty acids that are stored in many seeds as oils and are also needed for the waxy cuticle on the upper surface of the leaves.

> **DID YOU KNOW?**
>
> Maple syrup is prepared from phloem sap of maple trees. Farmers drill holes into the tree and insert tubes through which the sap flows and collects in buckets.

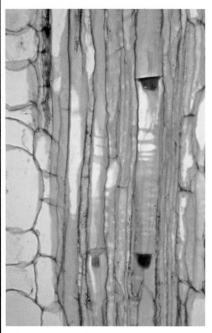

Phloem tubes are living tissues.

Aphids and phloem transport

Aphids are small insects that feed on the sap in the phloem of plants. They have piercing mouthparts called **stylets** which can be inserted through the surface of the stem into a single phloem tube. The pressure of the sap inside the phloem tube causes it to pass along the stylet and so provide the aphid with food.

Aphids can be used to study the transport of sugars in the phloem. If the stylet of a feeding aphid is cut, the sugary sap will pass out of the stylet for some time.

Isotopically-labelled sucrose can be injected into a leaf and samples collected from aphid stylets, that have been inserted into the stem above and below the leaf. The contents of the sap can be analysed for the isotope and the passage of the sucrose traced.

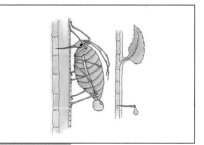

| **Figure 8.4.1** | Aphids are crop pests. They feed on phloem sap by using their piercing mouthparts. |

Transpiration and translocation

In plant transport, the part of the plant where a substance begins its journey is called a **source**. The part where it ends its journey is a **sink**.

Water and ions are absorbed by roots (the source) and travel to leaves, flowers and fruits (the sinks). Xylem vessels are columns (or pipes) of dead, empty cells. The movement of water in xylem vessels is *passive*, since it relies upon the evaporation of water vapour from the leaves producing a tension in the xylem. This pulls up water from the roots to give the transpiration stream. Transpiration is greatest on hot, dry and windy days.

Translocation is an *active* process which occurs in phloem. The leaves are the source of food substances and the sinks are the respiring plant tissues, regions of growth (root and shoot tips) and regions of storage such as the root cortex and the seeds. Phloem tubes are living cells that contain some cytoplasm. Movement in the phloem requires active transport of sucrose at the source. Water enters the phloem tubes to build up a head of pressure that forces the phloem sap to the sinks. Translocation is most active on sunny, warm days when plants are producing increased levels of sugar.

Black bean aphids. When there is plenty of food and conditions are favourable, aphids reproduce very rapidly. Farmers control them by spraying insecticide.

STUDY TIP

Movement of water and ions in the xylem is only one way – upwards from the roots. The transport of substances in the phloem is in both directions. In the growing season it is downwards from leaves to roots and upwards from leaves to stem tips, where growth occurs, and to flowers, fruits and seeds. Remember this when answering Summary question 3b.

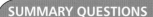

SUMMARY QUESTIONS

1 a Define the term *translocation*.

  b What substances are translocated in the phloem?

  c What are the materials transported in the phloem used for?

2 Draw a table to compare transpiration with translocation.

3 Substances are transported in plants from sources to sinks in the xylem and phloem.

	substances transported	source of substances in the plant	sink for substances in the plant
transport in the xylem			
transport in the phloem (translocation)			

  a Copy and complete the table by giving two substances that are transported in each tissue and one source and one sink in each case.

  b Explain how a root tuber can be a source and a sink at different times of the year.

4 Aphids feed on plants.

  a How are they used to investigate movement in phloem?

  b Suggest why farmers and growers control aphids by spraying insecticides.

KEY POINTS

1 Translocation is the movement of sucrose and amino acids from the leaves to regions of respiration, growth and storage.

2 Regions of production are called sources and regions of storage, respiration and growth are called sinks.

3 Transpiration is a passive process involving dead tissue. Translocation is an active process involving living tissue.

1 In which conditions would a plant lose most water by transpiration?

  A cool and dry   **C** hot and dry

  B cool and humid   **D** hot and humid

(Paper 1)   *[1]*

2 Which row best describes the substances transported in the phloem and xylem?

	phloem	xylem
A	amino acids and ions	sucrose and water
B	amino acids and sucrose	ions and water
C	sucrose and water	amino acids and ions
D	glucose and sucrose	ions and water

(Paper 1)   *[1]*

3 Which is the correct pathway taken by a molecule of water as it moves through a plant?

  A root cortex, xylem in root, xylem in stem, leaf mesophyll, atmosphere

  B root hair cell, root cortex, xylem, leaf mesophyll cells, air spaces in mesophyll, stoma

  C stoma, air spaces in leaf, xylem, root cortex, root hair cell

  D xylem, root cortex, leaf mesophyll cell, stoma, atmosphere

(Paper 1)   *[1]*

4 Which absorbs water from the soil?

  A cuticle   **C** root hairs

  B mesophyll   **D** xylem vessel

(Paper 1)   *[1]*

5 A sweet potato plant grows in a field in bright sunlight. Which row shows the correct movement of carbohydrate in the phloem?

	type of carbohydrate	source of carbohydrate	destination
A	glucose	leaves	tubers
B	glucose	tubers	leaves
C	sucrose	leaves	tubers
D	sucrose	tubers	leaves

(Paper 2)   *[1]*

6 A potted plant has wilted after being left without water for 6 days. Which combination of environmental conditions is likely to have caused it to wilt?

	light intensity	humidity	air temperature
A	high	high	low
B	low	low	high
C	high	low	high
D	low	high	low

(Paper 2)   *[1]*

7 Which statement explains the movement of water from the stem into leaf cells?

Water moves from:

  A mesophyll down a water potential gradient

  B phloem down a water potential gradient

  C xylem down a water potential gradient

  D xylem from low water potential to high water potential

(Paper 2)   *[1]*

8 Which statement best describes the pathway and the mechanism by which water is lost from leaves?

  A Diffusion of water vapour from intercellular air spaces through stomata into the atmosphere.

  B Evaporation of water and diffusion of water from mesophyll cells through stomata into the atmosphere.

  C Evaporation of water from mesophyll cells and diffusion of water vapour through the cuticle into the atmosphere.

  D Evaporation of water from mesophyll cells and diffusion of water vapour from air spaces through stomata into the atmosphere.

(Paper 2)   *[1]*

9 The diagrams show transverse sections of **A** root, **B** stem and **C** leaf.

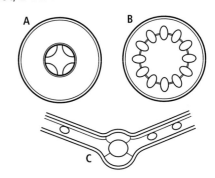

(a) Copy the diagrams and shade the areas that represent the positions of xylem and phloem in **A**, **B** and **C**. *[3]*

(b) Copy and complete the table to compare the functions of xylem and phloem. The term *sap* refers to the liquid transported in these two tissues.

feature	phloem	xylem
composition of the sap		
direction of flow in the stem		
destinations		

[6]

(c) Describe how you would discover the pathway taken by water as it travels through the parts of the plant that are above ground. *[3]*

(Paper 3)

10 A potometer (see Topic 8.3) was used to measure water uptake by a leafy shoot.

	conditions	time taken for the water to move 100 mm / min
A	cool, humid air in daylight	2
B	cool, dry air in daylight	6
C	warm, dry air, in daylight	1
D	warm, humid air, in daylight	3
E	warm, humid air, at night	60

(a) Calculate the rate of water movement in $mm\,min^{-1}$ for each of the conditions, **A** to **E**, and present your answers in a table. *[3]*

(b) Draw a bar chart to show the effect of the five conditions on the rate of water movement. *[5]*

(c) From the results, state three environmental conditions that affect the rate of water movement through the plant. *[3]*

(d) State two ways in which the air around the shoot would be affected if it was covered with a transparent plastic bag. *[2]*

(Paper 3)

11 A leafy shoot was cut from a plant and placed into a potometer (see Topic 8.3) which was placed on a balance. The table shows the rate of water uptake and the rate of transpiration by the leafy shoot during a hot, dry day.

time / h	rate of water loss / $g\,h^{-1}$	rate of water absorption / $g\,h^{-1}$
midnight	4	7
0300	6	9
0600	12	12
0900	18	14
1200	24	17
1500	22	19
1800	18	20
2100	8	14
midnight	6	10

(a) Plot a graph to show the data in the table. *[5]*

(b) Calculate the percentage change in the rate of water loss between 0600 hours and 1200 hours. Show your working. *[2]*

(c) Describe the changes in water loss and water absorption over the 24-hour period. *[4]*

(d) Explain how water moves from the cut end of the stem to the atmosphere. *[4]*

(Paper 4)

LEARNING OUTCOMES

- Describe the one-way flow of blood around the body
- Describe the heart as a pump for the flow of blood and the role of valves to permit only one-way flow
- Describe the single circulation of a fish
- Describe the double circulation of a mammal and explain its advantages

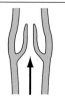

Valves open to let the blood flow towards the heart

Valves close to stop blood flowing backwards

Figure 9.1.1   How semi-lunar valves permit one-way flow of blood.

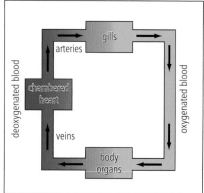

Figure 9.1.2   Fish have a single circulation in which blood flows through the heart once in a circuit around the body.

The circulatory system

Most animals have a circulatory system that consists of blood, some sort of pump and blood vessels. Circulatory systems in animals transport useful substances such as oxygen absorbed from a gas exchange surface and nutrients from the part of the gut that is adapted for absorption. Circulatory systems also transport waste, such as carbon dioxide, from all over the body to the gas exchange surface.

Blood circulation

The **heart** is a pump which circulates blood through blood vessels.

Blood flows in **arteries** away from the heart to the different organs of the body. Blood flows back to the heart in **veins**. The smallest blood vessels that connect arteries to veins are **capillaries**. Cells are very close to capillaries so they receive a good supply of oxygen and nutrients from these capillaries and can have their carbon dioxide and their other wastes removed efficiently.

The circulatory system allows a **one-way flow of blood** around the body. The heart pumps blood giving it pressure so that it flows inside arteries and this helps to maintain a one-way flow. This is good for getting blood to the capillaries, but high pressure blood will damage the delicate capillaries so small muscular blood vessels known as arterioles reduce the pressure before the blood enters capillaries (you can find out more about arterioles on page 107). When blood leaves capillaries the blood pressure is even lower.

There are **semi-lunar valves** in veins to make sure blood does not flow backwards away from the heart (see Figure 9.1.1). If this happens, veins swell and the blood is not circulated properly. Each valve has three of these pockets. The valves open when the pressure of the blood pushes against them, but they close when blood flows back to fill the pockets.

There are semi-lunar valves in the heart at the places where blood leaves the heart chambers to enter arteries. There are also valves between the chambers of the heart (see page 102).

A single circulation

Fish have a single circulation because blood flows through the heart once during a complete circuit of the body (see Figure 9.1.2). Blood from the organs flows into the heart from veins. This blood has little oxygen it and is called **deoxygenated blood**. The heart pumps blood into an artery that takes it to the gills to be oxygenated. The blood flows on from the gills in arteries to the body organs.

Supplement

A double circulation

Mammals have a double circulation. Blood flows through the heart twice during one complete circuit around the body. You can confirm this by following the pathway that blood takes around Figure 9.1.3.

The mammalian heart is divided into two halves: right and left. The septum is a thick wall of muscle that separates the two halves of the heart that stops blood in the right side mixing with blood in the left side.

Blood circulation has two functions:

• The right side of the heart pumps deoxygenated blood to the lungs and back to the heart again. The pressure required to force blood to the lungs is not very high since there is little resistance to flow in the lungs as they are a spongy tissue filled with air. Gas exchange occurs as blood flows through capillaries in the lungs. Blood absorbs oxygen and loses carbon dioxide. The **oxygenated blood** returns to the heart in a vein. This blood is bright red in colour and is shown in red on diagrams of the circulation.

• The left side of the heart pumps oxygenated blood to the rest of the body and back to the heart again. The pressure of blood leaving the left side is much greater than on the right side as there is much more resistance to flow than there is through the lungs. Gas exchange occurs as blood flows through capillaries in organs such as muscles, the gut, liver and kidneys. Oxygen leaves the blood and carbon dioxide enters. The deoxygenated blood that leaves capillaries and flows through veins is dark red in colour, but is always shown as blue.

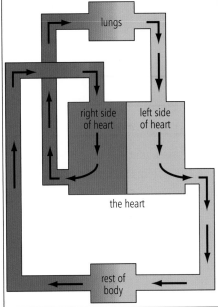

Figure 9.1.3 Mammals have a double circulation in which blood flows through the heart twice in a complete circuit of the body.

KEY POINTS

1 The heart and valves ensure a one-way flow of blood around the circulatory system.

2 Fish have a single circulation, so blood passes through the heart once during one circuit of the body.

3 Blood flows twice through the heart during one circuit of the body in a mammal.

4 The heart pumps blood at low pressure to the lungs and at high pressure to the rest of the body in a double circulation.

5 Blood flows in arteries away from the heart. Blood flows in veins towards the heart.

SUMMARY QUESTIONS

1 Copy and complete the sentences using these words (some may be used twice):

**oxygenated   body tissues   left   lungs   carbon dioxide
veins   deoxygenated   right   oxygen**

The right side of the mammalian heart pumps blood to the _____ . Here it absorbs _____ and is now called _____ blood.

The gas _____ is removed from the blood in the lungs. The blood returns to the _____ side of the heart and from here it is pumped to the _____ where it gives up its _____ and is now called _____ blood. It passes back to the _____ side of the heart in _____ .

2 Explain what is meant by the terms *single circulation* and *double circulation*.

3 Explain the following statements about the mammalian heart.
   a   Blood on the right side of the heart is at low pressure.
   b   Blood on the left side of the heart is at high pressure.
   c   Veins contain semi-lunar valves.

The heart

LEARNING OUTCOMES

- Describe the four chambers of the heart and locate the major blood vessels and valves associated with the heart
- State the sequence of events that take place during one heartbeat
- State that blood is pumped away from the heart into arteries and returns to the heart in veins
- Explain the importance of the septum in separating oxygenated and deoxygenated blood
- Locate and name the valves in the heart
- Explain why the walls of the ventricles are thicker than those of the atria and why the left ventricle is thicker than the right

STUDY TIP

When you look at a diagram or drawing of the heart you are viewing it from the front of the body – so the left side of the heart is on the right side of the drawing.

DID YOU KNOW?

Ibn Al-Nafis was an Arab physician of the 13th century. He was the first person to connect the functions of the heart and lungs. He suggested that blood was purified in the lungs.

Heart structure

The human heart consists of two pumps lying side by side. The right side of the heart is one pump and the left side of the heart is another pump. The heart consists almost entirely of cardiac muscle tissue (see page 24).

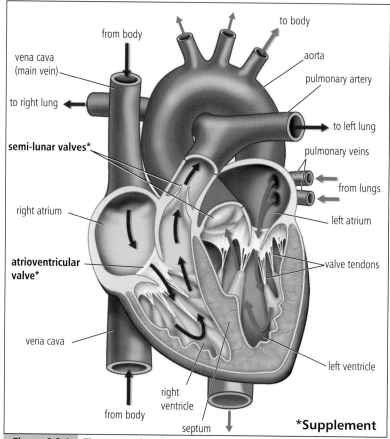

Figure 9.2.1 The arrows show how the blood moves through the heart.

The **septum** (wall of tissue) prevents deoxygenated blood on the right side mixing with oxygenated blood on the left side.

On each side of the heart there are two chambers. The upper chambers are called **atria** (singular: **atrium**). Blood empties into them from veins. When the atria contract they pump blood into **ventricles**, which have much more muscular walls than the atria. When the ventricles contract they pump the blood out into arteries at higher pressure.

Between each atrium and ventricle is a one-way **valve**. The valves prevent the blood flowing back into the right and left atrium when the right and left ventricles contract. The ventricles have more muscular walls than the atria because they have to pump the blood much further than the atria – either to the lungs or to the tissues of the body.

The left ventricle has a more muscular wall than the right ventricle because it has to pump blood around the body and has to overcome more resistance to flow.

Heart action

The heart pumps blood when its muscles contract. When the muscles contract, the chamber gets smaller and squeezes the blood out. After each chamber contracts it relaxes so it fills up with blood again.

The two sides of the heart work together. The atria contract and relax at the same time. The ventricles contract and relax at the same time.

During the relaxation phase, blood flows into the atria from the veins.

During the contraction phase:

• The atria contract and force blood into the ventricles. The valves between the atria and ventricles open due to the pressure of blood against them.

• Then the ventricles contract to force blood out into the arteries. The valves close to prevent blood flowing back into the atria.

The **right ventricle** pumps blood to the **lungs** in the **pulmonary artery**. The **left ventricle** pumps blood to the rest of the body in the **aorta** (main artery).

The semi-lunar valves occur at the base of the aorta and pulmonary arteries. They act like the ones in Figure 9.1.1. During the relaxation phase they shut, preventing back flow of blood into the ventricles. During the contraction phase they open allowing blood to leave to the lungs and the body.

Deoxygenated blood returns to the **right atrium** in the **vena cava** (main vein). Oxygenated blood returns to the **left atrium** in the pulmonary veins.

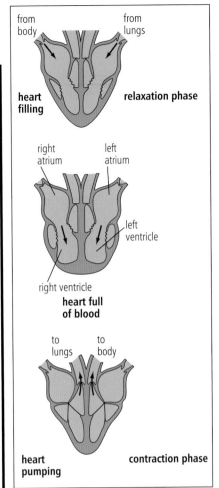

Figure 9.2.2 The cardiac cycle – the changes that occur in the heart during one heartbeat.

KEY POINTS

1 The heart consists of two muscular pumps divided by a septum.

2 Ventricles contract to force blood into arteries; valves make sure that blood flows in one direction.

3 The atria have less muscular walls than the ventricles, as they only force blood into the ventricles.

4 The right ventricle forces blood to the lungs, then the left ventricle is more muscular, as it has to pump blood at a pressure that overcomes greater resistances.

SUMMARY QUESTIONS

1 a What are the two upper chambers of the heart called?
  b What are the two lower chambers of the heart called?
  c What is the function of the septum?

2 Explain each statement.
  a The atria have less muscular walls than the ventricles.
  b The left ventricle has a much more muscular wall than the right ventricle.

3 In each of the following cases, which blood vessel carries:
  a blood to the rest of the body
  b blood from the lungs to the left atrium
  c blood from the body to the right atrium

Heart and exercise

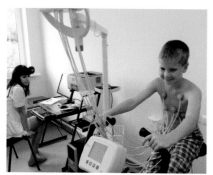

A boy using an exercise bicycle while having an ECG to monitor his heart function.

There are different methods by which the activity of the heart may be monitored. These include electrocardiograms (ECGs), pulse rate and listening to the sounds of the heart valves closing.

The electrocardiogram (ECG)

The control of the heartbeat depends upon electrical activity. A variety of heart disorders can produce irregularities in this activity. The electrocardiogram, or ECG, is a useful diagnostic tool that can detect these defects. Electrodes are taped at various positions on the body and the electrical activity of the heart is then displayed on a monitor. Figure 9.3.1 shows a normal ECG, while Figure 9.3.2 shows ECGs for a heart disorder.

The resultant trace can be compared with the normal ECG with its characteristic P, QRS and T waves. The P wave shows the atria contracting, the QRS 'spike' immediately precedes the contraction of the ventricles, and the T wave represents ventricles relaxing.

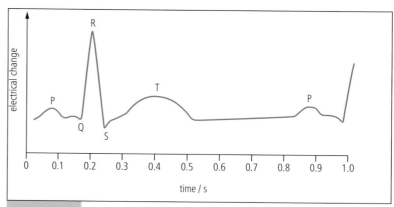

Figure 9.3.1   In this example one heart beats takes 0.8 seconds.

Heart sounds

The heart valves respond to pressure changes during a cardiac cycle. The noise of the blood when the valves open and close make the sound of your heartbeat – 'lub-dub'.

During the contraction phase, the muscular walls of the ventricles contract to force blood out of the pulmonary artery and aorta. The pressure of blood against the atrioventricular valves causes them to shut, preventing blood going back into the atria. This produces the first part of the heart sound – 'lub'.

During the relaxation phase, the ventricles relax. The blood under high pressure in the arteries causes the semi-lunar valves to shut, preventing the blood from going backwards into the ventricles. This produces the second part of the heart sound – 'dub'. Doctors are able to listen to these heart sounds with a stethoscope and identify any irregularity in the heartbeat.

ECG   A normal (ECG) trace

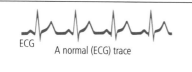

ventricular fibrillation
the contractions of the ventricles are extremely irregular

complete heart block
the atria and ventricles are beating independently

Figure 9.3.2   A normal ECG trace and two ECG traces for heart disorders.

The heart and exercise

During exercise, your muscles need more energy from respiration in order to contract. So the heart beats faster and arteries supplying muscles **dilate** (widen).

These changes increase the blood flow to muscles and result in:

- an increase in supply of oxygen and glucose
- an increase in removal of carbon dioxide.

You can detect the flow of blood through arteries as a 'pulse'. Each time the left ventricle beats, a wave passes along the arteries which you can feel. You can find your pulse by feeling an artery at your wrist. The number of pulses per minute is your **pulse rate** and this is the same as the heart rate.

The **resting pulse rate** gives a good indication of a person's fitness. The fitter you are, the lower your resting pulse rate (see table).

The same volume of blood is being passed out of the heart per minute but with fewer heartbeats.

A person having their pulse taken.

The table shows the general relationship between resting pulse rate and levels of fitness.

pulse rate	level of fitness
less than 50	outstanding
50–59	excellent
60–69	good
70–79	fair
80 and over	poor

PRACTICAL

Investigating the effect of exercise on pulse rate

Sit still and get a partner to measure your pulse rate at rest.

Try doing step-ups for 1 minute (light exercise).

As soon as you have finished, sit down and count your pulse rate.

Wait until your pulse returns to the resting rate.

Now do step-ups as quickly as you can for 3 minutes (heavy exercise). Then count your pulse rate.

Remember that to make this a valid investigation you should have the same person doing the exercise, do the same type of exercise each time, rest for 5 minutes between each exercise or until your pulse returns to its resting rate.

You can also repeat the tests and calculate an average to make your results more reliable.

You can explain the results in terms of supplying substances to your muscles and removing their wastes.

SUMMARY QUESTIONS

1 a State what happens to pulse rate when a person exercises.

b Explain how and why pulse rate changes with exercise.

2 a Draw a normal ECG trace.

b Explain what is happening to the heart at P, QRS and T waves.

3 Pulse rate is affected by certain risk factors. Explain how some of these can be the result of a person's lifestyle.

Blood vessels

LEARNING OUTCOMES

- Name the main blood vessels to and from the heart, lungs and kidney
- Describe the structure and function of arteries, veins and capillaries
- State the functions of arterioles, venules and shunt vessels
- Explain how the structure of arteries, veins and capillaries are adapted for their functions

Blood flows in arteries to all the organs of the body and flows away from them in veins. This table shows you some of the most important arteries and veins.

vessel	organs			
	heart	lungs	liver	kidneys
bringing blood to organ	vena cava to right atrium; pulmonary vein to left atrium	pulmonary artery	hepatic artery; hepatic portal vein (see page 144)	renal artery (see page 146)
taking blood away from organ	pulmonary artery from right ventricle; aorta from left ventricle	pulmonary vein	hepatic vein (see page 144)	renal vein (see page 146)

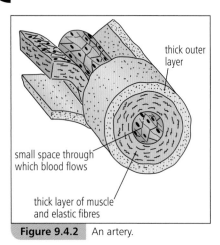

thick outer layer

small space through which blood flows

thick layer of muscle and elastic fibres

Figure 9.4.2   An artery.

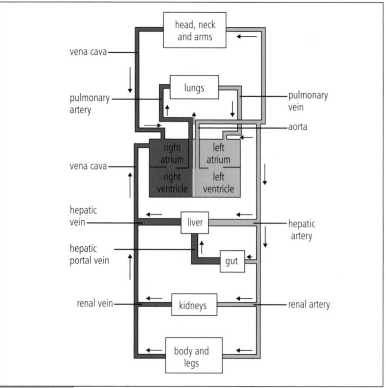

Figure 9.4.1   Main blood vessels of the body.

Structure and function of blood vessels

When your heart muscles contract they force blood into the arteries at a high pressure. Arteries have quite a narrow space in the centre for the blood to flow along. They have thick walls made of muscle and elastic fibres to withstand the pressure of blood.

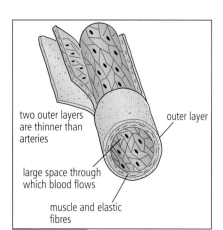

two outer layers are thinner than arteries

outer layer

large space through which blood flows

muscle and elastic fibres

Figure 9.4.3   A vein.

Veins have a wider space for blood to flow than arteries and are thinner, less muscular and have less elastic walls. The pressure inside veins is much lower than in arteries so they do not need such thick walls. There are semi-lunar valves at intervals along veins to ensure that blood flows in one direction – back to the heart (see page 100).

The arteries branch many times until the smallest branches form capillaries. Capillaries are very narrow; a red blood cell can only just squeeze through. The wall of a capillary is made from a single layer of very thin cells so it is easy for substances to pass to and from the blood.

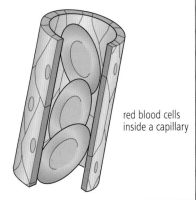

red blood cells inside a capillary

Figure 9.4.4 A capillary magnified ×2500 (the capillary is actually cylindrical, section removed to view red blood cells).

Supplement

How blood vessels are adapted to their functions

The elastic fibres in the walls of the arteries stretch and then recoil like a piece of elastic when you stretch it and then let it return to its original length. The recoil of elastic fibres helps to push blood along so maintaining its pressure. As a result the pressure at the end of an artery where it enters an organ is only a little less than when it left the heart.

It is estimated that there are over 80 000 km of capillaries in the human body. These tiny vessels provide a huge surface area for exchange between blood and cells. Blood flows through capillaries very slowly (about 1 mm per second) giving time for exchange of substances with cells. As the walls are made from one layer of thin cells the diffusion distance between blood and cells is very short. This makes it easy to supply oxygen and nutrients, such as glucose, and to remove carbon dioxide and other wastes.

As the pressure of blood in the veins is very low, it is squeezed along by pressure from the contraction of body muscles and other organs that surround the veins.

There are other vessels in the circulatory system.

Arterioles

Arterioles are the small subdivisions of arteries that carry blood to capillary networks. Like arteries, they have muscle in their walls. Apart from blood transport, arterioles are important in regulating blood pressure. They receive nerve impulses and respond to various hormones in order to regulate their diameter. This regulates blood flow through the capillaries.

Venules

Venules are small blood vessels whose function is to collect blood from the capillary beds. These thin-walled vessels then unite to form veins which transport deoxygenated blood back to the heart.

Shunt vessels

A shunt vessel is a blood vessel that links an artery directly to a vein, allowing the blood to bypass the capillaries in certain areas. Shunt vessels can control blood flow by constriction and dilation. In endotherms (warm-blooded animals), shunt vessels constrict in response to cold, thereby cutting off the blood flow to the extremities and reducing heat loss.

KEY POINTS

1 Arteries have thick muscular and elastic walls to withstand high blood pressure. Veins have much less muscle and elastic tissue and have thin walls. Blood pressure in veins is low.

2 Capillaries have walls that are one cell thick so that substances can pass easily in and out of the blood in body tissues.

SUMMARY QUESTIONS

1 Make a table to show the main differences between arteries and veins. Include three columns: feature, artery, vein.

2 In what ways are the capillaries well adapted for exchange of materials between the blood and body tissues?

Coronary heart disease (CHD)

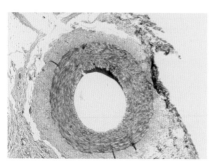

Cross-section of a healthy artery.

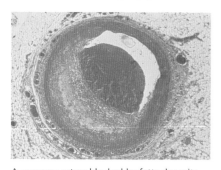

A coronary artery blocked by fatty deposits. A blood clot has formed to further narrow the artery. Compare with the healthy artery above it.

You need a healthy heart to pump blood around your body. Heart muscle needs glucose and oxygen to keep it contracting. These are transported to the heart in the **coronary arteries**. If these arteries get blocked then heart muscle could become starved of oxygen and die. This could cause a **heart attack**.

Slowing the flow

Healthy arteries have a smooth lining, letting blood flow easily. However, **cholesterol**, which is made in the liver, can stick to their walls. This can narrow the artery and slow down the flow of blood. This condition is called **atherosclerosis**.

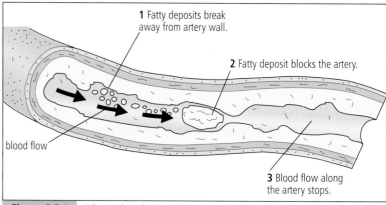

1 Fatty deposits break away from artery wall.

2 Fatty deposit blocks the artery.

blood flow

3 Blood flow along the artery stops.

Figure 9.5.1   Atherosclerosis.

The artery walls can become rough, which can cause the blood to clot and block the vessel. The blockage is called a **thrombosis**.

Narrowing of the coronary artery causes serious problems.

If the coronary artery gets partly blocked it can cause chest pains, especially if activity or emotion makes the heart work harder.

This is called **angina**. It is caused by not enough oxygen getting to the heart muscle.

Angina should act as a warning to the sufferer because it may lead to a heart attack.

A total blockage or thrombosis can cause a heart attack. When this happens, the supply of oxygen is cut off. It causes a severe pain in the chest and the affected part of the heart is damaged. The heart may stop beating altogether – this is called **cardiac arrest**. Death will follow unless the heart starts beating again within minutes.

Risk factors for CHD

These are factors thought to increase the chance of getting CHD:

- **eating a diet with too much saturated (animal) fat**, which can increase the concentration of cholesterol in the blood (see page 71). A good diet – eating more poultry and fish, which are less fatty; cutting down on fried foods; eating less red meat and more fresh fruit and vegetables – helps to reduce the risk.

- **being overweight**
- **taking little or no exercise**
- **smoking**
- **stress**.

People can do something about these risk factors by changing their way of life, but other risk factors cannot be avoided such as:

- **genes** inherited from parents – CHD tends to run in families. These genes control the metabolism of fat and cholesterol

- **age** – the chances of getting CHD increase with age

- **sex** – men are more likely to get CHD than women.

Figure 9.5.2 Three activities that can put you at risk of heart disease.

Supplement

Treatment of CHD

A **coronary artery bypass** can relieve patients with the symptoms of CHD. A blood vessel is taken from another part of the body, usually the arm or leg, and is attached to the coronary artery above and below the narrowed or blocked area. Depending on the severity of the CHD, a number of separate grafts may be needed.

If there is just one incidence of narrowing, or blockage, then a **coronary angioplasty** may be carried out. This is a less complex operation and involves inflating a small balloon inside the artery to widen it. A small metal tube is then inserted to help keep the artery open. This tube is called a **stent**.

Antiplatelet medicines, like **aspirin**, prevent blood clots forming in the arteries. Taking low-dose aspirin daily is part of the treatment for people with known CHD, stroke or other forms of heart disease, e.g. angina, heart attack, coronary stent and bypass surgery.

STUDY TIP

The coronary arteries branch from the base of the aorta just above the semi-lunar valves.

They transport oxygenated blood to the tissues of the atria and ventricles.

SUMMARY QUESTIONS

1 Define **a** coronary heart disease **b** thrombosis **c** angina **d** cardiac arrest

2 **a** If the heart is full of blood, then why does it need its own blood supply?
   b How is the heart muscle supplied with oxygenated blood?
   c Why does the heart muscle need this oxygenated blood?

3 There are risk factors that can increase the chances of a person getting heart disease.
   a List some of the factors that cannot be controlled.
   b List some of the factors that can be controlled.
   c What advice would you give to a person recovering from a heart attack?

KEY POINTS

1 Coronary heart disease is caused by blockage of the coronary arteries that supply the heart with glucose and oxygen.

2 Risk factors which we cannot control are genes, age and sex. Avoidable risk factors include diet, stress, lack of exercise and smoking.

3 CHD may be treated by aspirin and surgery (stents, angioplasty and bypass).

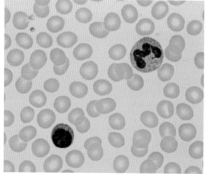

This blood smear seen with a light microscope shows red blood cells and two white blood cells with their nuclei stained blue.

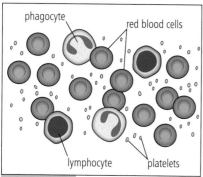

Figure 9.6.2 Red and white blood cells and platelets.

Blood composition

You have about 5 litres of blood in your body. Blood is made of cells and cell fragments suspended in a yellow liquid called **plasma**. The red colour of blood is due to the pigment **haemoglobin** in red blood cells.

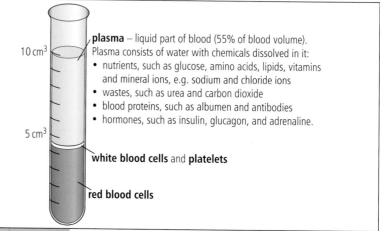

plasma – liquid part of blood (55% of blood volume). Plasma consists of water with chemicals dissolved in it:
- nutrients, such as glucose, amino acids, lipids, vitamins and mineral ions, e.g. sodium and chloride ions
- wastes, such as urea and carbon dioxide
- blood proteins, such as albumen and antibodies
- hormones, such as insulin, glucagon, and adrenaline.

white blood cells and **platelets**

red blood cells

Figure 9.6.1 When blood is spun in a centrifuge (or left to stand for 24 hours), it separates into three layers.

Blood cells

There are three main types of cell in the blood.

Red blood cells have no nuclei and have cytoplasm that is full of many thousands of molecules of haemoglobin for transporting oxygen.

White blood cells have nuclei. They look white when separated by spinning blood in a centrifuge (see Figure 9.6.1). Under a microscope they are colourless as they do not have haemoglobin or any other pigment. To see them, stain has to be added as in the blood smear shown in the photograph. One group of white blood cells searches out bacteria and takes them into vacuoles where they are digested. This process is known as phagocytosis (see Figure 9.7.1). Another group of white blood cells makes protein molecules called antibodies to protect us against different types of disease-causing organisms that invade the body.

Platelets are tiny fragments of cells which cause blood to clot, for example when you cut yourself. The **plasma** contains many soluble substances as listed in Figure 9.6.1. Among its functions are the transport of ions, nutrients, such as glucose and amino acids, carbon dioxide and hormones.

Red blood cells transport oxygen

Red blood cells are disc-shaped with the middle pushed in. They have a large surface area compared with their volume and this helps them to absorb oxygen.

Haemoglobin is a special type of protein that contains iron. You must have enough iron in your diet to make enough haemoglobin for your red blood cells (see anaemia on page 77). As the cells have no nuclei this makes more space for haemoglobin.

Haemoglobin combines easily with oxygen in the lungs to form **oxyhaemoglobin**:

haemoglobin + oxygen ⟶ oxyhaemoglobin

Look at the diagram below:

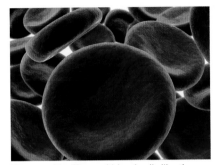

There are 5 million red blood cells like these in every cubic millimetre (mm³) of blood.

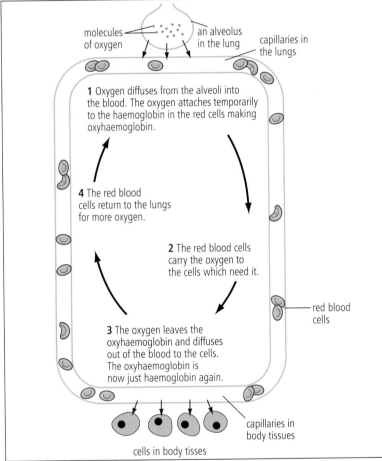

molecules of oxygen

an alveolus in the lung

capillaries in the lungs

1 Oxygen diffuses from the alveoli into the blood. The oxygen attaches temporarily to the haemoglobin in the red cells making oxyhaemoglobin.

4 The red blood cells return to the lungs for more oxygen.

2 The red blood cells carry the oxygen to the cells which need it.

3 The oxygen leaves the oxyhaemoglobin and diffuses out of the blood to the cells. The oxyhaemoglobin is now just haemoglobin again.

red blood cells

capillaries in body tissues

cells in body tisses

Figure 9.6.4   Transport of oxygen from the lungs to body tissues.

- Notice that oxyhaemoglobin forms while blood flows through capillaries in the alveoli in the lungs.
- Oxyhaemoglobin gives out oxygen as blood flows through capillaries in body tissues.

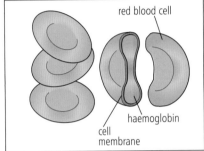

red blood cell

haemoglobin

cell membrane

Figure 9.6.3   Red blood cells magnified ×2500.

KEY POINTS

1 The main components of blood are plasma, red blood cells, white blood cells and platelets. Red and white blood cells are visible under the light microscope.

2 Red blood cells contain haemoglobin which combines with oxygen to form oxyhaemoglobin.

SUMMARY QUESTIONS

1 a   List four chemicals found in plasma.

  b   Give three ways in which a red blood cell differs from a white blood cell.

  c   Name the two main types of white blood cell.

2 a   How is the red blood cell adapted for oxygen transport?

  b i   What is haemoglobin?

   ii   Where does it combine with oxygen and what molecule is formed?

9.7

Blood in defence

- Describe how white blood cells such as lymphocytes and phagocytes protect the body from disease
- State the role of lymphocytes in producing antibodies
- Describe the process of blood clotting

This image was taken with an electron microscope and then coloured with a computer program. You can see two red blood cells and three white blood cells. The image below was taken in the same way. It shows a phagocyte ingesting a yeast cell.

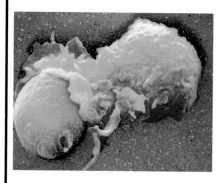

White blood cells

White blood cells defend us against disease. There are far fewer of them than red blood cells. The two groups of white blood cells described in Topic 9.6 are **phagocytes** and **lymphocytes**.

Phagocytes

These are white blood cells that ingest pathogens, such as bacteria.

They surround the pathogens, ingest them and take them into food vacuoles. Then they digest them by using enzymes and this kills them. The process is called **phagocytosis**.

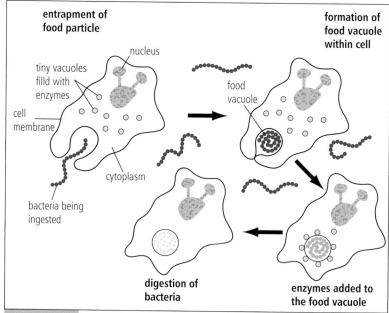

entrapment of food particle

formation of food vacuole within cell

nucleus

tiny vacuoles filld with enzymes

food vacuole

cell membrane

cytoplasm

bacteria being ingested

digestion of bacteria

enzymes added to the food vacuole

Figure 9.7.1   A white blood cell destroys some bacteria.

When pathogens invade the body, phagocytes move towards them. They can squeeze through capillary walls.

Find out about tissue rejection on page 151.

Lymphocytes

Lymphocytes are a type of white blood cell. When a bacterium or virus enters the body the lymphocytes recognise that it is 'foreign' and should not be there. Lymphocytes then make proteins called **antibodies**.

Like enzymes, antibodies are proteins with many different shapes. Enzymes have active sites that combine with their substrates. Similarly, each type of antibody has a binding site that combines with one type of pathogen.

Antibodies attack the pathogens in a number of ways:

- they make them stick together (agglutinate)
- they dissolve their cell membranes
- they neutralise the toxins (poisons) that some pathogens, such as some bacteria, produce.

There is a different type of antibody for each type of pathogen.

After you have had a disease, such as measles, lymphocytes are ready to produce more of the appropriate antibodies should the pathogen enter the body again. This makes you immune to that particular disease.

Blood clotting

When you cut yourself you bleed. Before long, platelets help the blood to thicken and the bleeding stops. The thickened blood has formed a **clot**. Without clotting, blood would be lost and pathogens would enter.

This is another way that the blood defends against disease.

Platelets are small fragments of cells. They are made in the bone marrow like the blood cells. When a blood vessel is damaged, platelets release substances to change the soluble protein **fibrinogen** in the plasma into the insoluble protein **fibrin**, which forms a meshwork of fibres. Red blood cells get trapped in these threads to make the clot. The clot hardens to make a scab which keeps the cut clean. With time the skin heals and the scab falls off.

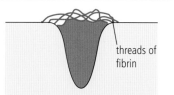

1 The skin is cut and blood starts leaking out of the body. However there is a protein in the blood called **fibrinogen**. When **platelets** come in contact with air they turn the fibrinogen into threads of **fibrin**. The threads make a net over the cut.

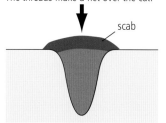

2 Red blood cells get caught in the net and make a blood clot which seals the cut. The clot dries to make a scab. New skin grows under the scab.

Figure 9.7.2   Blood clotting.

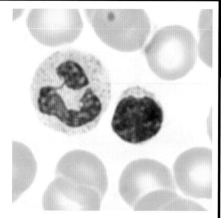

Like the blood smear on page 110, these two white cells have been stained. The larger white cell in this photograph of a blood smear is a phagocyte. The smaller white cell is a lymphocyte.

STUDY TIP

When writing about white blood cells, make sure you use the names 'phagocytes' and 'lymphocytes' when you are describing their functions – do not call them 'white blood cells'.

KEY POINTS

1 Phagocytes ingest pathogens and digest them using enzymes.

2 Lymphocytes make antibodies that protect against pathogens in the body.

3 After a person has had a disease, the lymphocytes remain to produce more antibodies for that pathogen if the disease is encountered again. This is called **immunity**.

4 Platelets help to convert fibrinogen to fibrin during blood clotting.

SUMMARY QUESTIONS

1 a State where white blood cells are made.

   b Explain how lymphocytes act to defend the body from pathogens.

2 Explain how a person can become immune to an infectious disease like measles.

3 a Explain, in detail, the role of platelets in the clotting of blood.

   b What would happen if a person's blood did not clot when exposed to the air at a cut?

   c Describe the process of phagocytosis.

LEARNING OUTCOMES

- Describe the exchange of materials between capillaries and tissue fluid
- Describe the functions of the lymphatic system
- Outline the lymphatic system as vessels and lymph nodes

DID YOU KNOW?

Laughter may protect our bodies from disease and boost our immune systems. Experiments have found that people watching humorous videos produced significantly more white blood cells and had reduced stress, than those watching non-humorous videos.

Tissue fluid

When blood reaches body tissues, some of the constituents of the plasma move out through small gaps in the capillary walls to form the **tissue fluid** that surrounds the cells. White blood cells, which can change shape, are able to squeeze out of the capillaries through these gaps. Red blood cells stay in the blood. Tissue fluid 'bathes' the cells providing a stable environment for them.

Tissue fluid helps substances to diffuse into and out of cells. Useful substances like glucose and oxygen pass from tissue fluid into cells. Carbon dioxide and waste chemicals like urea pass out of cells into the tissue fluid. Most of the tissue fluid then passes back into the blood capillaries. Fluid is constantly flowing from the plasma and back into the plasma, but some of it drains into our lymphatic system.

The lymphatic system

We all know about our blood circulatory system – we see blood when we cut ourselves, we can feel our heart beating and see our veins through the skin. Our other circulatory system – the lymphatic system – is not so obvious. But feel the glands beneath your lower jaw, think about adenoids and tonsils – these are the lymph nodes that are connected by lymphatic vessels. In Unit 7, page 85, you saw a lacteal in a villus. This is a thin-walled vessel that absorbs fat from the gut and is another part of the lymphatic system.

Not all of the tissue fluid returns to plasma in blood capillaries. About one-tenth of it enters a separate system of capillaries called **lymph capillaries**. Once inside these, the tissue fluid is called **lymph**.

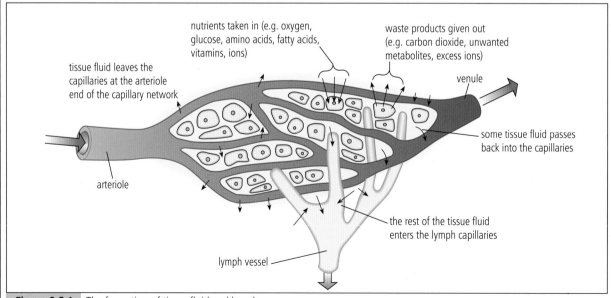

Figure 9.8.1 The formation of tissue fluid and lymph.

Lymph capillaries have tiny valves that allow the tissue fluid to enter but will not let the lymph pass out again. They join up to form **lymph vessels**. These have a structure similar to veins; they are thin-walled and contain semi-lunar valves to make sure lymph flows in one direction.

The lymphatic system has no pump so the flow of lymph is slow. The contraction of surrounding muscles helps to make it flow.

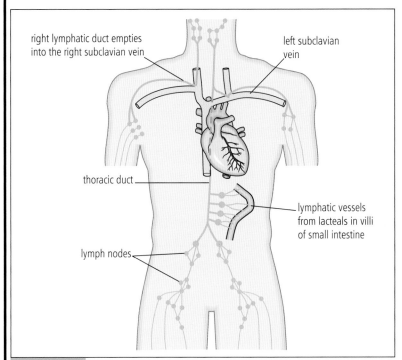

right lymphatic duct empties into the right subclavian vein

left subclavian vein

thoracic duct

lymphatic vessels from lacteals in villi of small intestine

lymph nodes

Figure 9.8.2 The lymphatic system.

Lymph flows from the tissues to the heart. The smaller lymph vessels join up to form two large lymph vessels which empty into the **subclavian veins**, under the collar bones. Here the lymph mixes with the blood before joining the **vena cava** just before it enters the heart.

At intervals along the length of the lymph vessels are **lymph nodes**. There are many **lymphocytes** in the nodes and during an infection they multiply by cell division and make antibodies. Some leave lymph nodes and circulate around the body in the blood, constantly patrolling for fresh invasions of pathogens. The lymphatic system is an important part of the body's immune system.

KEY POINTS

1 When tissue fluid enters the lymph capillaries it is termed *lymph*.

2 Lymph consists of plasma and white blood cells, but has no red blood cells or large plasma proteins.

3 The lymphatic system has a separate circulation which returns lymph to the blood.

4 Lymphocytes are in lymph nodes where they multiply during an infection and produce antibodies.

SUMMARY QUESTIONS

1 a What is tissue fluid and what are its functions?

   b What is lymph and what are its functions?

2 Explain, in as much detail as you can, why the flow of lymph is so slow and how the circulation is maintained.

3 What are lymph nodes, where are they found and what is their main function?

1 Which chamber of the heart pumps blood into the aorta?

   A left atrium

   B left ventricle

   C right atrium

   D right ventricle

(Paper 1)   *[1]*

2 The plan diagram shows the human circulatory system. Which is the pulmonary artery?

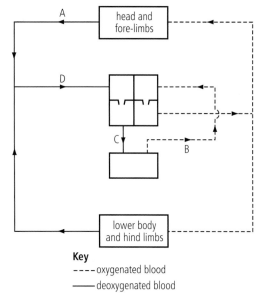

Key
- - - - oxygenated blood
———— deoxygenated blood

(Paper 1)   *[1]*

3 Which is the correct pathway taken by blood flowing through the human circulatory system?

   A body ⟶ vena cava ⟶ left atrium ⟶ left ventricle ⟶ lungs

   B left ventricle ⟶ aorta ⟶ body ⟶ vena cava ⟶ right atrium

   C lungs ⟶ pulmonary artery ⟶ left atrium ⟶ left ventricle ⟶ aorta

   D right ventricle ⟶ aorta ⟶ lungs ⟶ body ⟶ right atrium

(Paper 1)   *[1]*

4 Which row shows the correct functions of the components of the blood?

	plasma	platelets	red blood cells
A	antibody formation	oxygen transport	clotting
B	clotting	antibody formation	oxygen transport
C	transport of ions	clotting	oxygen transport
D	oxygen transport	clotting	antibody formation

(Paper 1)   *[1]*

5 In a single circulation of a fish, the blood flows through the heart once during a complete circulation of the body. What type of blood leaves the heart?

   A deoxygenated blood at high pressure

   B deoxygenated blood at low pressure

   C oxygenated blood at high pressure

   D oxygenated blood at low pressure

(Paper 2)   *[1]*

6 Arterioles are blood vessels that transport blood from:

   A arteries to veins

   B arteries to capillaries

   C capillaries to veins

   D capillaries to venules

(Paper 2)   *[1]*

7 Blood capillaries are exchange vessels. They permit the exchange of substances between:

   A cells and lymph

   B plasma and lymph

   C plasma and tissue fluid

   D lymph and tissue fluid

(Paper 2)   *[1]*

8 Prevention of coronary heart disease may involve:

A angioplasty

B eating more saturated fats

C putting stents into coronary arteries

D taking more exercise

(Paper 2) [1]

9 This is a drawing made of a blood smear as seen with a light microscope.

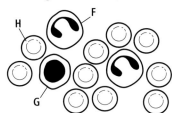

(a) (i) Name the cells **F**, **G** and **H**. [3]

(ii) State the functions of the cells **F** to **H**. [3]

(b) The magnification of the drawing is ×1100. Calculate the diameter of cell **H**. Show your working. [2]

(c) Some blood was spun in a centrifuge to separate the components. The diagram shows the results.

(i) Identify the cells that would be found in regions **K** and **L**. Give reasons for your answer. [3]

(ii) Name the fluid at **J** and describe its role in the body. [5]

(Paper 3)

10 Taking the pulse is one way to monitor the health of the heart.

(a) What is the pulse? [2]

(b) Describe what happens to the pulse rate during and after exercise. [4]

(c) State two other ways in which the health of the heart can be monitored [2]

(Paper 3)

11 The diagram shows the heart.

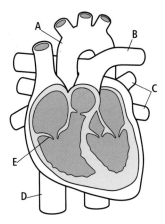

(a) (i) Name **A** to **E**. [5]

(ii) Give the letters from the diagram that indicate the blood vessels that contain oxygenated blood. [1]

(b) Describe what happens to blood as it flows through the capillaries in the lungs. [4]

(c) Explain why blood flowing to the lungs is at a lower pressure than blood flowing to the rest of the body. [3]

(Paper 3)

12 (a) Copy and complete the table to compare arteries, veins and capillaries.

feature	arteries	veins	capillaries
relative thickness of wall		thin	very thin
muscle tissue		some	none
elastic tissue		some	
direction of blood flow	heart to organs		arteries to veins

[5]

(b) Explain how the structure of an artery helps it to carry out its function. [3]

(c) Describe how substances are provided to muscle cells by capillaries. [4]

(Paper 4)

10 Diseases and immunity

10.1 Disease

LEARNING OUTCOMES

- Define the terms *pathogen* and *transmissible disease*
- Describe how pathogens can be spread from person to person
- Outline the body's defences against disease

Even though she is using a handkerchief, droplets of water containing pathogens from her nose and mouth are entering the air and may be inhaled by others.

These houseflies may have picked up bacteria on their bodies while feeding on rubbish and human wastes and are now transferring them to uncooked meat.

Pathogens are organisms that cause disease. Most of the organisms that cause disease in humans are bacteria and viruses. Other pathogens are fungi, protoctists and worms. The diseases that they cause are **transmissible diseases** because the pathogens are passed, or transmitted, from one person to another. They are also called infectious diseases.

The table shows different ways in which diseases are transmitted.

method of transmission	description	examples of diseases
through the air	pathogens are in tiny droplets of liquid from the airways and lungs of infected people (see photo opposite)	influenza, tuberculosis, common cold
contaminated food and drink	people preparing food do not wash their hands; foods are not cooked properly; human faeces contaminate water supplies; flies transfer pathogens on their bodies	cholera, typhoid
direct contact	uninfected people touch infected people, or items, that infected people have used	athlete's foot
insect vectors	insects, e.g. mosquitoes, feed on the blood of an infected person and then feed on an uninfected person	malaria, dengue fever
body fluids	blood from an infected person enters the blood of an uninfected person, e.g. in an unsterilised needle shared between drug addicts	HIV/AIDS, hepatitis
sexual activity	pathogens pass from infected person to sexual partner in blood, semen or vaginal fluid	HIV/AIDS, non-specific urethritis (NSU), chlamydia

Defences against disease

We have three different lines of defence against disease. The first line prevents pathogens entering the body; the second line destroys any pathogens that break through the first line and enters the blood. The third line produces antibodies that defend us against specific pathogens.

Barriers to infection

The defences we have against entry of pathogens are mechanical and chemical barriers.

Mechanical barriers:

- The dead outer layers of the skin form a barrier to entry (see Topic 14.8)
- The hairs in the nose trap larger particles that you breathe in.

Chemical barriers:

- The stomach makes hydrochloric acid that kills pathogens in food (see Topic 7.7).
- Cells that line the airways (trachea and bronchi) make mucus that traps small dust particles and microorganisms. The mucus is moved away from the lungs and up to the throat by cilia (see Topic 11.3). The mucus is swallowed and any pathogens are destroyed by stomach acid.

If any pathogens get through these barriers to enter the blood, then there is a second line of defence.

Blood defences

White blood cells form a line of defence against any pathogens that enter the body's tissues or the blood. **Phagocytes** engulf bacteria and viruses into vacuoles where they are digested and destroyed. The process of **phagocytosis** is described in Topic 9.7. **Lymphocytes** are white blood cells which produce **antibodies**. When activated during an infection by pathogens they produce antibodies which are proteins that have a variety of effects, such as stopping pathogens moving through the body and making it easier for phagocytes to engulf them. Vaccination is a way to make lymphocytes produce antibodies and give long-term protection against certain diseases.

All outdoor activities involve some risk of infection.

SUMMARY QUESTIONS

1 Copy and complete the following:

Pathogens are _____ that cause disease. Human diseases are caused by bacteria, _____ and _____. The airways have cells that make _____ to trap any pathogens that enter the body when people _____. The stomach lining makes _____ which kills pathogens in the _____. Dead cells on the surface of the _____ form a _____ barrier to pathogens.

2 Explain why you should:

  a always wash your hands before handling food

  b never share a towel with someone

  c always wash your hands after going to the toilet

  d use a handkerchief when sneezing

  e never let a dog lick your face

3 Explain how the climber in the photograph is at risk of infection.

KEY POINTS

1 A pathogen is a disease-causing organism.

2 Pathogens can be transmitted from one host to another through the air in droplets, by direct contact, in food and water and by animal vectors (e.g. mosquitoes).

3 The skin and hairs in the nose are mechanical barriers to entry of pathogens.

4 Mucus in the airways and acid in the stomach are chemical barriers.

5 Phagocytes are white blood cells that engulf pathogens and digest them; lymphocytes make antibodies to attack pathogens that enter the blood.

Defence against disease

STUDY TIP

Do not become confused between all the words that begin with 'anti-' in this Unit. An antigen is a chemical that stimulates lymphocytes to produce antibodies. Antibodies are proteins secreted by lymphocytes that help to destroy pathogens. Antitoxins are antibodies that make toxins harmless.

If pathogens get through our mechanical and chemical defences, then phagocytes in the blood may engulf and destroy them. However, phagocytes need help in order to find and engulf pathogens. That help is provided by the antibodies produced by lymphocytes.

All pathogens have chemicals on their surface called **antigens**. Many of these are made of protein. When you catch a disease your lymphocytes respond by making antibodies, which can 'lock on' to the antigens on the surface of the pathogens.

There are several ways in which antibody molecules attack pathogens.

- They cause bacteria to stick together in a group. This makes it much easier for phagocytes to find them and engulf them.
- Some bacteria have flagella which they use to move through the body. Antibodies can attach to the flagella and stop them moving. This helps phagocytes to destroy these bacteria.
- Bacteria such as those that cause tetanus and diphtheria release toxins (poisonous substances) into the blood. **Antitoxins** are a special group of antibodies that combine with the toxins so neutralising them and making them harmless.
- Some antibodies kill bacteria directly by 'punching' holes through their cells walls. This weakens the cell walls so water enters by osmosis causing the cells to burst.

Each type of pathogen has its own antigens which have specific shapes. The antibodies that are made against each type of antigen have a shape that fits around the antigen. The antibody has a shape that is complementary to the antigen.

If you have an infection, such as the common cold, you will be ill for a while and have a variety of symptoms. You will then begin to feel better and within a week or two you will be back to your normal, healthy self. If the strain of the virus that causes the common cold invades again, you will not get the same symptoms. You will not be ill. Lymphocytes are responsible for this and are activated during an **immune response**.

You are born with a very large number of different types of lymphocytes. These cells are specific to different antigens of the surfaces of bacteria, viruses and other pathogens. When a pathogen invades, some lymphocytes respond and divide to make more cells of the same type.

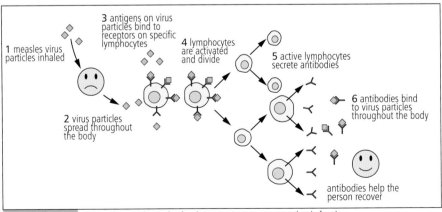

Figure 10.2.1 Active immunity – the body's response to a measles infection.

Many of these lymphocytes become bigger and fill many ribosomes which are attached to a very extensive network of endoplasmic reticulum. The antibody molecules are secreted into blood and lymph to be distributed throughout the body. The antibodies 'lock on' to the specific antigens on the pathogens that have invaded the body.

Other lymphocytes are activated to patrol the body looking for infected cells. When they find an infected cell they destroy it to stop it producing more pathogens. The reason HIV is such a serious infection is that it destroys the body's lymphocytes that are responsible for much of our defence against infectious diseases (see Topic 16.14).

The response to the first infection is very slow. This is why you have fallen ill. The response to the second infection by the same pathogen is much faster; so fast that you are unlikely to have any symptoms.

This process of defence against the pathogen by antibody production is called **active immunity**. Becoming immune after catching a transmissible disease and being ill is a natural way to gain this type of immunity.

During an immune response, many lymphocytes of the specific types are produced. Many of these cells develop into antibody-producing cells, but some do not. These other cells remain in the blood and the lymphatic system circulating throughout the body. These are **memory cells** which respond whenever there is another invasion by the same pathogen with the same antigens. Memory cells do not have a memory of the antigen in the same way that we store memories in the brain. During each immune response to specific antigens, more and more of these lymphocytes are produced to remain in the body, so that next time a pathogen invades the body there are far more lymphocytes with the ability to produce the right antibodies. The response to any subsequent infection is therefore much faster and much greater.

Vaccination

It is possible to promote active immunity without having to be ill. This is done by injecting a vaccine that contains live pathogens, dead pathogens or antigens taken from the surface of pathogens. Each vaccine stimulates immunity to a specific disease, such as measles, rubella and mumps. In some cases, the vaccine provides protection against a specific strain of that disease as is the case with influenza vaccines. Vaccination is an artificial way to gain active immunity.

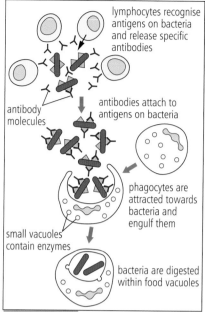

lymphocytes recognise antigens on bacteria and release specific antibodies

antibodies attach to antigens on bacteria

antibody molecules

phagocytes are attracted towards bacteria and engulf them

small vacuoles contain enzymes

bacteria are digested within food vacuoles

Figure 10.2.2 Lymphocytes secrete antibodies that make it easier for phagocytes to ingest bacteria and destroy them.

SUMMARY QUESTIONS

1 Explain what is meant by:
   a antigen   b antibody   c antitoxin
   d active immunity   e memory cells

2 a Outline how someone becomes naturally immune to a transmissible disease.

   b Explain why vaccination is a form of active immunity.

3 Describe how the second response to an antigen differs from the first response.

KEY POINTS

1 Antibodies 'lock onto' antigens leading to the destruction of pathogens.

2 Each pathogen has its own antigens which have specific shapes which fit the shapes of specific antibodies.

3 Active immunity is a defence against a pathogen by antibody production in the body.

4 Memory cells remain in the blood and lymph after an immune response to provide a much faster and greater response to subsequent infections.

5 Active immunity is gained after an infection by a pathogen or by vaccination.

Aspects of immunity

STUDY TIP

The mother's lymphocytes do not cross the placenta or pass to the baby in breast milk. The fetus and later the baby 'borrow' antibodies from the mother. It has to wait until the immune system has developed before its own lymphocytes can produce antibodies.

There are many advantages of breast-feeding – one of the most important is providing antibodies to protect the baby against infectious diseases.

So far we have looked at active immunity, the type of immunity that occurs during the course of an infection or that follows vaccination. During an infection the pathogen enters the body and antigens stimulate an immune response involving the production of many lymphocytes, some of which secrete antibodies and others which become memory cells.

Passive immunity

If protection is required in a hurry, then antibodies may be given by injection. This provides only temporary immunity, but should be sufficient to give protection. Immunity gained in this way is **passive immunity**.

Antibodies pass across the placenta and are present in breast milk, so giving a baby protection against diseases that are in the environment and which his/her mother will have immunity against. This is also only temporary as the baby's body treats the mother's antibodies as foreign and destroys them. The immunity is gained without exposure to antigens, so is much safer than active immunity. Amongst other diseases, babies are usually immune to tetanus and measles, both of which can cause deaths of children. The protection gained from passive immunity lasts until the time when they should receive their first vaccination against these diseases.

Passive immunity is used when people are involved in nasty accidents and are at risk of tetanus bacteria entering the body through open wounds. In these cases, health workers will inject tetanus antibodies as soon as possible to neutralise the toxins produced by these bacteria that can cause muscle paralysis.

The venom from snakes and other animals, such as spiders and scorpions, acts far too quickly for active immunity to be of any use. People who have been bitten by the venomous animals need an injection of antivenom quickly. Antivenom contains antibodies to the molecules in the venom. Specialist companies collect venom from these animals and inject small quantities into horses or sheep. This prompts an immune response and the antibodies are collected from the blood, processed and made available to hospitals in areas where people are at risk of being bitten by venomous animals.

No memory cells are produced in passive immunity, so this type of immunity is short-term, but it is instantaneous. As soon as the antibodies enter the body they are available to be used. There is no wait between the arrival of antigens and the production of antibodies as there is in active immunity.

Malfunction of the immune system

Sometimes the immune system does not work perfectly. When this happens it detects our own antigens as something foreign and this prompts an immune response. As a result the immune system attacks and destroys healthy tissues by mistake.

Malfunctions of the immune system cause a variety of diseases that are collectively known as **autoimmune diseases**. There are many of these diseases: examples are rheumatoid arthritis, multiple sclerosis and Type 1 diabetes.

Type 1 diabetes

This type of diabetes is caused when the immune system attacks and destroys the cells of the pancreas that make insulin. This usually happens rapidly and often before the age of 20.

Insulin helps to control the concentration of glucose in the blood. It is made in response to an increase in the glucose concentration in the blood usually while a meal is being absorbed. Insulin stimulates the liver and muscles to store glucose as glycogen (see Topic 14.7). This means that glucose is not stored for times when it will be needed by cells to release energy in respiration.

The symptoms of Type 1 diabetes are:

• weight loss – cells use protein and fat instead of glucose as sources of energy

• thirst – due to the increased concentration of glucose in the blood that lowers its water potential

• tiredness due to lack of glycogen that can be converted to glucose to provide energy between meals.

One indicator of diabetes is the presence of glucose in the urine.

People with diabetes are treated with regular injections of insulin. The drug is manufactured using techniques of genetic engineering (see Topic 20.4) and many different types are available so that diabetics can control their blood glucose concentrations efficiently.

Milking the venom from a dangerous snake. The venom will be used as an antigen to produce antibodies to give protection against the bites of this species of snake.

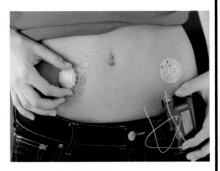

Many people with diabetes use programmable pumps to deliver insulin continuously into body fat through a tube known as a catheter.

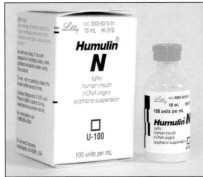

Figure 10.3.1 Human insulin made by cells genetically modified to make this human protein.

SUMMARY QUESTIONS

1 What is meant by the following terms?

   a passive immunity   b Type 1 diabetes

2 a Use an example to explain the advantages of passive immunity.

   b Explain why your example is only a short-term form of immunity.

3 a Explain the cause and describe the symptoms of Type 1 diabetes.

   b Suggest why insulin has to be injected or delivered by catheter and not taken orally (by mouth).

KEY POINTS

1 Passive immunity is a short-term defence against pathogens using antibodies from another individual.

2 No memory cells are produced in passive immunity.

3 Antibodies cross the placenta and are in breast milk. These protect a baby against the diseases its mother has had or been vaccinated against.

4 Type 1 diabetes is caused by the immune system destroying cells in the pancreas that make and release insulin to control blood glucose.

Controlling the spread of disease

LEARNING OUTCOMES

- Explain the importance of hygienic food preparation, good personal hygiene, proper waste disposal and sewage treatment in preventing the spread of disease
- Explain the role of vaccination in controlling the spread of disease

It is important that everyone learns to wash their hands with soap and water to prevent the spread of disease.

Chicken meat is often found to be contaminated by bacteria, such as *Salmonella*. It is safe to eat if it is cooked thoroughly.

Preventing infection

Personal hygiene is important in preventing the spread of some infectious diseases.

- People of all ages should wash their hands after going to the toilet to urinate or defecate and also before handling or eating food.
- Hair should be washed with shampoos to prevent dandruff and headlice.
- Everyone should wash themselves frequently, especially in hot weather.
- Dental hygiene is most important in fighting dental caries (see Topic 7.6).
- Cuts and bruises should be washed with an antiseptic and plasters applied to open wounds.

Hygienic food preparation

- Food should be covered to keep flies away.
- Kitchen surfaces should be cleaned with disinfectants to kill bacteria.
- Food should be cooked thoroughly to make sure any bacteria, such as *Salmonella*, are killed.
- Cooked food that is going to be eaten cold should be kept separate from raw food, especially meat.
- Water used for cooking and/or drinking should be boiled or sterilised by adding water purification tablets if it comes from sources that might be contaminated.

Proper waste disposal

- Household waste should be put into covered bins and collected at regular intervals, e.g. weekly.
- Garbage collected from houses and businesses should be disposed correctly so it is not a health hazard. It should be recycled, incinerated or buried in properly regulated landfill sites. Putting garbage onto rubbish tips where it is not buried carefully means it will attract rats and flies which spread disease. The effluent from rubbish tips may also contain harmful chemicals that cause pollution.

Sewage treatment

- Toilet waste is a serious health hazard if it is not disposed properly through drainage pipes to a sewage treatment works.
- Human wastes are broken down by microorganisms in sewage treatment works (see Topic 21.8).
- The pathogens that cause typhoid and cholera are transmitted through faeces and transmitted to people who drink food or water contaminated with raw sewage.

- Thorough sewage treatment breaks the transmission of typhoid and cholera so these diseases are unknown in countries with proper sanitation.

Supplement

The role of vaccination

Vaccination programmes are an important part of the health protection offered by governments to their citizens. Infants and children are vaccinated against diseases that used to be very common and were responsible for much ill health and many deaths.

Many of these diseases are now very rare in many parts of the world, e.g. the last case of polio in the Americas was in 1991, and in 1994 it was declared that transmission of polio had been successfully interrupted. But the disease still exists in other regions of the world and could be introduced into the Americas by travellers. In 2013, there were 93 cases of polio in Pakistan with more in neighbouring countries and in Syria, parts of West Africa, Somalia and Kenya. You can follow the progress of the campaign to eradicate polio from the whole world by searching online for 'polioeradication'.

During eradication programmes vaccination is used in two ways. Mass vaccination schemes attempt to give active immunity to everyone. Some people do not respond to vaccines, but they can be protected because the chances of them coming into contact with the disease are small as most people around them have immunity and will not transmit the disease. Careful surveillance by health workers identifies people who have infectious diseases and their spread can be limited or stopped by vaccinating all people in the neighbourhood who may have come into contact with infected people. For some diseases there are no vaccines (as of 2015); examples are HIV, Ebola and dengue fever.

Household garbage should be kept in covered bins and collected before overflowing and becoming a health hazard.

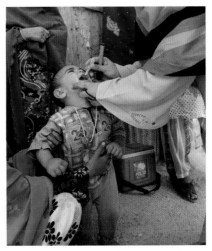

Vaccination against polio in Quetta, Pakistan in 2014.

DID YOU KNOW?

There may be as many as 100 million different lymphocytes that have the ability to become antibody secreting cells. Whatever foreign substance enters the body, there will be a lymphocyte with the right specificity to destroy it.

SUMMARY QUESTIONS

1 Suggest why it is important to:
  a teach children to wash their hands
  b put food waste into covered bins
  c connect toilets to the main drainage system
  d cook chicken thoroughly

2 The Americas were declared free of polio in 1994. Why do children throughout the Americas receive polio vaccinations?

3 Why is it important that everyone receives the vaccinations offered by their country's health service?

Supplement

KEY POINTS

1 The spread of transmissible diseases is prevented by good personal hygiene, hygienic food preparation, proper waste disposal and sewage treatment.

2 Vaccination programmes are used to prevent the spread of transmissible diseases.

3 Many diseases are controlled effectively by mass vaccination of children.

Supplement

1 A pathogen is an organism that:

  A feeds on dead and decaying organisms

  B lives in an organism and causes disease

  C lives inside a host organism without causing harm

  D spreads disease organisms from infected to uninfected people

(Paper 1)   *[1]*

2 Which is a transmissible disease?

  A cholera

  B coronary heart disease

  B rickets

  D scurvy

(Paper 1)   *[1]*

3 Which barrier to infection is most effective against cholera bacteria?

  A mucus

  B skin

  C stomach acid

  D phagocytosis

(Paper 1)   *[1]*

4 Which gives people long-term protection against transmissible diseases?

  A hand washing

  B garbage collection

  C sewage treatment

  D vaccination

(Paper 1)   *[1]*

5 An antigen is:

  A a disease-causing organism

  B a molecule produced by a lymphocyte

  C any substance that stimulates an immune response

  D a pathogen

(Paper 2)   *[1]*

6 During an immune response:

  A lymphocytes and specific phagocytes divide to make antibodies

  B all lymphocytes divide to produce cells that make antibodies

  C phagocytes engulf and completely digest bacteria

  D specific lymphocytes divide to produce cells that make antibodies

(Paper 2)   *[1]*

7 Passive immunity to snake venom is gained by:

  A being bitten by a venomous snake

  B being treated with antibodies to the venom of a spider

  C injecting a small quantity of snake venom

  D injecting snake antivenom with antibodies produced in a horse

(Paper 2)   *[1]*

8 The second immune response to an invasion by the measles virus many months after the first invasion is much faster because:

  A the concentration of antibodies specific to measles remains high from the first response

  B lymphocytes respond faster with age

  C lymphocytes specific to other viruses respond as well

  D memory cells are present in the blood and lymph

(Paper 2)   *[1]*

9 Some microorganisms, such as the bacteria that cause cholera, enter the body in our food and cause disease.

  (a) State the term given to a disease-causing organism.   *[1]*

  (b) State two ways in which food can become contaminated with microorganisms.   *[2]*

  (c) Explain how the stomach provides a defence against microorganisms.   *[2]*

  (d) Microorganisms may enter the blood. Explain how the body prevents their spread in the blood.   *[4]*

  (e) Some diseases are transmitted in food and water. State four other ways in which infectious diseases can be transmitted.   *[4]*

(Paper 3)

10 (a) Explain the importance of the following in maintaining good health:

    (i) personal hygiene

    (ii) sewage treatment

    (iii) proper waste disposal   *[6]*

(b) Rubella is a transmissible disease caused by a virus. Many children are vaccinated against rubella. Explain how vaccination provides protection against disease.   *[3]*

(Paper 3)

11 A child received a first vaccination against measles at four months of age and then a booster at eight months. The concentration of antibodies to measles is shown in the graph.

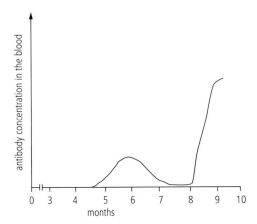

(a) Explain why no antibodies were present in the blood for the first week.   *[3]*

(b) The response to the two infections of the vaccine is different. Use the information in the graph to describe how the response to the booster at eight months differs from the response to the first injection at four months.   *[4]*

(c) Explain why the response to the second injection is different to the response to the first injection.   *[4]*

(d) Suggest why further boosters of this vaccine may be given.   *[2]*

(Paper 4)

12 The diagram shows four different antigens and an antibody molecule.

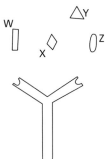

(a) The antibody forms a complex with one of the antigens.

    (i) Which one?   *[1]*

    (ii) Explain your answer.   *[1]*

There is no vaccine for chicken pox. A child has been vaccinated against several diseases including measles, tetanus and rubella. The child catches chicken pox and has the symptoms of the disease.

(b) Use the information in the diagram to explain why it is possible to be immune to many diseases, but still be ill with another disease, such as chicken pox.   *[4]*

(c) Explain how vaccinations can be used to eradicate a disease from a country.   *[3]*

(d) Suggest why vaccination against diseases must continue to be carried out even if there are no cases of the disease for many years.   *[3]*

(Paper 4)

13 (a) The lymphatic system consists of lymphatic vessels and lymph nodes.

Describe the functions of **(i)** lymphatic vessels, and **(ii)** lymph nodes   *[4]*

(b) The immune system consists of phagocytes, lymphocytes and antibodies. Explain how the immune system defends us against infection by bacterial pathogens.   *[5]*

(Paper 4)

11 Gas exchange in humans

11.1 The gas exchange system

Structure of the gas exchange system

We need to breathe air into our lungs in order to get oxygen. We breathe air out of our lungs to get rid of carbon dioxide. The lungs form part of the **gas exchange system**.

The lungs are spongy organs found inside the chest (**thorax**) and are surrounded and protected by the ribs and the sternum (breastbone). The **diaphragm** is a sheet of fibrous tissue and muscle that separates the thorax from the abdomen. Its movement up and down changes the volume of the lungs to move air when you breathe out and in.

The **intercostal muscles** between the ribs move the ribs during breathing, especially during deep breathing.

Air enters the mouth or nose and passes through the throat to the **larynx** (voice-box). It then enters the **trachea** (windpipe), which connects the throat to the lungs. It branches to form two **bronchi** which enter each lung. These continue to divide to form many small **bronchioles**, which end in tiny air sacs called **alveoli**. It is here that gas exchange takes place.

The tubes through which air moves are often called airways.

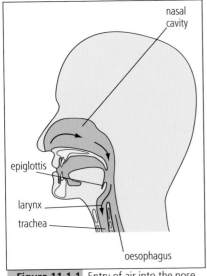

An X-ray of healthy lungs which appear black.

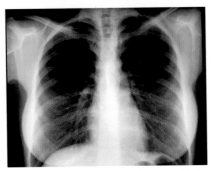

Figure 11.1.1 Entry of air into the nose and trachea.

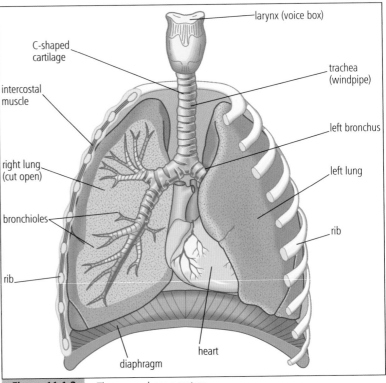

Figure 11.1.2 The gas exchange system.

The larynx contains the vocal cords. When air passes over these you make sounds. You cannot breathe and swallow at the same time. This is because when you swallow, a flap called the **epiglottis** moves to cover the opening to your larynx. This stops any food from going down your trachea (see page 82).

The trachea is kept open by C-shaped rings of cartilage. The 'arms' of the C are joined by muscle at the back of the trachea. The cartilage prevents the trachea from collapsing as you breathe in when the air pressure decreases.

Deeper into the lungs

There is a very large number of alveoli in the lungs to give a huge surface area for diffusion (see page 28 to remind yourself about diffusion). Each alveolus is surrounded by a network of blood capillaries for efficient gas exchange into and out of the blood.

The alveoli have thin walls made of a single layer of cells, so there is a short distance for diffusion. They are moist so oxygen dissolves in this watery fluid before diffusing through the walls into the blood.

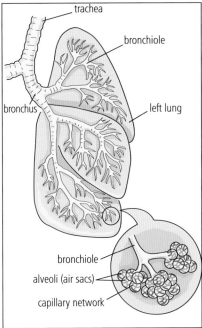

Figure 11.1.3   Structure of a lung.

SUMMARY QUESTIONS

1 Describe the route taken by inspired air (air we breathe in) from the nasal cavity to the alveoli in the lungs.

2 The diagram shows the human breathing system:

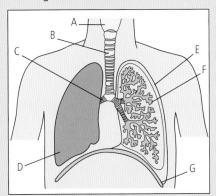

a   Name structures A to G.
b   Match parts A to G with these descriptions:
   i   sheet of muscle forming the floor of the thorax
   ii   one of these enters each lung
   iii   contains the vocal cords
   iv   flexible tube kept open by rings of cartilage
   v   where exchange of gases takes place
   vi   slippery membranes
   vii   an organ made of spongy tissue and found in the chest

KEY POINTS

1 The inspired air passes down the larynx, trachea, bronchi, bronchioles to the alveoli.

2 Cartilage keeps the trachea open when air passes through it.

3 Gas exchange occurs between the air in the alveoli and the blood in the surrounding capillaries.

STUDY TIP

See page 28 to remind yourself about diffusion and concentration gradients.

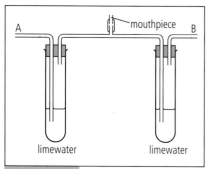

Figure 11.2.1

STUDY TIP

Beware! We can say that alveoli and capillaries have <u>walls made of cells</u>, but not <u>cell walls</u> – remember only plant cells have cell walls, animal cells do not.

Breathing out carbon dioxide

Carbon dioxide is a waste gas made in respiration. It can become toxic if it builds up in cells.

We breathe out in order to get rid of the carbon dioxide.

gas	composition / %	
	inspired air (breathing in)	expired air (breathing out)
oxygen	21	16
carbon dioxide	0.04	4
nitrogen	78	78
water vapour	variable	saturated

The table shows that we breathe out more carbon dioxide than we breathe in. We breathe in more oxygen than the air we breathe out, and the air that we breathe out has a lot more water vapour.

The following practical demonstration confirms that we breathe out more carbon dioxide.

PRACTICAL

Carbon dioxide in inspired and expired air

1 Set up the apparatus as shown in Figure 11.2.1.

2 Then breathe gently in and out of the mouthpiece several times.

- When you breathe in the air comes in through A.
- When you breathe out the air goes through B.
- The limewater turns cloudy first in B, showing that there is more carbon dioxide in expired air than in inspired air.
- The limewater in A will also go cloudy if you breathe through the apparatus long enough, showing there is a lower concentration of carbon dioxide in atmospheric (inspired) air.

Gas exchange surfaces

Gas exchange surfaces, such as the gills of a fish and alveoli (air sacs) of a human, have features in common that adapt them for efficient exchange of oxygen and carbon dioxide:

- a very large surface area for the diffusion of gases
- moist surfaces so that gases can dissolve before diffusion
- a thin surface (only one cell thick in each alveolus) so the gases do not have to diffuse very far
- a good blood supply so that lots of oxygen is removed quickly and lots of carbon dioxide is supplied quickly. This maintains the **concentration gradients** for these gases

- ventilation of the lungs ensures that the air in the air passages is changed; this helps maintain the gas concentration gradients between air in the alveoli and that in the blood.

Gas exchange at the alveolus

When inspired (breathed in) air reaches the alveoli (air sacs) it contains a lot of oxygen. Oxygen dissolves in the water lining each alveolus. It then diffuses through the wall of the alveolus and through the capillary wall into the blood. Although this involves diffusing through two cells, the distance is very small.

Each alveolus has a network of capillaries around it. Oxygen molecules from the alveolus diffuse into the red blood cells and combine with haemoglobin. The blood cells can then transport this oxygen to the body tissues.

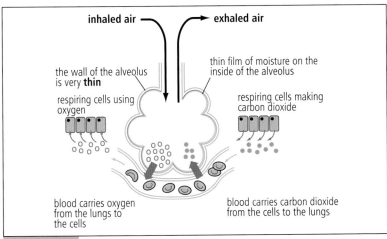

inhaled air ──→ ←── exhaled air

the wall of the alveolus is very **thin**

thin film of moisture on the inside of the alveolus

respiring cells using oxygen

respiring cells making carbon dioxide

blood carries oxygen from the lungs to the cells

blood carries carbon dioxide from the cells to the lungs

Figure 11.2.2 Gas exchange at an alveolus.

There is a lot of carbon dioxide in the capillary. It has been carried there from the respiring tissues in the blood plasma. It diffuses in the opposite direction, through the capillary wall across the alveolar wall into the space inside the alveolus. From here it is breathed out.

Alveoli are surrounded by elastic tissue. This stretches when you breathe in and recoils when you breathe out to help remove air from the lungs.

The air we breathe out is saturated with water vapour that has evaporated from the moist walls of the alveoli.

KEY POINTS

1. There is more carbon dioxide, less oxygen and more water vapour in expired air than in inspired air.

2. Gas exchange surfaces are moist, thin, have a large surface area and a good blood supply.

3. Alveoli are well adapted for gas exchange as they have these features and are surrounded by many capillaries.

SUMMARY QUESTIONS

1 List four features of the alveoli of the lungs that adapt them for efficient exchange of gases.

2 Look at the apparatus used in this experiment:

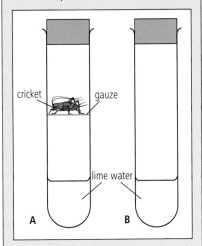

cricket   gauze

lime water

A   B

After an hour the tubes were examined.

a What result would you expect?

b Explain your answer.

c Explain why tube B is included.

3 Give an explanation for each of the following.

a When you breathe out on to a cold surface, water droplets form.

b Expired air turns limewater cloudy quicker than inspired air.

c Alveoli (air sacs) are very thin, have a large surface area and have walls with elastic fibres.

LEARNING OUTCOMES

- Describe how volume and pressure changes lead to ventilation of the lungs
- Explain the role of goblet cells, mucus and cilia in protecting the gas exchange system from pathogens and dust particles

PRACTICAL

A model of the chest

1 Obtain a model like the one in the diagram. The rubber sheet represents the diaphragm.

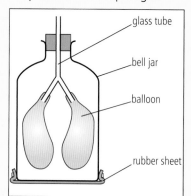

2 Pull the rubber sheet down and then push it up.

3 Do this a few times and watch what happens to the balloons.

This model shows how movement of the diaphragm changes the volume and pressure in the thorax represented by the bell jar. As the diaphragm contracts and moves downwards, the lungs inflate as a result of an increase in volume and decrease in pressure in the space around the balloons.

When you breathe it feels as if the lungs are expanding to push your ribs outwards. This is not so. When you breathe in, movements of your diaphragm and ribcage move the chest cavity and pull on the lungs so they occupy a larger volume. This decreases the air pressure inside the lungs to below the pressure of atmospheric air so that air moves in through your nose and/or mouth.

When you breathe out, the diaphragm and intercostal muscles relax and this causes the chest cavity to decrease in volume. The air pressure in the lungs increases to above atmospheric pressure so air is forced out. You can do this just by movement of your diaphragm alone (try breathing without moving your ribs).

This is fine for quiet breathing, but during deep breathing and when you exercise the intercostal muscles contract to move the ribs. (Intercostal means between the ribs.) There are two layers of these: the external muscles that are closer to the skin, and the internal muscles that are deeper into the chest wall (see Figure 11.3.3). The diaphragm is made of tough fibrous tissue in the centre and strong muscles that connect to the backbone, the lower ribs and the sternum (breastbone).

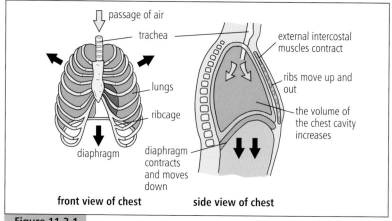

Figure 11.3.1

Inspiration (breathing in)

- The external intercostal muscles *contract* and the internal intercostals muscles *relax*, raising the ribs upwards and outwards.
- At the same time, the diaphragm contracts and flattens.
- Both of these actions *increase* the volume inside the thorax, causing the pressure inside the thorax to *decrease*.
- Since atmospheric pressure is greater, air moves *into* the lungs and they inflate.

Expiration (breathing out)

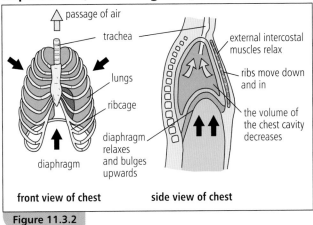

front view of chest   side view of chest

Figure 11.3.2

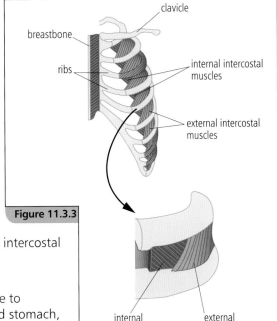

Figure 11.3.3

- The internal intercostal muscles *contract* and the external intercostal muscles *relax*.
- This lowers the ribs downwards and inwards.
- The muscle in the diaphragm *relaxes* and it bulges up due to pressure from the organs below, for example the liver and stomach, and contraction of the abdominal muscles.
- Both of these actions *decrease* the volume inside the thorax, causing the pressure inside the thorax to *increase*.
- Elastic recoil of the alveoli helps to force air *out* of the lungs.

Cleaning the air

As you breathe in through your nose, air is warmed and moistened by evaporation of water from the lining. Hairs inside the nose filter the air, removing particles and some pathogens.

The trachea, bronchi and bronchioles are lined with ciliated epithelial cells and goblet cells which secrete mucus. Dust particles and pathogens become trapped in the slimy mucus. The **cilia** beat to carry a stream of mucus up to your nose and throat, removing the particles and pathogens which you then swallow.

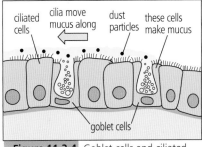

Figure 11.3.4 Goblet cells and ciliated epithelial cells keep the lungs free of dust and pathogens.

Rate and depth of breathing

LEARNING OUTCOMES

- Investigate and describe the effects of physical exercise on the rate and depth of breathing
- Describe how the effects of exercise increase carbon dioxide concentration in the blood and how this affects the rate and depth of breathing

Performing step-ups.

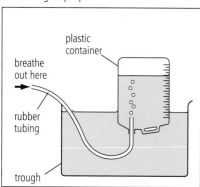

Figure 11.4.1 This apparatus can be used to measure your vital capacity (see top of next page).

Breathing and exercise

At rest you probably take between 12 and 16 breaths per minute. When you exercise this breathing rate changes.

PRACTICAL

The effects of exercise on breathing rate

You can measure your breathing rate by counting the number of times that you breathe out in a minute.

You could record your results in a table like this:

	breathing rate / breaths per min⁻¹		
	at rest	light exercise	heavy exercise
1			
2			
3			
average =			

- Try doing step-ups for 1 minute (light exercise).
- As soon as you have finished, sit down and count your breathing rate.
- Wait until your breathing rate returns to its resting value.
- Now try doing step-ups as quickly as you can for 3 minutes (heavy exercise).
- Then count your breathing rate when you have finished.
- You will notice that your breathing rate and depth of breathing change after doing exercise.
- Repeat this twice and calculate averages.

The rate and depth of breathing increase with exercise. When they are working hard, your muscles need more oxygen for respiration. They also produce more carbon dioxide in respiration.

Increasing the rate and depth of breathing gets more oxygen into the blood and gets rid of more carbon dioxide from the blood. Muscles continue to respire quite fast after exercise finishes. They still need a good supply of oxygen and they still have carbon dioxide to be removed. Your pulse rate remains high after exercise because your heart is beating fast to deliver plenty of blood to your muscles so they gain this extra oxygen and have their carbon dioxide removed.

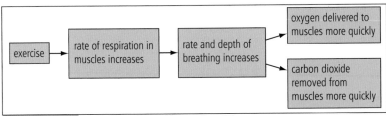

Figure 11.4.2 The effect of exercising on breathing.

Depth of breathing

An adult can take in about 5 litres of air in their deepest breath. This is their **vital capacity** – the maximum volume of air that is breathed out after breathing in as much air as possible. At rest, about half a litre of air is breathed in and out. During exercise approximately four and a half litres of extra air can be taken in during a deep breath.

STUDY TIP

See page 28 to remind yourself about diffusion. Compare your results with the results you got when investigating pulse rate (page 105). Summarise the effects of exercise on the body including the muscles, heart and lungs. You will be able to add some more about body temperature and blood glucose in Unit 14.

Supplement

Control of breathing

You don't think about your breathing – most of the time it just happens. So what makes you breathe faster and deeper when you exercise? The brain has a special part for controlling breathing. When you exercise the tissues respire more quickly and make *more carbon dioxide* which *lowers* the *pH* in the tissues and the blood (i.e. they become more acidic). They may also make lactic acid which does the same (see page 140).

Your brain detects this rise in carbon dioxide and the lowering of the pH of the blood reaching it. The brain sends nerve impulses to the diaphragm and to the intercostal muscles so they contract faster and further to increase the rate and depth of breathing. By breathing deeper and more rapidly, you lower the concentration of carbon dioxide in the blood and this raises the blood pH back to normal. In turn this has the same effect on the tissues because the blood flows through them. This is an example of **homeostasis** – see page 166.

STUDY TIP

Remember the muscles! They are the organs doing the hard work. They require the extra oxygen and glucose and need to have their wastes taken away. Start any answer on this topic with the changes that occur in the muscles during exercise.

Supplement

SUMMARY QUESTIONS

1 When you exercise, describe what happens to:
   a   the rate of breathing   b   the depth of breathing

2 Describe what happens to the concentration of carbon dioxide in the blood when you exercise.

Supplement

3 a   Explain how the brain monitors and controls the rate of breathing when you exercise.
   b   Explain what happens to the following when you exercise:
      i   the intercostal muscles   ii   the diaphragm
      iii   the heart rate

KEY POINTS

1   Exercise increases both the rate and depth of breathing.

2   An increase in the concentration of carbon dioxide in the blood reaching the brain triggers these changes in breathing.

Supplement

135

1 Which is a pathway that air takes from the atmosphere to the gas exchange surface in the lungs?

A   mouth → larynx → bronchus → bronchiole → alveolus

B   mouth → nose → trachea → bronchus → alveoli → bronchiole

C   nose → larynx → trachea → bronchus → bronchiole → alveolus

D   nose → trachea → bronchus → bronchiole → alveolus

(Paper 1)   [1]

2 A student filled a conical flask with lime water and put a drawing of a cross beneath it. The student then breathed into the flask of lime water to find out how long it took before the cross became invisible. The student repeated the test after doing three minutes of strenuous exercise.

Which describes the time taken for the cross to become invisible in the test after exercise?

A   The cross became invisible in a shorter time.

B   The cross took three times as long to become invisible.

C   The cross took twice as long to become invisible.

D   The time was the same.

(Paper 1)   [1]

3 The diagram shows apparatus to investigate the production of carbon dioxide during respiration of insects. What would you expect to see in tubes **A** and **B** five minutes after setting up the apparatus?

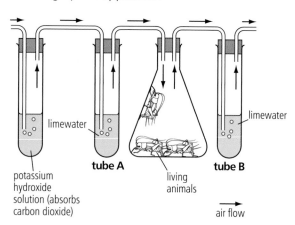

potassium
hydroxide
solution (absorbs
carbon dioxide)

living
animals

tube A   tube B

→
air flow

	tube A	tube B
A	cloudy	cloudy
B	cloudy	no change
C	no change	cloudy
D	no change	no change

(Paper 1)   [1]

4 What are the features of a gas exchange surface?

A   thick-walled, large surface area, many capillaries

B   thick-walled, small surface area, few capillaries

C   thin-walled, large surface area, many capillaries

D   thin-walled, small surface area, few capillaries

(Paper 1)   [1]

5 Which is the function of the cartilage in the trachea?

A   cleans the air as it moves to the lungs

B   contracts to move air into the lungs

C   increases resistance to air flowing to the lungs

D   prevents the trachea collapsing

(Paper 2)   [1]

6 The diaphragm and the intercostal muscles act together during expiration at the end of strenuous exercise. Which of the following occurs?

	diaphragm		intercostal muscles	
			internal	external
A	contracts	moves down	relax	contract
B	contracts	moves up	relax	contract
C	relaxes	moves down	contract	relax
D	relaxes	moves up	contract	relax

(Paper 2)   [1]

7 The photograph shows some cells from the gas exchange system.

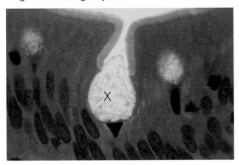

What is cell X?

A alveolar cell

B ciliated cell

C goblet cell

D muscle cell

(Paper 2) [1]

8 The percentage composition of air changes as it passes through the lungs because:

A carbon dioxide and water vapour are absorbed from the air

B carbon dioxide is absorbed and oxygen excreted

C oxygen is absorbed and carbon dioxide excreted

D oxygen is absorbed and nitrogen excreted

(Paper 2) [1]

9 Some students investigated the effect of exercise on the rate and depth of breathing. Here are the results for one of the students.

activity	volume of each breath / cm³	number of breaths taken per minute
rest	500	18
20 step-ups per minute	750	25
50 step-ups per minute	1200	34

(a) (i) Calculate the change in the volume of each breath when the student changed from 20 to 50 step-ups per minute. [1]

(ii) Describe the effect of exercise on the volume of each breath. [2]

(b) (i) Calculate the percentage change in the volume of each breath when the student changed from rest to 50 step-ups per minute. [2]

(ii) Describe the effect of exercise on the rate of breathing. [2]

(c) The students decided to investigate the effect of this exercise on the boys and girls in the class. State two features of the investigation that they would have to keep the same so that they could make valid comparisons between the boys and girls.

(Paper 3) [2]

10 The diagram shows the ribs and muscles of the thorax from the side.

(a) Name **A** to **C**. [3]

(b) (i) Describe how the structures shown bring about breathing movements during inspiration and expiration. [6]

(ii) Explain how the movements you describe cause air to move into and out of the lungs. [5]

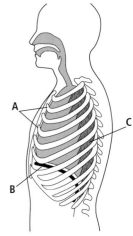

(Paper 3)

11 Look at the table of results in question 9.

(a) Describe the effect of exercise on the student's breathing. [5]

(b) Explain why the student's breathing changed during exercise. [6]

(Paper 4)

LEARNING OUTCOMES

- Define *aerobic* respiration and state the word equation
- State that respiration involves enzymes catalysing reactions in cells
- State the uses of energy in the human body
- Investigate the uptake of oxygen by respiring organisms
- State the balanced chemical equation for aerobic respiration

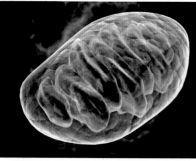

A computer generated image of a mitochondrion. Most of the ATP in cells is made in these cell structures. Each is about 0.002 mm long.

STUDY TIP

Do not confuse respiration with breathing. Respiration is a chemical process consisting of lots of chemical reactions, each one catalysed by its own enzyme. What you do when you breathe in and out is not respiration!

Aerobic respiration

When we burn a fuel such as petrol, oxygen is used up and carbon dioxide is made. The flame gives off a lot of light and heat.

Every cell in every living organism needs energy. A fuel used to provide energy in the cells is called glucose, but unlike burning petrol, the energy is released very gradually in a series of small, enzyme-controlled reactions. **Respiration** involves a number of chemical reactions that break down nutrient molecules, such as glucose, in living cells to release energy.

In **aerobic respiration** oxygen is used in the breakdown of glucose:

glucose + oxygen $\longrightarrow$ carbon dioxide + water + energy released

Cells also respire fats and proteins to provide energy, but you only have to know about the respiration of glucose, which is a carbohydrate.

Using energy

Your body needs energy for many different things. Energy is used in the following processes:

- muscle contraction
- cell division
- absorption of nutrients in the gut by active transport
- sending impulses along nerves
- protein synthesis for making enzymes, some hormones and antibodies
- making new cell membranes and cell structures like the nucleus during growth
- keeping the body temperature constant. Some of the energy released is in the form of heat.

Respiration takes place in *all* living cells *all of the time* because cells need a constant supply of energy to stay alive. The balanced chemical equation for aerobic respiration is:

$$C_6H_{12}O_6 + 6O_2 \longrightarrow 6CO_2 + 6H_2O + \text{energy released}$$

Most of the energy released in aerobic respiration is released in **mitochondria**. Cells that require a lot of energy have many mitochondria. For instance, insect flight muscle has numerous mitochondria located between the muscle fibre. The mitochondria provide the energy for the muscles to contract during flight. Liver cells have a high metabolism and have many mitochondria. The epithelial cells of the small intestine absorb glucose and other molecules by active transport, so have large numbers of mitochondria.

PRACTICAL

Investigating the uptake of oxygen by respiring seeds

The apparatus in Figure 12.1.1 is a respirometer and it can measure how quickly germinating seeds use up oxygen. The seeds take up oxygen during aerobic respiration and release carbon dioxide, which is absorbed by the soda lime. The decrease in volume and air pressure causes the red liquid to move along the capillary tube. The rate of movement can be measured by the scale – the faster the rate of respiration, the quicker the liquid will move. Why is it important to keep the apparatus at a constant temperature?

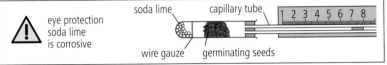

Figure 12.1.1 A simple respirometer.

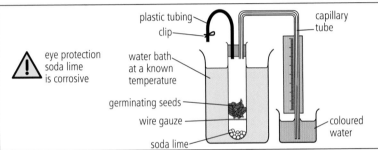

Figure 12.1.2 A respirometer for investigating the effect of temperature on respiration.

SUMMARY QUESTIONS

1 Copy and complete the sentences using these words:

**aerobic glucose oxygen mitochondria
energy carbon dioxide cells**

Respiration takes place in all living _____ . A fuel for respiration is _____ . It is broken down to release _____ Also produced is water and the waste gas _____ . When it takes place in the presence of _____ we call it _____ respiration.

These reactions take place in every living cell, inside tiny structures called _____ .

2 a Write down the word equation for aerobic respiration.

b Energy is released during aerobic respiration.
Make a list of five processes that use this energy.

3 a Describe how a simple respirometer works.

b What is the function of soda lime in the apparatus in Figure 12.1.1?

c What do you think will happen to the rate of respiration of the germinating seeds if the temperature **i** decreases to 5 °C or **ii** increases to 50 °C?

PRACTICAL

Investigating the effect of temperature on the rate of respiration of germinating seeds

The apparatus shown in Figure 12.1.2 is a more accurate respirometer.

1 Place some germinating seeds into the boiling tube.

2 Set the rest of the apparatus up as shown. Be very careful not to touch the soda lime as it is corrosive.

3 Close the clip and find out how much the coloured liquid rises in the capillary tube in 30 minutes.

This investigation works on the same principle as the previous one. It can be repeated at a number of temperatures by changing the temperature of the water and so compare the rate of respiration of the seeds.

KEY POINTS

1 In aerobic respiration, glucose is broken down to release energy in the presence of oxygen, forming carbon dioxide and water.

2 Respiration involves the action of enzymes to catalyse reactions in cells.

3 The word equation for aerobic respiration is:

glucose + oxygen $\longrightarrow$ carbon dioxide + water

4 The balanced chemical equation for aerobic respiration is:

$C_6H_{12}O_6 + 6O_2 \longrightarrow 6CO_2 + 6H_2O$

5 Respirometers can be used to measure the rate of respiration under different conditions.

LEARNING OUTCOMES

- Define *anaerobic* respiration
- State the word equations for anaerobic respiration in muscle during vigorous exercise and in yeast
- State that anaerobic respiration releases much less energy per glucose molecule than aerobic respiration
- State the balanced chemical equation for anaerobic respiration in yeast
- Describe the build-up of lactic acid in muscle to cause an oxygen debt and how it is removed during recovery from exercise

Yeast cells under an electron microscope.

Hurdlers build up lactic acid in their muscles.

Muscles without oxygen

Your muscles need oxygen and glucose to respire aerobically. These are brought to your muscles by your blood system. During vigorous exercise your heart and lungs cannot get enough oxygen to your muscles quickly enough. When this happens your muscles start to carry out anaerobic respiration.

Anaerobic respiration

It is possible to release energy from glucose without using oxygen. **Anaerobic respiration** is the release of a little energy from each molecule of glucose in the *absence* of oxygen.

During strenuous exercise, not enough oxygen may reach the body muscles for aerobic respiration to supply all the energy the muscles need. Muscle tissue respires anaerobically to release energy. Most of the enzyme-catalysed reactions of aerobic respiration do not happen without oxygen. As a result the glucose is not broken down to carbon dioxide and water, but to **lactic acid** instead:

glucose $\longrightarrow$ lactic acid + energy released

Other tissues, such as cardiac muscle in the heart, do not normally respire anaerobically as this would not release enough energy to keep the heart beating properly. Some bacteria also respire anaerobically to make lactic acid in the same way, for example those that we use to make yoghurt.

Other microorganisms, such as yeast, respire anaerobically when oxygen is absent or in short supply in their surroundings, to make alcohol and carbon dioxide:

glucose $\longrightarrow$ alcohol + carbon dioxide + energy released

$$C_6H_{12}O_6 \longrightarrow 2C_2H_5OH + 2CO_2 + \text{energy released}$$

Plant roots respire anaerobically when land is flooded and soils become saturated with water so little or no oxygen is available.

Far less energy is released from each molecule of glucose in anaerobic respiration compared to aerobic respiration. This is because glucose is not completely broken down and a lot of energy remains stored in lactic acid and in alcohol in the form of chemical bond energy. You can see this if you set alcohol alight.

	energy released (kJ / g^{-1} glucose)
aerobic respiration	16.1
fermentation by yeast	1.2
anaerobic respiration in muscle	0.8

Lactic acid

Lactic acid can slowly poison muscles and cause cramps, so it must be removed from the body. The build up of lactic acid in the muscles and blood, during vigorous exercise, causes the **oxygen debt**.

Our oxygen debt builds up after we exercise hard, and it has to be 'paid back' straight away. We carry on breathing faster and deeper after vigorous exercise in order to supply more oxygen for aerobic respiration to break down lactic acid. The oxygen debt is removed during recovery by aerobic respiration of lactic acid in the liver.

After exercise the heart continues to beat at a fast rate in order to transport lactic acid in the blood from the muscles to the liver. Sprinters build up lactic acid in their muscles. They often hold their breath during a 100 metres race. Afterwards they need about $7\,dm^3$ of oxygen to get rid of the lactic acid. That is why they breathe deeply after the race in order to 'repay their oxygen debt'.

The enzyme-catalysed reactions that release most of the energy in aerobic respiration occur in mitochondria. These cell structures do not function without a supply of oxygen. During anaerobic respiration, glucose is partly broken down in the cytoplasm and alcohol or lactic acid is produced from the end product of this partial breakdown.

This runner is repaying her oxygen debt by breathing deeply after her training.

SUMMARY QUESTIONS

1. a Define the term *anaerobic respiration*.
   b Give two examples of anaerobic respiration.
   c Compare the energy released from a molecule of glucose during aerobic and anaerobic respiration.

2. Describe and explain the differences between the energy released in aerobic and anaerobic respiration. (See the table opposite for some figures to use in your answer.)

3. The table below shows the units of lactic acid produced in the leg muscles of an athlete during a race:

time / minutes	0	10	20	30	40	50	60	70	80
units of lactic acid	0	1	7	12	9	6	3	1	1

   a Draw a graph using the data.
   b When did the lactic acid reach a maximum?
   c Explain why the muscles produced lactic acid.
   d Suggest why the muscles produced less lactic acid towards the end of the race.

4. Explain why we carry on breathing heavily after we finish a session of hard exercise.

5. Explain how an oxygen debt develops and how it is repaid.

KEY POINTS

1. Anaerobic respiration is the chemical reaction in cells that break down glucose to release energy without using oxygen.

2. In anaerobic respiration, glucose is broken down to release energy in the absence of oxygen, forming lactic acid in muscle tissue, or alcohol, plus carbon dioxide in yeast and plants.

3. Far more energy is released from each molecule of glucose in aerobic than in anaerobic respiration.

4. The build up of lactic acid in muscles during hard exercise produces an *oxygen debt* that has to be paid back by breathing in extra oxygen.

1 Which process does *not* require energy from respiration?

   A diffusion of oxygen across the alveolar wall

   B contraction of muscle during exercise

   C passage of nerve impulses

   D synthesis of protein molecules

(Paper 1) [1]

2 When yeast respires without oxygen, the cells produce:

   A alcohol

   B carbon dioxide and alcohol

   C lactic acid

   D lactic acid and carbon dioxide

(Paper 1) [1]

3 During aerobic respiration, enzymes catalyse chemical reactions to breakdown nutrient molecules to:

   A produce glucose

   B produce lactic acid

   C produce oxygen

   D release energy

(Paper 1) [1]

4 During vigorous exercise, muscle respires without oxygen. The result of this respiration is the release of

   A carbon dioxide

   B carbon dioxide and water

   C little energy from each molecule of glucose

   D much energy from each molecule of glucose

(Paper 1) [1]

5 Which does not have mitochondria?

   A ciliated epithelial cell

   B liver cell

   C muscle cell

   D red blood cell

(Paper 2) [1]

6 The equations for respiration show the overall change that occurs to glucose as a result of enzyme-catalysed reactions. Which term describes the role of glucose for the first enzyme in respiration?

   A activator

   B active site

   C product

   D substrate

(Paper 2) [1]

7 After vigorous exercise, a student's heart rate is high. Which is *not* a reason for this?

   A carbon dioxide from anaerobic respiration needs to be removed from the muscles

   B lactic acid from anaerobic respiration needs to be removed from the muscles

   C oxygen is required in the liver for metabolism of lactic acid

   D waste products of respiration need to be removed

(Paper 2) [1]

8 The oxygen debt is the additional volume of oxygen absorbed by the lungs after exercise. This oxygen is required for

   A anaerobic respiration in muscle tissue

   B metabolism of lactic acid in the liver

   C removal of carbon dioxide

   D respiration of muscle tissue to release energy for contraction

(Paper 2) [1]

9 (a) Write out and complete the word equation for aerobic respiration:

   ____ + oxygen $\longrightarrow$ ____ + ____ +____

[4]

A student set up the apparatus shown opposite to find the rate of respiration of crickets.

The student took several readings from the apparatus. After each reading the clip was opened for several minutes before being sealed again.

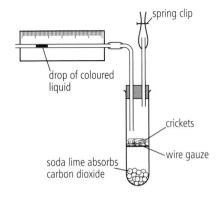

spring clip

drop of coloured liquid

crickets

wire gauze

soda lime absorbs carbon dioxide

(b) State the direction in which the coloured drop will move as the crickets respire in the apparatus. Explain your answer. [4]

(c) Explain why the clip is opened after taking the readings. [2]

(d) State one factor that should be kept constant during the investigation. [1]

(Paper 3)

10 Copy and complete the table with ticks and crosses. The first row has been completed.

feature	photosynthesis	respiration
occurs only in plant cells	✓	✗
produces oxygen		
consists of a series of enzyme-catalysed reactions		
produces carbon dioxide		
releases energy		
takes place all the time (night and day)		

(Paper 3) [5]

11 The apparatus shown in question 9 was used to determine the effect of temperature on the rate of respiration of Pinto beans.

The apparatus was put into a water bath at 27 °C with the clip open. After ten minutes the clip was closed and the position of the droplet recorded over time. The results are as follows.

time / minutes	0	5	10	15	20	25	30	35
position of droplet / mm	0	0	0	31	65	95	130	162

(a) Explain why
  (i) a water bath was used
  (ii) the apparatus was left for ten minutes before closing the clip. [2]

(b) The diameter of the capillary tube is 0.8 mm. Use the results in the table to calculate the rate of oxygen uptake in mm³ per hour. [3]

(c) Explain how the results would differ if the investigation was repeated at 17 °C and at 50 °C. [4]

(Paper 4)

12 (a) Name the structures inside animal cells that carry out aerobic respiration. [1]

The graph shows the oxygen consumption of an athlete who exercised for 10 minutes after a short period of rest.

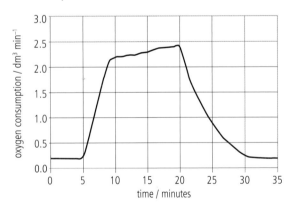

(b) Describe the change in oxygen consumption over the 35 minutes that recordings were taken. [4]

(c) Explain the changes in oxygen consumption between
  (i) 10 and 20 minutes [6]
  (ii) 20 to 30 minutes. [5]

(Paper 4)

13 Excretion

13.1   Excretion

LEARNING OUTCOMES

- Define the term *excretion*
- Name the excretory organs and excretory products
- Describe the role of the liver in the assimilation of amino acids
- Define *deamination*

Excretion is the removal from the body of the waste products of metabolism, toxic materials and substances in excess of requirements. The waste substances are removed from the body by the excretory organs, which are the lungs, liver and kidneys (see Figure 13.1.1).

Excretory products

Humans have two main excretory products.

- **Carbon dioxide** is made in body tissues during respiration. It is transported to the lungs in the blood plasma. Here, it diffuses out of the blood into the air in the alveoli and is breathed out (see page 130).
- **Urea** is made in the liver from excess amino acids. It is carried to the kidneys in the plasma where it is filtered out and leaves the body dissolved in the form of urine.

Excretion is a vital process in the metabolism of the body. Without it, substances like carbon dioxide and urea would be allowed to build up in the tissues. These substances would reach a toxic level and destroy the tissues.

Do not confuse 'excretion' with 'egestion'. Substances that are **excreted** are produced by body cells in metabolism. Substances that are **egested** have been eaten and have passed through the alimentary canal without being digested and absorbed into the bloodstream. Fibre is an example.

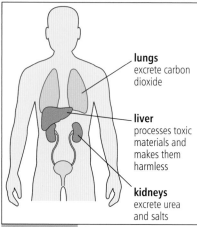

lungs
excrete carbon dioxide

liver
processes toxic materials and makes them harmless

kidneys
excrete urea and salts

Figure 13.1.1 The excretory organs.

Assimilation

Assimilation means that the food molecules that have been absorbed now become part of the cells or are used by the cells.

The liver carries out a number of important functions as part of assimilation:

- stores glucose by removing it from the blood and storing it as glycogen. This helps to regulate the concentration of glucose in the blood (see page 166)
- uses amino acids to make proteins, such as plasma proteins, e.g. fibrinogen, involved with blood clotting
- breaks down excess amino acids
- converts fatty acids and glycerol into fat which is stored around the body, e.g. under the skin
- produces cholesterol from fats (see page 71).

Supplement

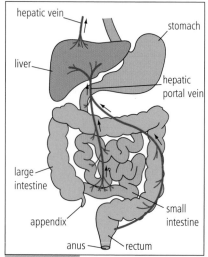

hepatic vein

stomach

liver

hepatic portal vein

large intestine

small intestine

appendix

anus   rectum

Figure 13.1.2 The position of the hepatic portal vein.

Roles of the liver

The liver carries out over 200 different roles in the body. As you can see from Figure 13.1.2, all the blood from the digestive system flows to the liver first before going into the rest of the circulation. The amino acids produced in the alimentary canal by the digestion of protein are absorbed into the blood in the small intestine. Your body cannot store the amino acids that it cannot use and so each molecule is broken down into two parts. One of these parts is made into urea.

Other roles of the liver include making bile, which neutralises acid as it enters the small intestine, and emulsifies fat (see page 84). The liver breaks down hormones that have circulated in the blood for a while and harmful substances, such as alcohol and drugs (see page 178).

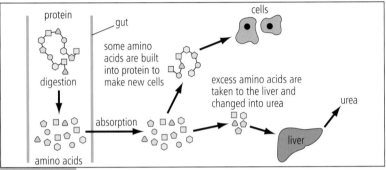

Figure 13.1.3 This shows what happens to amino acids in the body.

Amino acids cannot be stored so the liver breaks down amino acids that are not required for making proteins. Each amino acid molecule is broken down into two by the process of **deamination**. One molecule is converted to carbohydrate or fat and used as a source of energy. The other molecule is ammonia (NH_3) which combines with carbon dioxide to form the excretory product **urea**. This is carried to the kidneys in the blood where it is filtered out and is excreted in urine (see page 146). The part of the amino acid molecule that contains nitrogen is known as the amino group – this is why the process is called deamination because this group is removed.

The liver breaks down toxins, for example drugs such as alcohol and paracetamol.

KEY POINTS

1 Excretion is the removal from the body of waste chemicals made in tissues during metabolism.

2 Excretory products include carbon dioxide and urea.

3 The lungs, liver and kidneys are excretory organs. Carbon dioxide is excreted by the lungs.

4 The liver makes urea from excess amino acids. The kidneys excrete urea in urine.

5 The liver breaks down toxins such as **alcohol** and **drugs.**

6 The liver uses amino acids to make proteins such as plasma proteins and fibrinogen.

7 Deamination is the breakdown of excess amino acids in the liver to form ammonia, which is made into urea.

SUMMARY QUESTIONS

1 a Explain how each of these excretory products are produced and then removed from the body:
   i carbon dioxide   ii urea

2 Distinguish between excretion and egestion.

3 Describe what happens during deamination.

The kidneys

The kidneys make up part of the **urinary system**. They are responsible for the excretion of urea and excess salts from the body. They also control the water and ion content of the blood.

Your kidneys are at the top of your abdominal cavity just underneath the diaphragm. They are protected by the backbone, lower ribcage and the fat that surrounds them.

Waste chemicals like urea diffuse from cells into the blood and are transported to the kidneys. Blood enters the kidneys through the **renal arteries**. Inside each kidney is a complex network of filtering units called kidney tubules. As the blood flows through these units small molecules, such as glucose, salts, water and urea, are forced out of the blood plasma to form a fluid known as filtrate. As the filtrate passes along the tubule the useful substances (glucose and salts) are reabsorbed into the blood.

The volume of water reabsorbed depends on the body's water content. If you are dehydrated (short of water) perhaps because you have been sweating a lot during exercise, or because it is very hot, then the kidneys will take back as much as they can so you produce concentrated urine and conserve water. If you have taken in a lot of water, or it is a cold day and you have not been sweating, the kidneys reabsorb less water and you will produce dilute urine.

At the end of each tubule the urine is released and flows into the **ureter** and to the **bladder** where it is stored. From here it leaves the body, at intervals, through a shorter tube called the **urethra**. The **renal vein** carries blood with a low concentration of waste chemicals away from the kidney.

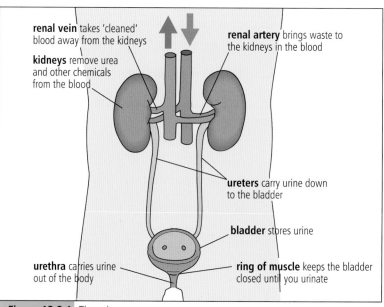

Figure 13.2.1 The urinary system.

Inside the kidneys

If you cut open a kidney lengthways, you can see three areas: the **cortex** – a brown outer area, the **medulla** – a reddish inner area, and the **pelvis** – a white area.

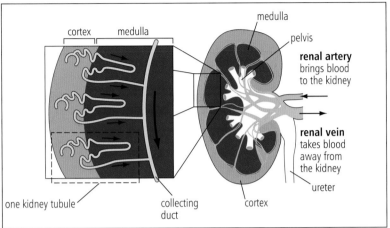

Figure 13.2.2 A vertical section of a kidney (right hand side). Detailed view of part of the kidney (left hand side).

Inside each kidney are thousands of tiny tubes called **kidney tubules** which you can see only with a microscope. The function of the kidney tubules is to filter the blood and remove waste chemicals and determine how much water is excreted. The filtering is carried out in the cortex. The waste chemicals and excess water are removed from the body in the urine which flows from the kidneys down the ureter and is stored in the bladder.

A kidney with part of the ureter attached.

A kidney cut open to show the cortex, medulla, pelvis and ureter.

SUMMARY QUESTIONS

1 Match each structure in the urinary system in the left column with one of the functions in the right column.

renal artery	stores urine
ureter	filters urea and other waste chemicals out of the blood
kidney	carries blood with a high concentration of urea
bladder	carries urine down to the bladder
renal vein	carries blood with a low concentration of urea

2 Sketch a vertical section of the kidney and label:

A cortex **B** medulla **C** renal artery **D** renal vein **E** ureter

3 a Distinguish between excretion and egestion.

b What is the main nitrogenous waste product excreted by the kidneys and where is it produced?

c Distinguish between the ureter and the urethra.

d Name two substances found in the blood plasma which are not found in the urine of a health person.

KEY POINTS

1 The urinary system consists of the kidneys, ureters, bladder and urethra.

2 The internal structure of a kidney consists of an outer cortex and an inner medulla.

3 The kidneys filter materials out of the blood, take back useful substances into the blood, e.g. glucose, control the quantity of water lost in the urine and excrete urea in the urine.

4 The ureters carry urine to the bladder from the kidneys.

5 The volume and concentration of urine depends on the intake of water and how much water has been lost in sweat during exercise and in hot weather.

LEARNING OUTCOMES

- Describe the structure of a kidney tubule
- Describe the role of the kidney tubule in filtration, removal of urea and excess water and the reabsorption of glucose and some salts

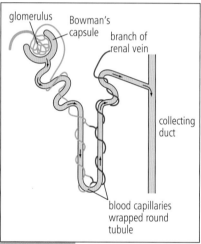

Figure 13.3.1 Kidney tubule.

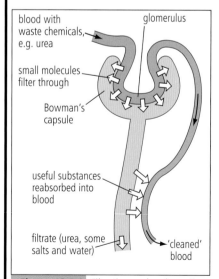

Figure 13.3.2 Filtration and reabsorption.

Structure of a kidney tubule

Blood containing waste chemicals flows into the kidney in the renal artery. Inside the kidney, the renal artery branches many times to give arterioles. Each of these arterioles supplies blood to a closely packed group of capillaries called a **glomerulus**.

After filtration, blood flows out of the glomerulus into another arteriole. Blood then flows into capillaries around the rest of the tubule, which joins to form the renal vein. The kidney tubule itself consists of a Bowman's capsule at one end. This is formed by the end of the tubule being pushed in like the finger of a rubber glove. The glomerulus is found inside the Bowman's capsule.

Figure 13.3.1 shows that the tubule forms a loop which goes deeper into the medulla and back to the cortex (see page 147). The end of the kidney tubule drains into a **collecting duct** which goes through the medulla and empties urine in the pelvis and then flows into the ureter.

DID YOU KNOW?

People on survival courses are taught to assess their level of dehydration by looking at the colour of their urine. The darker yellow the urine, the more dehydrated the person.

Filtration and reabsorption

Filtration

Look at Figure 13.3.2 showing how a kidney tubule works.

The kidneys are close to the heart so the blood pressure in the renal artery is high. You can see that the blood vessel entering the glomerulus is wider than the one leaving it. So there must be more blood entering the glomerulus than there is leaving it. This causes pressure to increase inside the glomerulus (it's rather like a bottleneck situation). It is this pressure that causes the blood to be filtered.

The lining of the capillaries is like a net with tiny holes in it. Blood cells and large molecules, like blood proteins, are too big to pass through the capillary lining and so stay in the blood. Small molecules, like urea, glucose, salts and water, pass out of the glomerulus and into the Bowman's capsule. It is in this space that the filtrate collects.

Reabsorption

All of the glucose, some salts and much of the water are needed by the body. They are **reabsorbed** back into the blood from the kidney tubule. This reabsorption involves active transport (see page 35).

If you look at the structure of the cells in the wall of the tubule, you will see adaptations associated with active transport:

- **microvilli** provide a large surface area for absorption.
- numerous mitochondria provide energy for active transport.

After the process of reabsorption, what is left is urea and excess salts dissolved in water. As this fluid flows on through the tubule, some water may be reabsorbed if the body is low in water. The fluid that enters the collecting ducts is urine. It flows down the collecting duct and then to the ureter and the bladder. Urine collects and is stored in the bladder. From here it is expelled through the urethra at intervals. The blood leaving the kidney in the renal vein has a much lower concentration of waste chemicals.

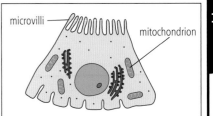

Figure 13.3.3 An epithelial cell from the lining of the kidney tubule.

STUDY TIP

Reabsorption of glucose and salts occurs by diffusion and active transport. Water is reabsorbed by osmosis. Expect to explain reabsorption in the kidney by referring to these processes (see pages 30 and 34).

SUMMARY QUESTIONS

1 Sketch a diagram of the kidney tubule and label:

  a glomerulus   b Bowman's capsule   c collecting duct
  d branch of the renal vein   e branch of the renal artery

2 Describe what happens in kidney tubules to form urine.

3 The table shows the concentrations of five substances in three body fluids.

substance	concentration / g dm^{-3}		
	blood in renal artery	filtrate in kidney tubule	urine
urea	0.2	0.2	20.0
glucose	0.9	0.9	0.0
amino acids	0.05	0.05	0.0
salts	8.0	8.0	16.5
protein	82	0	0

Use the information in the table to:

  a state the substances that pass from the blood into the kidney tubule,

  b explain how the substances you have named pass into the kidney tubule,

  c state the substances that are reabsorbed into the blood from the kidney tubule and explain why they are reabsorbed,

  d explain the results for protein.

KEY POINTS

1 Blood is filtered as it passes through the glomerulus removing urea, salts, glucose and water.

2 The useful materials such as glucose, some salts and most of the water are reabsorbed into the blood capillaries along the tubule.

3 Urine, consisting of urea, some salts and water passes out of the kidneys, along the ureters, to be stored in the bladder until urination.

Supplement

LEARNING OUTCOMES

- Describe the process of dialysis and explain how it controls the concentration of glucose, urea and salt in the blood
- Describe how a kidney dialysis machine works
- Describe the advantages and disadvantages of kidney transplants compared with dialysis

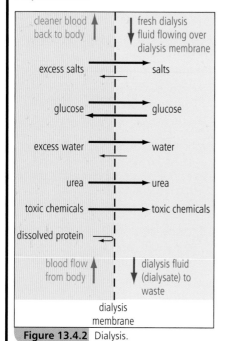

Dialysis machine.

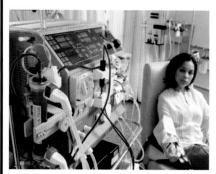

Figure 13.4.2 Dialysis.

The kidney dialysis machine

A person can survive on one kidney. However, if both kidneys become diseased or damaged, it can be fatal unless a medical procedure known as **kidney dialysis** is available. A common indicator that the kidney is damaged is the presence of protein in the urine. This happens when the glomeruli are damaged so that large molecules of protein pass out of blood plasma into the filtrate.

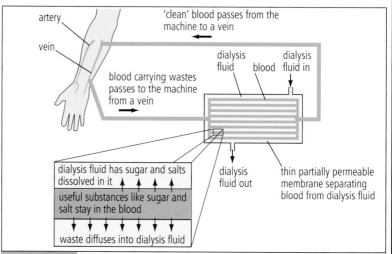

Figure 13.4.1 Renal dialysis.

This is what happens when someone has dialysis treatment.

- A tube is connected to one of the patient's veins.
- The blood flows along the tube and into the machine.
- Inside the machine the blood is pumped over the surface of a dialysis membrane, which separates the patient's blood and the dialysis fluid.
- Urea diffuses out of the blood, across the dialysis membrane and into the dialysis fluid.
- The dialysis fluid already has glucose and salts in it, so there will be no overall loss of glucose and salts from the blood by diffusion into the fluid.
- Urea and other waste chemicals leave the machine in the dialysis fluid.
- The patient's 'cleaner' blood passes back into the vein.

The patient has to use the dialysis machine for several hours, two or three times a week. This may involve attending a special ward in hospital or it can be done at home. Between visits to the dialysis ward of the hospital, patients eat a restricted diet without too much protein and salt. If they eat too much protein, they will produce too much urea which becomes toxic and would mean more frequent dialysis treatment.

Dialysis

Dialysis involves the separation of two solutions by a partially permeable membrane. This **dialysis membrane** has tiny pores in it which allow water and small solutes to diffuse through, but not larger molecules such as proteins. In the case of kidney dialysis the blood is pumped over the surface of a dialysis membrane which contains a fluid (the **dialysate**) which has the normal concentrations of salts and glucose and is the same water potential as normal plasma.

During dialysis, solute molecules such as salts and glucose diffuse through the membrane. As the dialysis fluid has the same water potential as normal blood plasma, the concentrations of solutes, such as blood proteins, reach equilibrium and are restored to normal. Excess salts, urea and other toxic chemicals are not present in the dialysis fluid, so they diffuse out of the blood and across the membrane into the dialysate.

Kidney transplants

A person with failed kidneys may have a kidney transplant. This involves replacing the diseased kidney with a healthy one from a donor. A donor may be a close relative or friend or someone who has just died.

Before a kidney transplant is carried out, the blood type and tissue type of donor and recipient (the person receiving the kidney) must be matched. Blood typing is the same as for blood transfusion. Tissue typing reduces the chances that the recipient's immune system will attack the new kidney and reject it. To reduce the chances of rejection occuring, drugs are given to the patient after the operation to suppress the immune system. They remain on these for life, which means they are more likely to catch certain infections and to have those infections for longer.

After recovering from the operation the patient is able to lead a normal life, including a normal diet, and does not need to have dialysis treatment in hospital or at home. Transplants replace all the functions of a kidney, not just some of them as with dialysis. Kidneys continuously remove urea from the blood rather than at intervals. They also make a hormone that stimulates red blood cell production in bone marrow. However, it is difficult to find a donor kidney that is a suitable match and there is always the danger of tissue rejection. Some wealthy people have resorted to paying healthy, poor people to donate a kidney to them.

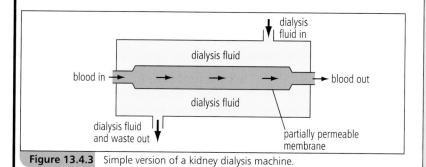

Figure 13.4.3 Simple version of a kidney dialysis machine.

SUMMARY QUESTIONS

1 Explain why the presence of protein in the urine may indicate kidney damage.

2 Figure 13.4.3 shows a simple version of a kidney dialysis machine.

   The patient's blood is kept separate from the dialysis fluid by a partially permeable membrane.

   a Name the process by which urea passes from the blood into the dialysis fluid.

   b Explain how concentrations of salts and glucose are returned to normal while on dialysis.

3 Make a table to show the advantages and disadvantages of kidney dialysis and kidney transplants.

1 Which substances are found in the urine of a healthy person?

A glucose and proteins

B glucose and water

C proteins and salts

D salts and water

(Paper 1) [1]

2 The urinary system makes urine, stores it and removes it from the body. Which organ is *not* part of the urinary system?

A bladder

B kidney

C liver

D ureter

(Paper 1) [1]

3 Which is used to make urea?

A excess amino acids

B excess fatty acids

C excess glucose

D excess hormones

(Paper 1) [1]

4 Which is the blood vessel with the lowest concentration of urea?

A aorta

B renal artery

C renal vein

C vena cava

(Paper 1) [1]

5 Which process reabsorbs glucose from the filtrate in the renal tubules?

A active transport

B cohesion

C diffusion

D osmosis

(Paper 2) [1]

6 Which is *not* an excretory function of the liver?

A breakdown of alcohol

B deamination

C production of urea

D protein synthesis

(Paper 2) [1]

7 The conversion of amino acids to plasma proteins, such as albumen and fibrinogen, occurs in which organ?

A heart   **C** liver

B kidneys   **D** lungs

(Paper 2) [1]

8 As part of an investigation, samples of urine were collected from a person throughout a 24 hour period. The urine sample collected early in the morning was more concentrated than the samples collected later in the day. Which is the most likely explanation for this?

A less urea was produced during the day

B less water was reabsorbed from the urine in the renal tubules at night

C the blood glucose concentration decreased during the day

D the concentration of the blood plasma increased at night

(Paper 2) [1]

9 The diagram shows the urinary system and its blood supply.

(a) (i) Name parts **A** to **D**. [4]

     (ii) State the functions of structures **A** to **D**. [4]

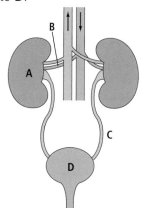

(b) (i) The following is a list of components of the blood. Identify those that are found in the urine of healthy people.

     blood proteins, urea, glucose, platelets, red blood cells, salts, water [1]

     (ii) State three factors that affect the volume and concentration of urine. [3]

(Paper 2)

10 Ammonia is produced by all animals as a waste product of their metabolism.

(a) Ammonia produced by humans is converted to urea. Explain why this is necessary. *[3]*

The table shows the composition of blood plasma in the renal artery, filtrate in Bowman's capsule and urine.

substance	concentration / g dm⁻³		
	blood plasma in renal artery	filtrate in Bowman's capsule	urine
urea	0.2	0.2	20.0
glucose	0.9	0.9	0.0
amino acids	0.05	0.05	0.0
mineral ions	8.0	8.0	16.5
protein	82	0.0	0.0

(b) Explain why
  (i) four of the substances shown in the table are present in the filtrate, but protein is not *[2]*
  (ii) glucose and amino acids are not present in the urine *[2]*
  (iii) the concentration of urea and mineral ions is higher in the urine than in the filtrate *[3]*
  (iv) the blood in the renal artery has a higher concentration of urea than blood in the renal vein. *[2]*

(c) Explain how filtrate is formed in the kidney. *[4]*

(Paper 4)

11 A man with kidney failure received regular dialysis treatment for 17 days. The graph shows how the concentration of urea in his blood changed over the 17 days of his treatment and for 10 days afterwards.

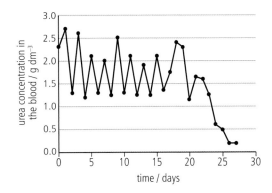

(a) State how many times the person received dialysis treatment; give the evidence from the graph to support your answer. *[2]*

(b) Calculate the decrease in the concentration of the man's urea in the blood from the start of dialysis until day 27. Show your working. *[2]*

(c) (i) Describe the changes that occur in the urea concentration in the blood shown in the graph. *[4]*
  (ii) Explain the changes that you have described. *[3]*

(d) Kidney transplants are the most common organ transplants. Describe the advantages of having a kidney transplant compared with continuing with dialysis treatment. *[3]*

(e) Before a kidney is transplanted from a living donor, it is important that the donor and the recipient have compatible blood types. If a kidney from a deceased person is used then it is also tissue typed to match the recipient as closely as possible. Explain why these procedures are necessary. *[3]*

(Paper 4)

14 Coordination and response

Nervous control in humans

Supplement

LEARNING OUTCOMES

- Describe nerve impulses as electrical signals that pass along nerve cells
- Describe the structure of the human nervous system
- Distinguish between voluntary and involuntary actions

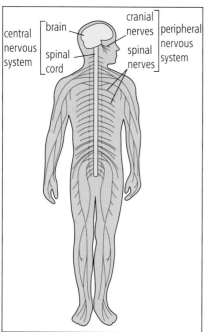

Figure 14.1.1 The main parts of the human nervous system.

STUDY TIP

Remember: nerve cells transmit impulses, not 'messages'.

All living organisms are sensitive to changes in their environment; the changes in daylight, wind speed, temperature, movement, availability of food that they detect are called **stimuli** (singular: stimulus). **Receptors** are the cells that detect the stimuli; for example, cells in your taste buds detect chemicals in your food. **Effectors** are the organs, such as glands and muscles, that bring about responses. Glands secrete useful substances and muscles contract to bring about movement. The actions they take as a result are called **responses**.

Animals have nervous systems which at their simplest are networks of specialised nerve cells, as in jellyfish and sea anemones. At their most complex, they are organs with powerful processing skills, such as the human brain. Two organ systems are involved – the sensory system for detecting changes in our environment and the nervous system for coordinating our responses and initiating actions. Here we discuss the nervous system. The nervous system is involved in the coordination and regulation of body functions. You will find information about the sensory system in Topic 14.4.

Human nervous system

The human nervous system consists of two parts: the **central nervous system** (CNS) and the **peripheral nervous system** (PNS). The central nervous system is made up of only two organs: the **brain** and the **spinal cord**. The brain is protected by the skull and the spinal cord is protected by the vertebral column.

The central nervous system is connected to different parts of the body by nerves that make up the peripheral nervous system. **Cranial nerves** and **spinal nerves** are in pairs on either side of the brain and spinal cord respectively. Cranial nerves link the brain with all the organs in the head and also some in the thorax and abdomen. Spinal nerves link the brain to the arms, thorax, abdomen and legs. Spinal nerves leave the spinal cord in pairs through spaces between the vertebrae.

Each nerve is made up of lots of nerve cells or **neurones** surrounded by a protective fibrous tube. Nerves are visible to the naked eye, but you need microscopes to see neurones. Neurones transmit information in the form of nerve impulses which are like pulses of electricity passing along a wire.

Voluntary actions are under conscious control and involve the brain in decision making. A footballer carries out a voluntary action when he or she kicks a ball. Turning over the pages of a book is an example of a voluntary action. These actions are under the conscious control of the brain.

Involuntary and voluntary actions

We make two types of action. Some happen automatically without us having to think about them. These are **involuntary actions** that occur unconsciously. Swallowing, blinking, breathing and the beating of the heart are four such actions that we are not conscious of most of the time, but we can choose to control occasionally. Movement of food in the ileum (see page 82) and of urine in the ureter (see page 146) are other involuntary actions you have studied that we can't control by thinking about them. **Voluntary actions** are those we choose to make and the decisions to make them occur in our brains.

When a person sits down on a sharp object such as a pin without realising it, there is a very quick, automatic response that the person does not need to think about.

This is an example of a **simple reflex** which is an involuntary action.

The stimulus is the pin and the receptors are pain sensors in the skin. The effectors are the muscles in the legs that cause the person to get up quickly.

The sequence of events is:

stimulus ⟶ receptor ⟶ coordinator ⟶ effector ⟶ response

The **coordinator** is the part of the body that connects information about the stimulus to the effector. This always happens inside the central nervous system to allow connections with other parts of the body and for the brain to be aware of stimuli. Inside the spinal cord, neurones from pain receptors connect with neurones to the muscles. Some reflexes involve the brain as we will see in the section on the eye (see page 160).

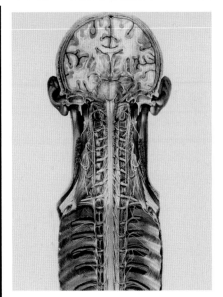
Front view of the top part of the nervous system.

STUDY TIP

You can see more about the neurones that control simple reflexes on page 157. They are arranged into reflex arcs involving peripheral nerves and the CNS. All responses are coordinated through the CNS.

SUMMARY QUESTIONS

1 Copy and complete the sentences using these words:

cranial   peripheral   spinal   neurones   central

The brain and spinal cord make up the _____ nervous system. All the nerves of the body make up the _____ nervous system. _____ nerves are connected to the brain and _____ nerves are connected to the spinal cord. The nervous system is composed of many specialised cells known as _____.

2 a Put the following words into the correct sequence:

coordinator   stimulus   effector   receptor   response

b Define each of the above words.

3 a Explain the difference between an involuntary action and a voluntary action.

b Give two examples of involuntary actions.

c Give two examples of voluntary actions.

KEY POINTS

1 The human nervous system is made up of the central nervous system (brain and spinal cord) and the peripheral nervous system (cranial and spinal nerves).

2 Neurones or nerve cells transmit nerve impulses.

3 Some actions are involuntary (not under conscious control). Other actions are voluntary and involve conscious control by the brain.

Neurones and reflex arcs

- Describe and distinguish between the three types of neurone found in a reflex arc
- Describe the functioning of these neurones in a reflex action

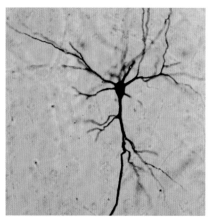

This neurone has a long axon.

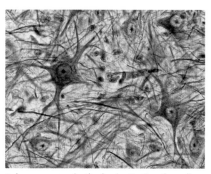

Relay neurones in the brain. Notice the cell bodies with their nuclei.

The axons of sensory neurones and motor neurones may be over a metre long. Think about the neurones in giraffes or the great whales!

Neurones

Neurones are highly specialised cells. Their structure allows them to transmit information as nerve impulses over long distances. The cell body contains the nucleus and almost all of the cytoplasm, but there are thin extensions of cytoplasm that may be quite short or extend a long way through the body.

Sensory neurones transmit impulses from sense organs to the brain and spinal cord.

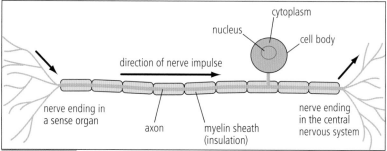

Figure 14.2.1 A sensory neurone.

Motor neurones transmit impulses away from the brain and spinal cord to effector organs – muscles and glands.

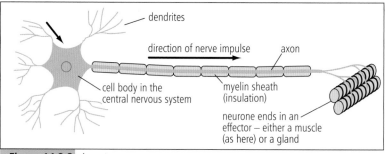

Figure 14.2.2 A motor neurone.

The diagrams above show the sensory and motor neurones surrounded by insulation. This insulation is known as **myelin**, formed by separate cells that grow around the neurones to form a layer rich in fat. This insulation makes impulses travel very quickly. Some myelinated neurones, which are used in our fast reflexes, transmit impulses at 100 metres per second.

A third type of neurone is the **relay (connector) neurone**. These are short and pass on impulses from sensory neurones to motor neurones inside the brain and spinal cord.

When any two neurones meet, they do not actually touch each other. There is a tiny gap between them called a **synapse**. When an impulse reaches a synapse a **chemical transmitter substance** is released from the first neurone. It diffuses across the synapse and triggers an impulse in the second neurone. Since the chemical transmitter is only produced on one side of the synapse, it ensures that impulses travel in one direction through the nervous system. Many drugs produce their effects by acting at synapses (see page 159).

Reflex arcs

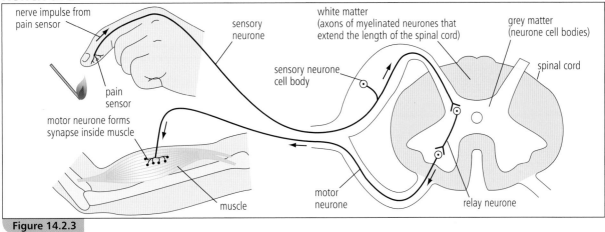

nerve impulse from pain sensor

sensory neurone

white matter (axons of myelinated neurones that extend the length of the spinal cord)

grey matter (neurone cell bodies)

sensory neurone cell body

spinal cord

pain sensor

motor neurone forms synapse inside muscle

muscle

motor neurone

relay neurone

Figure 14.2.3

If you touch a hot object, you remove your hand very quickly and automatically without thinking. This is an example of a simple reflex. Your brain receives impulses about the pain but not until you have already moved your hand. If you relied on the brain to make a decision you would be burnt in the time it took for the impulses to travel to the brain and back.

Look at the diagram above and trace the pathway of the impulse along the neurones.
This pathway is called a **reflex arc**.

The **stimulus** in the example is the hot flame. The **receptor** is the heat sensor in the skin. When it is

stimulated the heat sensor generates an impulse. The impulse travels to the spinal cord along a **sensory neurone**.

Inside the grey matter of the spinal cord the impulse passes across a synapse to a **relay (connector) neurone**. The relay neurone passes the impulse across a second synapse to a **motor neurone**.

The motor neurone transmits the impulse to a muscle in the arm. The muscle is the **effector** and it contracts to remove the hand from the hot object. This action is the **response**.

Many reflexes are protective; they happen very quickly, so you do not harm yourself.

SUMMARY QUESTIONS

1 Copy and complete the sentences using these words:

**synapse   grey   arc   relay   motor
sensory   effectors   transmitter**

In a reflex _____ nerve impulses are transmitted to the spinal cord by _____ neurones. Inside the _____ matter of the spinal cord the impulses are passed on to _____ neurones. The impulses leave the spinal cord along _____ neurones to go to _____ . There is a gap between two neurones called a _____ where chemical _____ substances are released to pass the impulse to the next neurone.

2 Look at the diagram of the motor neurone to the right:
   a   Name the parts labelled **A** to **D**.
   b   Give the functions of parts **B**, **C** and **D**.

3 a   Define each of the following:
      i   synapse
      ii   chemical transmitter substance
      iii   myelin   iv   relay neurone
   b   Explain the function of **i – iv** in the nervous system.

4 Make a drawing of a relay (connector) neurone.

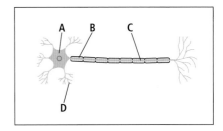

A   B   C

D

LEARNING OUTCOMES

- Define the term *synapse*
- Describe how an impulse passes across a synapse
- Describe how many drugs, e.g. heroin, affect nerve transmission at synapses

A **synapse** is a junction between two neurones. Where two neurones meet, they do not touch. There is a small gap between them, about 20 nm wide. The gap is called the **synaptic gap**. The neurone that carries the impulse to the synapse is called the **presynaptic neurone**. The neurone that carries the impulse away from the synapse is called the **postsynaptic neurone**.

How does a nerve impulse transfer across the synapse from one neurone to the next? Chemicals known as **neurotransmitters** are released by the presynaptic neurone and diffuse across the synaptic gap to trigger an impulse in the postsynaptic neurone.

Structure of the synapse

The axons of neurones end in swellings called **synaptic bulbs**. The surface of the synaptic bulb is called the **presynaptic membrane**. It is separated by the synaptic gap from the **postsynaptic membrane** of the cell body or axon of the next neurone.

If you look at Figure 14.3.1 you will see that the synaptic bulb has many vesicles containing neurotransmitter molecules. It also contains many mitochondria which suggests that energy is required in synaptic transmission. The postsynaptic membrane has a number of large protein molecules on its surface which act as receptor sites for the neurotransmitter substance.

Synaptic transmission

1 When an impulse arrives at the synaptic bulb it causes vesicles containing the neurotransmitter to move towards the presynaptic membrane.

2 The vesicles fuse with the presynaptic membrane, releasing the neurotransmitter into the synaptic gap.

3 The neurotransmitter diffuses across the synaptic gap and attaches to specific receptor sites on the postsynaptic membrane. These receptor sites have a complementary shape to the neurotransmitter, but the binding is only temporary.

4 The binding of the neurotransmitter triggers an impulse in the postsynaptic neurone. Once this has happened the neurotransmitter is broken down by an enzyme in the synaptic gap.

5 The mitochondria provide energy to reform the neurotransmitter.

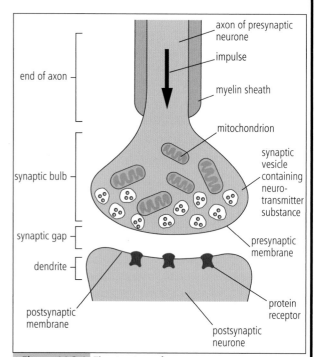

Figure 14.3.1 The structure of a synapse.

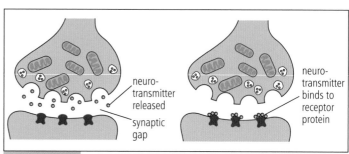

Figure 14.3.2 Action of a neurotransmitter at a synapse.

In a reflex arc synapses ensure that nerve impulses can only travel in one direction i.e. from presynaptic neurone to postsynaptic neurone. This happens because vesicles of neurotransmitter are only present in the presynaptic bulb and the receptors are only found on the postsynaptic membrane.

The effect of drugs on synaptic transmission

Most drugs that affect the nervous system have their effects at synapses. These drugs influence the transmission of impulses by influencing the release of neurotransmitters or by interacting with receptors either by stimulating them or by inhibiting them. **Amphetamines** are a type of **excitatory drug** that increases the activity of the nervous system. Amphetamines stimulate the release of neurotransmitters in the brain, so increasing brain activity and making a person more alert. They also have the effect of suppressing the appetite.

Heroin and **beta-blockers** are examples of **inhibitory drugs** that reduce the effects of neurotransmitters. Heroin acts on presynaptic neurones to reduce the release of neurotransmitters and this reduces the sensation of pain. Beta-blockers are drugs that block receptors for neurotransmitters and are taken by people to reduce blood pressure and heart rate.

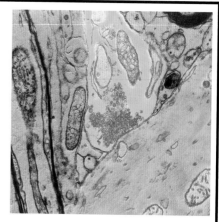

In this image the cytoplasm of the neurones is coloured orange. The image shows the synapse between a presynaptic neurone (above) and a postsynaptic neurone (below). The synaptic gap is the tiny space between them. You can see lots of vesicles clustered together at the bottom left of the presynaptic neurone. ($\times 14\,000$).

DID YOU KNOW?

Estimates for the number of neurones in the adult human brain reach 86 billion. As each neurone has many synapses with other neurones, the estimate for the number of these is therefore far greater: 0.15 quadrillion (0.15×10^{15}).

SUMMARY QUESTIONS

1 What is meant by each of these terms?

  a synapse b neurotransmitter c receptor molecules

2 Explain the function of each of the following in the transmission of an impulse across synapses:

  a synaptic vesicles b neurotransmitter

  c protein receptor molecules d mitochondria

3 a What is meant by the following terms?

    i excitatory drug

    ii inhibitory drug

  b Give an example of each different type of drug.

  c What effect do i heroin and ii amphetamines have on the transmission of nerve impulses across the synapse?

4 Do some research into the effects that the following drugs have on synaptic transmission:

  a caffeine b ecstasy c cocaine

KEY POINTS

1 A synapse is a junction between two neurones.

2 An impulse triggers the release of a neurotransmitter, which diffuses across the synaptic gap to bind with the receptor molecules on the postsynaptic membrane and trigger an impulse.

3 In a reflex arc, synapses ensure that impulses travel in only one direction.

4 Many drugs, e.g. heroin, have their effects by interacting with the receptor molecules at synapses.

Sense organs

- Define sense organs as groups of receptor cells responding to particular stimuli
- Describe the structure and function of the eye
- Explain the pupil reflex in terms of light intensity and pupil diameter

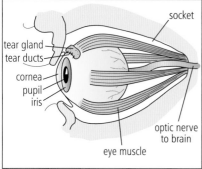

Musicians use a variety of their senses when performing.

Figure 14.4.1 The eye in its socket.

We are aware of our surroundings and use our sense organs to detect stimuli. We have receptor cells that are specialised to detect certain stimuli, such as light and chemicals. These cells generate nerve impulses when stimulated and these impulses travel along sensory neurones to the central nervous system.

Some receptors are found concentrated in groups in certain organs. The eye and the ear have receptors arranged in special tissues. They also have other tissues to modify the information from the environment before it reaches the receptors. In the eye there are structures for focusing light rays; in the ear there are structures for amplifying sound.

Sense organs are groups of receptor cells that respond to specific stimuli.

- The eyes respond to light rays and give us our sense of sight.
- The nose responds to chemicals in the air and gives us our sense of smell.
- The tongue responds to chemicals in our food and drink and gives us our sense of taste.
- The ears respond to sound vibrations and give us our sense of hearing. The ear also detects movement and position of the body, so provides information about our balance.
- The skin responds to pressure and gives us our sense of touch. It also detects pain and temperature (see page 168).

Eye structure

Your eye sits inside a socket in the skull and is moved about by three pairs of eye muscles as you can see in Figure 14.4.1. The eye is a complex organ which is composed of different tissues, such as fibrous tissue, muscle tissue, sensory tissue and blood. Each tissue has a function to help with the detection of light. Figure 14.4.2 shows the structures in the eye and their functions.

At the front of the eye is the transparent **cornea**, through which light enters the eye. Light then passes through the **pupil**, which is a hole in the centre of the pigmented **iris**, through the lens to the **retina** at the back of the eye. The retina is the tissue with light-sensitive receptor cells.

At the very front of the eye is a delicate, transparent layer, which provides protection. It is kept moist by the tear glands that make the tears that wash your eye clean every time you blink. Tears also contain **lysozyme**, an enzyme that kills bacteria.

In humans, the eyes are set side by side at the front of the head as opposed to at the side, like a rabbit. What you see in front of you is called your visual field. As the eyes are a few centimetres apart, each eye views the same visual field but from a slightly different angle. Just try shutting your eyes alternately to experience this. The two fields of view overlap to give stereoscopic vision which gives us the ability to judge distances accurately. Stereoscopic vision is an essential adaptation for tree-dwelling animals, such as monkeys, and predators, such as lions and tigers, which must judge distances accurately to survive.

The pupil reflex

Look in a mirror and shine a bright light near your eyes and then turn it off. When you turn the light on your pupils decrease in size. This lets less light into your eyes and protects the retina from damage. When you turn the light off, your pupils increase in size to let more light into the eyes so you can see clearly. This is another example of a simple reflex. Receptor cells in the retina send impulses along sensory neurones in the optic nerve. These connect with motor neurones that pass impulses to the muscles in the iris to stimulate them to contract.

When a bright light is shone into the eye the pupil constricts to protect the retina from damage.

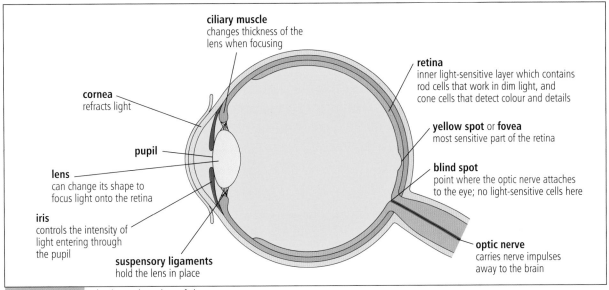

ciliary muscle
changes thickness of the lens when focusing

cornea
refracts light

pupil

lens
can change its shape to focus light onto the retina

iris
controls the intensity of light entering through the pupil

suspensory ligaments
hold the lens in place

retina
inner light-sensitive layer which contains rod cells that work in dim light, and cone cells that detect colour and details

yellow spot or **fovea**
most sensitive part of the retina

blind spot
point where the optic nerve attaches to the eye; no light-sensitive cells here

optic nerve
carries nerve impulses away to the brain

Figure 14.4.2 A horizontal section of the eye.

- Explain the pupil reflex in terms of light intensity and the antagonistic action of muscles in the iris

- Explain how the eye focuses on near and far objects

- State that rods and cones are light sensitive cells in the retina

- Describe the distribution of rods and cones and the way they function

- State the position of the fovea

STUDY TIP

Here is a way to remember this: the three Cs: circular, contract, constraint.

STUDY TIP

In Unit 2 we dealt with cells, tissues, organs and organ systems. The eye is a very good example of an organ where you can identify tissues (e.g. muscle tissue) and cells (e.g. rod and cone cells). The eye is part of an organ system: the sensory system.

Control of the pupil reflex

The size of the pupil is controlled by a pair of antagonistic muscles in the iris. The **circular muscles** are arranged around the pupil. The **radial muscles** run outwards from the pupil like the spokes in a wheel. These muscles are described as antagonistic because they have opposite effects: the circular muscles contract to reduce the size of the pupils while the radial muscles contract to increase the size of the pupils.

The diagrams show what happens in dim light when the light intensity is low and in bright light when it is much higher and has the potential to damage the retina.

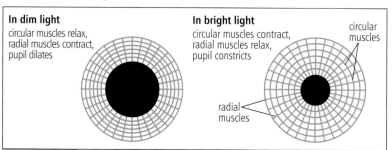

In dim light
circular muscles relax, radial muscles contract, pupil dilates

In bright light
circular muscles contract, radial muscles relax, pupil constricts

circular muscles

radial muscles

Figure 14.5.1 The iris muscles control the diameter of the pupil.

Focusing

Figure 14.5.2 shows how light is focused in the eye. Light enters the eye through the transparent cornea, passes through the lens to be focused on the retina. Notice that the image on the retina is inverted (upside down) because of the way the light rays reflected from the top and bottom of the object enter the eye and cross over behind the lens. Information about this inverted image goes to the brain where it is interpreted the right way up. We learn how to do this in the first few months of life.

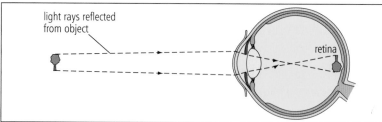

light rays reflected from object

retina

Figure 14.5.2 This shows light rays refracted by the cornea and the lens.

Rods and cones

Rods and cones are the light-sensitive receptor cells in the retina. Rod cells are sensitive to light of low intensity and send impulses only when it is dark, for example at night. They are responsible for our night vision. Cone cells are only sensitive to light of high intensity.

There are three types of cone cell, each sensitive to different wavelengths of light. The brain interprets the impulses from cone cells to give us our colour vision. The fovea in the centre of the retina contains cones and no rods. Each cone has its own neurone to the brain so this area in the middle of our visual field gives us a very detailed image. The rest of the retina contains rods and few cones. This area gives us our peripheral vision which is not as detailed. The position of the fovea is in the centre of the retina on the same horizontal level as the blind spot. The fovea in your right eye is to the right of your blind spot.

Accommodation is the term used to describe the changes that occur in the eye when focusing on far and near objects. As light enters the eye it must be refracted (bent) so that we can see the image clearly. About 60% of the refraction of the light rays is done by the cornea and the rest is done by the lens. The lens is surrounded by elastic tissue which can be stretched and can recoil. The shape of the lens is controlled by the **ciliary muscles**.

If you are looking at a distant object:

• the ciliary muscles relax,
• the pressure inside the eye pulls the suspensory ligaments tight (or taut) so the lens is pulled into an **elliptical** (thin) shape. Light rays are refracted as they pass through the lens and focused on the retina. The distant object is in focus.

If you are looking at a near object:

• the ciliary muscles contract to counteract the pressure inside the eye,
• the suspensory ligaments are not pulled and become slack so the elastic tissue around the lens recoils so the lens becomes more **spherical** (fatter). Light rays are refracted more than they were when looking at the distant object. The near object is in focus.

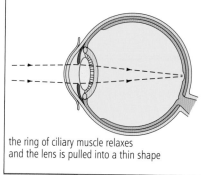
the ring of ciliary muscle relaxes and the lens is pulled into a thin shape

Figure 14.5.3 A distant object.

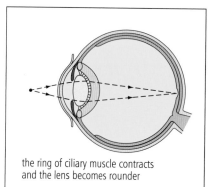
the ring of ciliary muscle contracts and the lens becomes rounder

Figure 14.5.4 A near object.

SUMMARY QUESTIONS

1 a When the eye views a distant object, explain what happens to:
   i the ciliary muscles ii the suspensory ligaments
   iii the shape of the lens
   b Explain what happens to each of these structures when the eye views a near object.

2 a State what happens to each of the following in bright light:
   i circular muscles of the iris ii radial muscles of the iris
   iii the size of the pupil
   b State what happens to each of the above in dim light.
   c Explain how the pupil reflex is controlled.
   d Suggest why doctors shine a light in people's eyes when they are unconscious.

3 Describe the distribution of rods and cones in the retina and explain their functions.

KEY POINTS

1 The pupil reflex involves the coordination of radial muscles and circular muscles in the iris to control the intensity of light entering the eye through the pupil.

2 The ciliary muscles, suspensory ligaments and the lens are all involved in accommodation (focusing) in the eye.

3 In the retina, rod cells are distributed around the periphery and respond to light of low intensity. Cone cells respond to high light intensity and detect colour. The fovea is made entirely of cones.

Hormones

LEARNING OUTCOMES

- Identify on a diagram the adrenal glands, the pancreas, the testes and the ovaries
- State the hormone released by each of these four glands
- Describe the effects of adrenaline on the heart, lungs and pupils
- Explain that adrenaline increases supply of glucose and oxygen to muscles during activity
- Compare the endocrine and nervous systems

STUDY TIP

The term 'secretion' means that cells make a substance that is useful for the body and release it into the blood or some other part of the body, e.g. the gut. Do not confuse secretion with 'excretion', which is the removal of metabolic waste (see page 144).

Adrenaline is released during times of excitement, fear or stress.

The endocrine system

A **hormone** is a chemical substance produced by an **endocrine gland** and transported in the blood. Endocrine means that the gland produces hormones that are secreted into the bloodstream rather than into a duct. Hormones travel throughout the body but only certain organs or tissues recognise each hormone and respond to it. These are referred to as **target organs**. Each hormone alters the activity of its target organs.

Hormones coordinate the activities of the body but in a different way to the nervous system. The main endocrine glands, the hormones that they secrete and the actions are show in Figure 14.6.1.

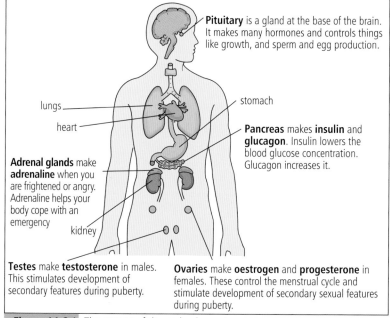

Pituitary is a gland at the base of the brain. It makes many hormones and controls things like growth, and sperm and egg production.

lungs

heart

stomach

Pancreas makes **insulin** and **glucagon**. Insulin lowers the blood glucose concentration. Glucagon increases it.

Adrenal glands make **adrenaline** when you are frightened or angry. Adrenaline helps your body cope with an emergency

kidney

Testes make **testosterone** in males. This stimulates development of secondary features during puberty.

Ovaries make **oestrogen** and **progesterone** in females. These control the menstrual cycle and stimulate development of secondary sexual features during puberty.

Figure 14.6.1 The organs of the endocrine system.

Adrenaline

Adrenaline is secreted by the adrenal glands. If you realise that you are in a situation that puts you at risk of injury, or even death, you will immediately become much more aware and ready to take action. The responses necessary to survive in these 'fight or flight' situations are coordinated by adrenaline, which gets you ready for action by:

- increasing the breathing rate
- increasing the heart (pulse) rate
- widening the pupils so that more light enters the eyes.

Adrenaline also stimulates an increase in the rate of chemical reactions within the body. To provide enough energy for increased levels of activity, adrenaline stimulates cells in the liver to convert glycogen to glucose, which diffuses into the blood. This increases the concentration of glucose in the blood, so that there is more available to the muscles as a source of energy for muscle contractions needed for sudden action.

Adrenaline coordinates the increased uptake of oxygen and changes in the distribution of blood so that oxygen and glucose reach the muscles that need them for respiration.

• The air passages (trachea, bronchi and bronchioles) widen to allow more air into the alveoli in the lungs. This increases the volume of oxygen that can be absorbed.

• The breathing rate increases to increase the uptake of oxygen and excrete carbon dioxide at a faster rate.

• Arterioles in the brain and muscles dilate (become wider) to deliver more glucose and oxygen for respiration in these organs.

• Arterioles in the gut and other organs constrict (become narrower) to divert blood to the muscles.

If too much adrenaline is released, as a result of prolonged stress and/or anxiety, constant high blood pressure and heart disease may result. Beta-blockers are drugs that act at synapses (see Topic 14.3). They also act on cell surface membranes to block the receptors for adrenaline and so reduce its effects. Beta-blockers help to reduce blood pressure and decrease the heart rate (see Topic 14.3).

Comparing the endocrine and nervous systems

The endocrine system and the nervous system act to control and coordinate the activities of the body.

feature	nervous system	endocrine system
structures	nerves	secretory cells in glands
forms of information	electrical impulses	hormones (chemicals)
pathways	along neurones	in the blood
speed of information transfer	fast	slow
longevity of action	short-lived, e.g. muscle contracts for a short time	usually slow and longer lasting
target area	only the area at the end of a neurone	whole tissue or organ
response	muscle contraction or secretion by glands	e.g. conversion of glucose to glycogen, protein synthesis, rate of respiration

KEY POINTS

1 Hormones are chemical substances that are secreted by endocrine glands and travel in the blood to alter the activity of target organs.

2 Adrenaline increases breathing and pulse rates and stimulates widening of the pupils in 'fight or flight' situations.

3 Adrenaline increases supply of oxygen and glucose to muscles during activity.

SUMMARY QUESTIONS

1 What is meant by the following terms?
   a endocrine system
   b hormone
   c target organ

2 State the gland which secretes each of the following:
   a testosterone   b insulin
   c oestrogen

3 Suggest four situations in which the secretion of adrenaline increases.

4 Explain how adrenaline prepares the body for action.

5 State four ways in which the endocrine system differs from the nervous system.

Supplement

LEARNING OUTCOMES

- Define the term *homeostasis*
- Explain the concept of negative feedback
- Describe the control of blood glucose concentration

STUDY TIP

Although we say that homeostasis is maintaining <u>constant</u> conditions in the body, they do change slightly all the time. It is more precise to say that homeostasis maintains <u>nearly constant</u> conditions or keeps them <u>within set limits</u>.

Controlling conditions

During metabolism, many chemical reactions occur inside the body. In cells, some molecules are built up, such as proteins, and some are broken down, such as glycogen. Enzymes control these reactions and work best when conditions, such as temperature, pH and concentration of water, are kept suitable and constant. See Unit 5 to see how temperature and pH influence enzyme activity.

Any slight change in the conditions in the body can slow down or stop enzymes from working. **Homeostasis** is the maintenance of constant internal conditions in the body, so that enzymes control metabolism efficiently.

The brain has overall control of our body processes. The control of body temperature is an example of homeostasis.

In humans, the normal temperature of the blood is 37 °C. When blood reaching the brain is warmer than this temperature it sends impulses along nerves to the skin to promote heat loss. For example, our sweat glands produce sweat and this helps us to cool down, returning body temperature to normal.

When we feel cold our blood temperature falls below 37 °C. Then the brain will stimulate the production of more heat in the muscles by shivering and by increasing the rate of respiration in the liver. It also stimulates the body to conserve heat by decreasing the production of sweat by sweat glands.

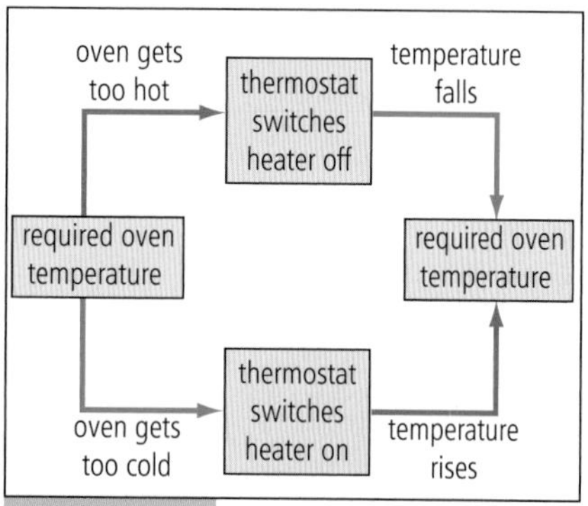

Negative feedback

The control of body temperature is an example of **negative feedback**. This is the same sort of control system as in an oven to keep the temperature constant at the pre-set temperature. Negative feedback acts to ensure that the actual temperature is as close to the pre-set temperature as possible. In our body, if the temperature increases above or below the normal 37 °C, then it responds by taking actions that will bring it back to normal. It may not be exactly 37 °C all the time, but will be close to it, within set limits. There are also control systems for blood glucose concentration, the water content of the blood, blood pH and the oxygen and carbon dioxide concentrations of the blood.

Figure 14.7.1 Controlling the temperature of an oven is an example of negative feedback.

Controlling blood glucose

Cells need glucose for energy and therefore need a constant supply from the blood.

When you eat a high carbohydrate meal your blood glucose concentration can increase by a factor of 20. However, it does not stay high because cells in the **pancreas** detect the high glucose concentration in the blood. These cells secrete the **hormone** called **insulin** into the blood. Insulin stimulates liver cells to convert glucose

Supplement

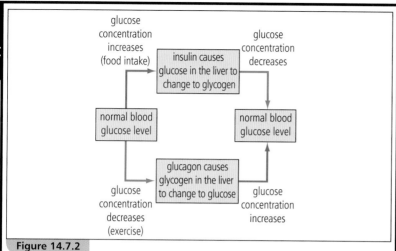

Figure 14.7.2

into the storage compound **glycogen** (see page 38). This stimulates the liver cells to absorb lots of glucose from the blood so the concentration decreases and returns to normal.

When you run a race or carry out some other type of exercise, your muscles take up lots of glucose from the blood to provide the energy you need. Your blood glucose concentration decreases, but it does not keep decreasing. Other cells in the pancreas detect this decrease and secrete the hormone **glucagon** into the blood. Glucagon stimulates liver cells to break down glycogen to glucose, which diffuses into the blood so the blood glucose concentration increases to normal.

Blood glucose concentration is controlled by the pancreas, which *monitors* the blood glucose concentration and releases hormones to instruct liver cells to remove glucose molecules from the blood or add more.

This is an example of negative feedback because the pancreas:

- makes insulin to stimulate a decrease in blood glucose concentration, or
- makes glucagon to stimulate an increase in blood glucose concentration,

to make sure the concentration stays within limits and does not increase or decrease too much (see treatment of Type 1 diabetes in Topic 10.3).

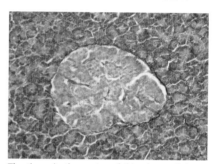

The tissue in the centre of this photo of the pancreas secretes insulin and glucagon.

KEY POINTS

1 Homeostasis is the maintenance of constant internal conditions, such as body temperature.

2 Homeostasis involves negative feedback. It operates to maintain constant conditions of glucose, water, carbon dioxide and oxygen in the blood as well as its pH.

3 The pancreas controls the concentration of glucose in the blood by producing the hormones insulin and glucagon which stimulate the liver.

SUMMARY QUESTIONS

1 a Define the term *homeostasis*.
  b Explain why is it important that conditions in cells are kept constant.

2 a Explain the term *negative feedback* using controlling blood glucose as an example.
  b Explain the roles of the pancreas and the liver in maintaining a constant blood glucose concentration.

Controlling body temperature

Polar bears have a thick layer of insulating fat to conserve their body heat.

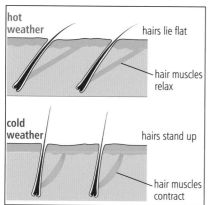

Figure 14.8.2 Hairs are involved in maintaining body temperature.

Birds and mammals maintain a constant body temperature. They are very successful animals as they can live in cold environments as well as hot ones and be active at night as well as during the day. The skin provides ways to adjust the body temperature in different conditions.

The skin

The skin is the largest organ in the body and has several important functions:

- protects the body from damage
- stops pathogens from entering
- prevents too much water loss
- detects changes in temperature
- detects pressure (touch) and pain
- loses heat by conduction, convection, radiation and evaporation.

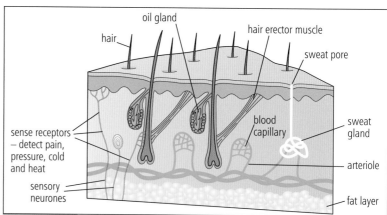

Figure 14.8.1 The structure of human skin.

Many of the structures in the diagram help in controlling body temperature. We lose and absorb heat through our skin – mostly by radiation. We use our skin for heat loss when we are hot and to conserve heat when we are cold.

Mammals have fur or hair, which traps a layer of air close to the skin. Air is a poor conductor of heat. So air pockets in the fur or hair reduce the loss of heat to the atmosphere by keeping convection currents of air away from the warm surface of the skin. When it is cold, hair erector muscles *contract* so that hairs stand erect to trap a thick layer of air so reducing heat loss.

When it is hot, hair erector muscles *relax* so that hairs lie flat. Less air is trapped close to the skin so more heat can be lost as convection currents flow closer to the surface of the skin.

Mammals also have a layer of fatty tissue beneath their skin, which is a good insulator as it does not conduct heat well. Mammals that live in cold environments have very thick layers of fat.

Controlling body temperature

No matter what the weather is like, your body temperature stays very close to 37°C unless you have a fever. There are receptors in the brain that detect changes in the temperature of the blood. When the brain detects any change in the blood temperature, it coordinates the action of effectors to restore the blood temperature to normal.

In the heat, the brain detects an increase in the blood temperature. The brain sends impulses to the skin to increase the rate of sweating so that more heat is lost by evaporation from the skin surface.

In the cold, the brain detects a decrease in the blood temperature. The sweat glands stop producing sweat and the hair erector muscles contract to raise the hairs. The body starts to shiver as some of the body muscles contract spontaneously and release heat from respiration. Blood flows through the muscles and is warmed by this heat.

Supplement

A very effective way to regulate body temperature is to change how much blood flows through the capillaries near to the surface of the skin. Arterioles supply these capillaries with blood. The muscle around the arterioles can contract and relax to change the blood flow. To lose heat, these muscles in the walls of the arterioles relax. The vessels widen and there is an increase in blood flow through the capillaries. This is known as **vasodilation**. More heat is now lost to the surroundings by convection and radiation.

In the cold, the muscles in the walls of the arterioles contract so the vessels become narrower. This reduces the blood flow through the capillaries and the blood is diverted to stay beneath the fat layer in the skin. This is known as **vasoconstriction**.

The control of body temperature is by negative feedback as shown in Figure 14.8.3. The brain coordinates the responses by effector organs to lose, conserve or gain heat in order to maintain a body temperature that stays within set limits.

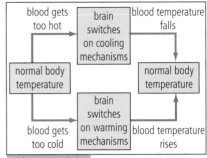

Figure 14.8.3

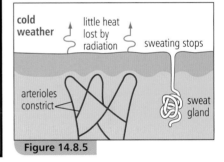

Figure 14.8.4

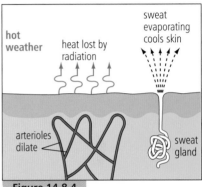

Figure 14.8.5

Supplement

SUMMARY QUESTIONS

1 Explain how the brain controls body temperature.

2 Describe what happens in the skin:
   a to lose heat when we are hot,
   b to conserve heat when we are cold.

3 How do we produce heat in the body if we are cold?

4 Explain how fur and fat tissue beneath the skin help in temperature control in cold environments.

5 Explain how vasodilation and vasoconstriction are involved in maintaining body temperature.

KEY POINTS

1 The brain controls body temperature by using the many structures in the skin.

2 Insulation conserves heat, sweating loses heat and shivering generates heat.

3 Vasoconstriction and vasodilation control how much heat is lost from the blood in the skin.

LEARNING OUTCOMES

- Define the terms *gravitropism* and *phototropism*
- Explain the chemical control of plant growth by auxins
- Explain the effect of synthetic plant hormones used as weedkillers

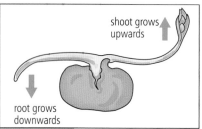

Figure 14.9.1 Which way is up?

STUDY TIP

A seed germinates. A root and then a shoot appear. Which way is up? Gravitropic responses ensure the survival of the seedling as it establishes itself in competition with other plants nearby.

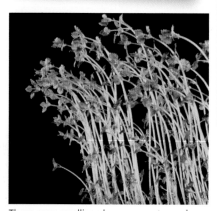

These cress seedlings have grown towards the light.

Tropisms

Plants respond to stimuli such as light and gravity. Different parts of the plant either grow towards or away from the different stimuli. These growth responses are much slower than the responses coordinated by the nervous systems in animals. They are called **tropisms**.

Gravitropism

Whichever way a seed is planted, the root always grows downwards in the direction of gravity. The shoot always grows upwards against the direction of gravity.

A growth response to gravity is called a **gravitropism**.

Roots are **positively gravitropic** because they grow in the same direction as the stimulus, gravity. They grow downwards into the soil to absorb water and mineral ions.

Shoots, on the other hand, are **negatively gravitropic**. They will always grow in the opposite direction to the direction of gravity. This ensures that shoots grow upwards getting the leaves into sunlight.

Auxin

A plant hormone called **auxin** controls this growth. Auxin in shoots stimulates growth by causing the cells to elongate, but in roots auxin inhibits growth by slowing down cell elongation.

If a seedling is put on its side the auxin builds up on the lower side of the shoot and root. In the shoot the auxin builds up on the lower side and stimulates the cells there to elongate and therefore to grow more on the lower side. This causes the shoot to bend upwards.

In the root the auxin also builds up on the lower side. But auxin slows down growth in a root by inhibiting cell elongation. As a result, cells on the upper side of the root elongate more than those on the lower side so the root bends downwards.

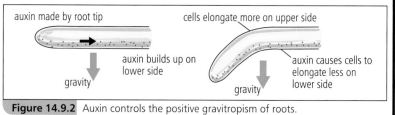

Figure 14.9.2 Auxin controls the positive gravitropism of roots.

Phototropism

The cress seedlings in the photo have been put in a window for some time and their shoots have grown towards the light.

This sort of growth response is called a **phototropism**. Because shoots grow *towards* the light they are **positively phototropic**.

The advantage of this tropic response is that the leaves expose the maximum surface area to the light. They can then absorb light more efficiently for photosynthesis.

This response is controlled by auxin made in the shoot tip. The auxin moves away from the shoot tip down the stem. When a shoot receives light only from one side the auxin is unequally distributed so most is found on the shaded side of the stem. The auxin stimulates cells on the shaded side to elongate more than the cells on the side in direct light. The shoot grows more on the shaded side and bends towards the light. Roots are not sensitive to light.

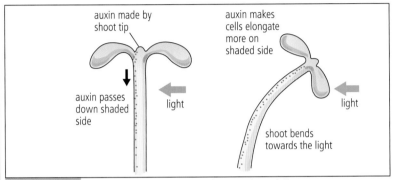

Figure 14.9.3 Auxin controls the positive phototropism of shoots.

Synthetic plant hormones

Synthetic auxins are very effective as **selective weedkillers**. Important cereal crops like wheat and barley are narrow leaved (monocotyledonous) whereas most of the weeds that compete with them for water, nutrients and light are broad leaved (dicotyledonous). Herbicides like 2,4-D are sprayed on the crop but tend to run off the leaves of the cereal plants because they are thin and point upwards. The weedkillers are more likely to be absorbed by the broad-leaved weeds.

The auxins increase the growth rate of the affected weeds, possibly by increasing the rate of cell division. The weeds cannot provide enough food from photosynthesis to maintain this rate of growth and soon die.

STUDY TIP

Cells become longer, or elongate, when they absorb water by osmosis. The vacuole swells, increasing the turgor pressure inside the cell which causes the cell wall to stretch. Auxins work by making it easier for the walls to stretch.

This plant has been sprayed with a hormone weedkiller that has made it grow in a twisted fashion, causing it to die.

KEY POINTS

1 A gravitropism is a growth response to the stimulus of gravity. Roots are positively gravitropic and shoots are negatively gravitropic.

2 A phototropism is a growth response to the stimulus of light. Shoots are positively phototropic and grow towards the light.

3 Auxins are plant hormones that control growth. Synthetic auxins are used as weedkillers by making the weeds grow too fast.

SUMMARY QUESTIONS

1 Define each of the following:

a a tropic response b phototropism c gravitropism

2 State what growth response will happen in each of the following.

a Some seedlings are placed in a window, in sunlight for 6 hours.

b Some germinating seeds are placed with their roots lying horizontally for a few hours.

3 Explain the observed growth responses in question 2 in terms of the action of auxin.

4 Explain how synthetic auxins are able to act as weedkillers.

1 In which pathway do nerve impulses travel in a spinal reflex?

A effector ⟶ sensory neurone ⟶ relay neurone ⟶ sensory cell ⟶ motor neurone

B motor neurone ⟶ relay neurone ⟶ sensory neurone ⟶ effector

C sensory cell ⟶ sensory neurone ⟶ effector ⟶ relay neurone ⟶ motor neurone

D sensory cell ⟶ sensory neurone ⟶ relay neurone ⟶ motor neurone ⟶ effector

(Paper 1)   *[1]*

2 Where are relay neurones situated?

A effector   **C** spinal nerve

B spinal cord   **D** receptor

(Paper 1)   *[1]*

3 A nerve impulse travels from a fingertip to the spinal cord in 0.02 s. This is a distance of 1.4 metres. The speed of the impulse is:

A $0.7\,\mathrm{m\,s^{-1}}$ **C** $70\,\mathrm{m\,s^{-1}}$

B $7\,\mathrm{m\,s^{-1}}$ **D** $170\,\mathrm{m\,s^{-1}}$

(Paper 1)   *[1]*

4 Which shows the correct match between the hormone and its effect?

A adrenaline – decrease in heart rate

B insulin – increase in blood glucose concentration

C oestrogen – decrease in the volume of urine

D testosterone – stimulates development of sperm

(Paper 1)   *[1]*

5 Which occur when in a hot environment?

	arterioles in the skin	sweat glands	hair erector muscles
A	constrict	active	relax
B	constrict	not active	contract
C	dilate	active	relax
D	dilate	active	contract

(Paper 2)   *[1]*

6 The diagram shows the response of a plant shoot to receiving light from one side only.

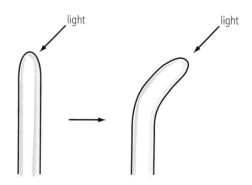

Which statement explains how this response is controlled?

A Auxins stimulate a gravitropic response.

B Auxins stimulate a phototropic response.

C Auxins stimulate cells on the side closest to the light to grow faster than the cells on the opposite side.

D Auxins stimulate cells on the side away from the light to grow faster than the cells on the opposite side.

(Paper 2)   *[1]*

7 Which are symptoms of Type 1 diabetes?

A bleeding gums and poor wound healing

B constant alertness and insomnia

C thirst and unexplained weight loss

D tiredness and unexplained weight gain.

(Paper 2)   *[1]*

8 A man injures his arm in an accident. Afterwards, he can feel objects touching his hand, but he cannot move his hand away from them. The cause of this could be that:

A all nerves between his hand and the spinal cord are damaged

B all nerves between his spinal cord and the effectors in his arm are cut

C all nerves between the receptors in his hand and his spinal cord are cut

D all the receptors in his hand are damaged

(Paper 2)   *[1]*

9 The diagram shows a horizontal section of the eye.

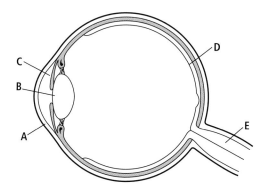

(a) Name the parts labelled A to E. *[5]*

(b) Explain why the eye is a sense organ. *[2]*

(c) A person walks from a dark room into a brightly lit place.

   (i) Describe the change that you would see happening to the person's pupils. *[1]*

   (ii) Suggest the advantage of the change you described in (i). *[1]*

(Paper 3)

10 (a) Copy and complete the flow chart that shows part of the control of the concentration of glucose in the blood.

> increase in blood glucose concentration
> ↓
> detected by …
> ↓
> insulin secreted
> ↓
> liver cells respond by …

[2]

(b) Write out a similar flow chart to show the body's response to a decrease in blood glucose concentration. *[3]*

(c) Explain the principle of negative feedback, using the control of glucose in the blood as an example. *[5]*

(Paper 4)

11 A scientist investigated the distribution of rods and cones across the retina of a mammal as shown in the diagram. The graph shows the distribution of cones in the retina.

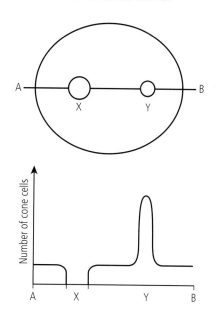

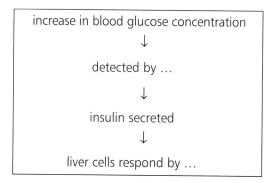

(a) Name the parts of the retina X and Y. Give reasons for your answers. *[4]*

(b) Copy the graph and show on it the distribution of rods in the retina. *[3]*

(c) (i) State an involuntary action involving the eye. *[1]*

   (ii) Explain how the control of an involuntary action differs from the control of a voluntary action. *[3]*

(d) Explain the advantages to an animal of having two coordination systems: the nervous system and the endocrine system. *[3]*

(Paper 4)

12 (a) Describe the events that occur at a synapse when a nerve impulse arrives at the end of a sensory neurone. *[4]*

(b) There are huge numbers of synapses in the spinal cord and the brain. Explain why this is so. *[3]*

(Paper 4)

15 Drugs

15.1   Drugs

LEARNING OUTCOMES

- Define the term *drug*
- Describe the use of antibiotics in the treatment of bacterial infection
- State that some bacteria are resistant to antibiotics which makes them less useful
- State that antibiotics kill bacteria but do not affect viruses
- Explain the strategies that can be used to reduce the development of antibiotic-resistant bacteria
- Explain why antibiotics kill bacteria but do not affect viruses

Supplement

STUDY TIP

There are many ways to define the term 'drug'. Learn the definition at the top of this page, which covers all the types of drugs we describe, so you can write it from memory.

Many medical drugs are dangerous if taken in an uncontrolled way and are available only on prescription from a doctor.

What are drugs?

A drug is any substance that is taken into the body that alters or influences chemical reactions in the body.

There are several different types of drug. Some are the active ingredients of medicines and are used to treat and cure people of disease. Others are mood-enhancing which alter sensory perception. Aspirin, paracetamol, morphine and antibiotics are medical drugs designed to suppress pain, counteract the symptoms of 'flu', soothe muscular pain or kill pathogenic bacteria. Alcohol, nicotine and caffeine are socially acceptable drugs that people take for their pleasurable effects, to help them relax or concentrate.

Drugs that tend to be illegal because of their more severe effects are sometimes called **recreational drugs**. Hallucinogens, like Lysergic Acid Diethylamide or LSD, cause psychedelic visions for the drug user. Stimulants, like cocaine, ecstasy and amphetamines are all mood-enhancing drugs which give the user a short-lived feeling of well-being and energy. Heroin, although it affects the body in a different way, also gives a feeling of well-being.

Drugs act upon the human body in various ways. Some drugs, such as nicotine and heroin, interfere with the way that the nervous system works and these are often abused by people. Many of the mood-enhancing drugs, such as heroin and nicotine, act at synapses in the nervous system and change the way in which neurones send impulses. They do this by combining with protein molecules on the cell membranes of the neurones.

The liver is the site of the breakdown of alcohol and other toxins. Drugs are broken down in the body by enzymes and the products are excreted. The breakdown products can be detected in the urine, and this is why urine tests are carried out to see if people have been taking drugs. Athletes taking part in competitions and car drivers who have been in accidents are routinely tested for drugs.

Many drugs have the potential to be **addictive**. If the body comes to rely on the drug, a person can become addicted and feel the need to take it regularly. In many countries some or all of the non-medicinal drugs are illegal and possession or dealing in them results in severe punishment.

The body's metabolism may become used to the drug. The liver may produce more enzymes to break it down so that the dose of the drug has to increase to have the effect that the user first experienced.

Antibiotics

Antibiotics are a group of chemicals made by microorganisms (bacteria and fungi) that are used to kill pathogens or stop their growth. Antibiotics are prescribed by doctors and vets to treat and cure human and animal diseases caused by bacteria and fungi. Antibiotics can be injected or be taken by mouth. Usually a course of antibiotics has to be taken over a certain period of time.

Penicillin was the first antibiotic to be discovered and mass-produced. Penicillin acts on bacteria by inhibiting cell wall formation, leading to a breakdown of the cell wall and the leakage of cell contents. Some antibiotics stop substances crossing cell membranes and others prevent enzymes catalysing important reactions, such as making proteins. Antibiotics have no effect on viruses. There is no point in taking antibiotics to treat viral diseases.

A big problem with antibiotics is that many bacteria have become resistant to them, so doctors and vets need to know which antibiotic to prescribe for a particular disease.

MRSA and other antibiotic-resistant bacteria are being controlled in hospitals by use of antibacterial gels which this doctor is using.

Supplement

Viruses are not cells. They do not carry out their own metabolism, but rely entirely on the cells of their host. This means that to control viruses we would have to use antibiotics that inhibit our own metabolism and that is not possible. This explains why antibiotics cannot be used to control diseases caused by viruses. There are anti-viral drugs available, such as AZT which is used to treat HIV/AIDS, but they work differently to antibiotics. There are far fewer anti-viral drugs than there are antibiotics.

Bacterial resistance to antibiotics has been caused by overuse of antibiotics, especially in hospitals. Overuse was responsible for the development of methicillin-resistant *Staphylococcus aureus* or MRSA. This has caused deaths in hospital patients with suppressed immune systems, such as those who have had organ transplants. The number of cases of MRSA in the USA and UK has decreased in recent years as a result of specific strategies. Prescribing antibiotics only when necessary and ensuring that people complete their courses of antibiotic treatment by taking all their pills are ways to control bacterial resistance. See more on antibiotic resistance on page 239.

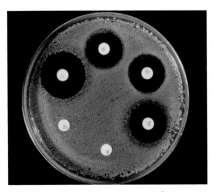

Before prescribing, a doctor may ask for a sensitivity test to be carried out in a hospital laboratory to find out which antibiotic to use against the pathogen that has infected a patient. The white discs on this agar plate contain different antibiotics. Here, four antibiotics would be effective because they have killed the bacteria, whereas two would not.

SUMMARY QUESTIONS

1 a Define the term *drug*.
  b Name two drugs that are used as medicines.
  c Name two mood-enhancing drugs.
  d Name two drugs that can be harmful, but are not illegal in many countries.

2 Explain how antibiotics act to kill pathogens like bacteria and fungi.

3 Explain why antibiotics can kill bacteria but not viruses.

Supplement

KEY POINTS

1 A drug is a substance taken into the body, which alters or influences chemical reactions in the body.

2 Antibiotics destroy pathogens by disrupting cell wall formation, inhibiting protein synthesis and metabolism in the pathogen cell.

3 Antibiotics cannot kill viruses since they have no cell wall but rather live inside host cells, taking over their metabolic processes.

Supplement

LEARNING OUTCOMES

- Describe the effects of the abuse of heroin
- State that injecting heroin can cause infections such as HIV
- Describe the problems of addiction to heroin, severe withdrawal symptoms and problems such as crime and HIV infection

- Explain how heroin affects the nervous system, limited to its effect on the function of the synapses

STUDY TIP

Heroin and alcohol are both depressants. You should use this term when writing about their effects on the body.

STUDY TIP

You should be able to apply these ideas of tolerance and dependence to heroin and alcohol.

Biology of heroin

Morphine and codeine are two compounds extracted from opium poppies and used as painkillers in medicine. **Heroin** is a compound modified from morphine. Heroin has only limited medical uses as it is highly addictive and so is rarely used.

Heroin is a powerful **depressant** that slows down the nervous system. Heroin has a chemical structure that is similar to **endorphins**, a group of chemicals that are found in the brain. Endorphins are made naturally in the brain and provide relief when the body experiences pain or stress. They work by flooding the synapses in the brain and preventing neurons from transmitting impulses from pain receptors, so producing pain relief. When a person takes heroin, the heroin molecules bind to the endorphin-receptor sites on the post-synaptic membrane of the synapse, blocking nerve transmission. This mimics the function of natural endorphins (see more on synapses in Topic 14.3). When people take heroin for the first time they usually feel a warm rush and a feeling of contentment and intense happiness. This state is known as **euphoria**.

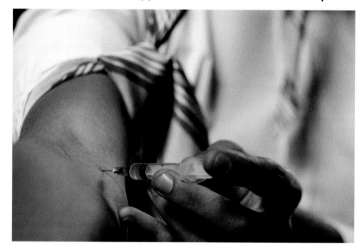

An addict injecting heroin into a vein.

Addiction to heroin

Heroin becomes part of the body's metabolism and the body quickly gets used to the drug. This includes the nerve cells where it has its effect. More painkillers are needed to prevent them sending impulses to the brain. But the body does not produce more of its own natural painkillers. The feeling of pain becomes unbearable and addicts have to take heroin to reduce the pain. This is how the body develops a **tolerance** to the drug, and it has to be taken in ever greater quantities in order to feel euphoria or just to deaden the pain.

People who start taking heroin want to repeat the feelings they first experienced, but they quickly get 'hooked' or addicted as they are dependent on the drug. This can happen within a matter of just two to three weeks of misuse of heroin.

Heroin is taken by smoking, sniffing or injecting it into a vein. Injecting is highly dangerous as veins can collapse and tissues surrounding injection points start to die, giving the condition known as gangrene. Long-term addicts can run out of suitable places to inject it. Some addicts share needles and syringes. The danger associated with this is the transmission of the human immunodeficiency virus (HIV) and the virus that causes hepatitis (see page 211).

Social problems

Heroin addiction is a serious problem for many countries. The drug often dominates the lives of addicts and they think only about how they can get their next 'fix'. The drug can be expensive, which drives many people to turn to crime in order to fund their habit. It can often become difficult to keep a job, and addiction often leads to family breakdown and homelessness. Addicts tend to shun friends and families, preferring the company of other addicts, and so they become isolated from society.

Heroin is produced in Afghanistan, which is the largest producer of heroin in the world, and in the Golden Triangle (Laos, Vietnam, Burma and Thailand). It is then trafficked illegally into Europe, North America and elsewhere. Networks of criminal gangs help process and smuggle the drugs into developed countries and distribute drugs to drug dealers who sell it on the streets. Most of the heroin-producing countries have harsh penalties, such as life imprisonment and the death penalty, to stop trafficking. A user who stops taking the drug experiences very unpleasant **withdrawal symptoms**. These can include sleeplessness and hallucinations (imagining terrible things are happening to them), muscle cramps, sweating, vomiting and nausea.

A great deal of willpower and support from others is needed to overcome heroin addiction. Often the only way to overcome the habit is to go into **rehabilitation**. This may involve staying in a treatment centre. In some countries addicts are given another drug, methadone, which gives some of the same effects as heroin.

Drugs are often the cause of many people becoming involved with crime. Support for drug addicts who want to give up their habit is available in rehabilitation centres.

Heroin addiction is a serious form of addiction, but far more people put their lives at risk from nicotine addiction.

Supplement

KEY POINTS

1 Heroin is a highly addictive depressant drug.

2 Heroin addicts may turn to crime to obtain money for their next dose. Shared needles have resulted in the spread of hepatitis and HIV amongst addicts who inject heroin.

3 Heroin interacts with synapses in the nervous system.

SUMMARY QUESTIONS

1 Define each of the following in relation to drug use:
   a tolerance b dependence
   c addiction d rehabilitation

2 a Why is heroin described as a highly addictive, depressant drug?

   b If a heroin addict stops taking the drug, describe what happens to them.

3 Heroin addiction can lead to crime, social and family problems and the spread of viral diseases.

   Explain how these consequences can affect society in a negative way.

Alcohol and the misuse of drugs in sport

Supplement

LEARNING OUTCOMES

- Describe the effects of excessive alcohol consumption
- Describe the long-term effects of alcohol on the body and on society
- Discuss the use of hormones in sport

type of drink	volume / cm³	alcohol by volume / %	units
alcopop	400	5	2
extra strong lager	500	8	4
cider (glass)	500	4	2
cocktail	125	40	5
wine	175	12.5	2.2
spirit (vodka) and mixer	35	40	1.4

alcopop   extra strong   cider
                     lager

cocktail   wine   spirit and
                                     mixer

Figure 15.3.1 Some alcoholic drinks.

Biology of alcohol

Alcohol is a socially acceptable drug in many countries in the world. In some, its supply and consumption are very tightly regulated and in others it is banned entirely.

When alcohol is consumed, it is absorbed into the blood very quickly since it is a small molecule that does not need to be digested. It is also soluble in cell membranes, so it is absorbed very quickly through the wall of the stomach and small intestine. The presence of food in the stomach slows down its absorption.

Alcohol gets distributed throughout the body in the blood. Some is lost through the lungs and the kidneys. It is absorbed by liver cells and broken down by enzymes so that its concentration in the blood decreases gradually. This breakdown happens more quickly in men than in women because men have more of these enzymes and their enzymes tend to be more active. They also have more water and less fat in their bodies than women, which tends to cause the concentration of alcohol in the blood to decrease more quickly.

Like heroin, alcohol is a depressant. It affects the brain by slowing down the transmission of nerve impulses. In small quantities, this has the effect of removing inhibitions so people find it easier to socialise with others. With larger quantities this lack of inhibitions leads to

- loss of coordination, judgment and control of fine movements
- slower reaction times
- loss of self-control.

Unlike heroin, alcohol does not lead to addiction as quickly if at all. Many people drink alcohol in small to moderate quantities throughout their lives without becoming addicted.

The alcohol content of drinks is measured in units. One unit is 8 grams of alcohol; this is the quantity broken down in the liver in one hour. The UK government recommends that men should not drink more than 3 to 4 units per day and women 2 to 3 (but see page 200 for advice about drinking during pregnancy).

Addiction

Some people become dependent upon alcohol and are sometimes referred to as alcoholics. They develop a tolerance as more enzymes that metabolise alcohol are made in the liver. They therefore need to take greater quantities of alcohol to get the same effect. They feel tense and irritable and find it hard to cope with everyday problems without a drink. Alcoholics can cause their families pain and misery. They can become aggressive after drinking and spend a lot of money on drink.

Social problems

The misuse of alcohol is a factor in crime, family disputes, marital breakdown, child neglect and abuse, absenteeism from work, vandalism, assault and violent crime including murder. In some countries the damage done by alcohol is considerable and often difficult for governments to measure.

Alcohol and other drugs are involved in many road accidents. Alcohol has adverse effects on people's concentration, while at the same time making them feel more confident. In some countries it is against the law to drink alcohol and then drive a vehicle or operate machinery. In most of Europe the legal limit of alcohol in the blood is 50 mg per 100 cm^3. There are severe penalties for driving with more than the legal limit.

Long-term effects of alcohol

Drinking large quantities of alcohol over a number of years can have serious effects on health. It can lead to stomach ulcers, heart disease and brain damage. The liver is the part of the body that breaks down alcohol and other toxins. Drinking large quantities of alcohol interferes with the metabolism in the liver so that fat is stored and builds up. This condition is known as fatty liver. If this continues the liver tissue is damaged and replaced by fibrous scar tissue.

If heavy drinking continues then the liver becomes full of nodules. This is the condition known as **cirrhosis**. The liver becomes less able to carry out its job of removing the toxins from the blood. This condition is not reversible and is fatal unless the person stops drinking. Long-term alcohol consumption also leads to brain damage, with changes in personality and behaviour.

The misuse of drugs in sport

Anabolic steroids are substances similar to the male sex hormone testosterone. They work by mimicking the protein-building effects of this hormone. The result is muscle growth, which gives the athlete increased strength and endurance.

There are a number of harmful side-effects resulting from the excessive use of anabolic steroids. In men, use of these drugs can bring about increased aggression, impotence, baldness, kidney and liver damage and even the development of breasts. In women, there is a development of male features, facial and body hair and irregular periods.

Alcohol and other drugs are thought to be the cause of many road accidents. Alcohol slows reaction times so it takes longer to respond to stimuli, such as traffic lights.

Lance Armstrong was banned from competitive cycling for life for doping offences in 2012.

Smoking and health

LEARNING OUTCOMES

- Describe the effects on the gas exchange system of carbon monoxide, nicotine and tar in tobacco smoke

- State that tobacco smoking can cause chronic obstructive pulmonary disease (COPD), lung cancer and coronary heart disease

- Discuss the evidence for the link between smoking and lung cancer

STUDY TIP

The chemicals in tobacco smoke have effects on the heart and circulatory system as well as on the gas exchange system. You may be asked about this in questions on transport in humans.

The actor Humphrey Bogart died at 57 from cancer brought on by smoking.

Tobacco plants make nicotine as a way to protect against insect attack. It is a natural pesticide. Nicotine happens to have a molecular structure that allows it to interact with our nervous system and so fits the definition of the term drug.

Biology of tobacco smoke

Nicotine is absorbed very quickly through the alveoli to enter the bloodstream. Like heroin, it is a drug that interacts with nerve cells at synapses. But unlike heroin it is a **stimulant**, not a depressant. Nicotine makes the heart beat faster and narrows the arterioles, which increases blood pressure. It also increases the stickiness of blood platelets that promote blood clotting.

Tar is the black sticky material that collects in the lungs as smoke cools. It does not pass into the bloodstream. Tar irritates the lining of the airways and stimulates them to produce more mucus. This tends to accumulate because the cilia which normally help to remove mucus are damaged by smoking, resulting in narrowing of the airways. Smokers cough to make this material move to the back of the throat giving rise to 'smoker's cough'. There are about 4000 different chemical compounds in tar. Some of these can cause cancer and are described as carcinogenic.

Carbon monoxide is a poisonous gas. It combines permanently with haemoglobin so reduces the volume of oxygen that blood can carry by as much as 10%. This puts an added strain on the heart that may already be beating harder and faster in response to nicotine. At a time when it needs more oxygen for its respiration, there is less delivered by the blood as it flows through the coronary arteries to supply heart muscle. If a woman smokes during pregnancy there may not be enough oxygen in the blood for the fetus to develop properly. Due to this, the baby may have a smaller birth weight and is sometimes premature (see page 200).

Diseases caused by smoking

Chronic obstructive pulmonary disease (COPD) is the term given for a collection of lung diseases including chronic bronchitis and emphysema. The effect of hot tobacco smoke on the airways is to reduce the efficiency with which they keep the lungs clean. Tobacco smoke irritates the lining of the airways. Mucus-secreting cells produce more mucus in response, but cilia on the cells lining the air passages stop beating. This means that mucus and the dust, dirt and bacteria that stick to it are not removed from the lungs. It accumulates and bacteria multiply. This stimulates the body's immune system to send phagocytes to the places where this happens – particularly the bronchi where particles settle out of the air.

Large amounts of **phlegm** (a mixture of mucus, bacteria and white blood cells) are produced, which people attempt to cough up. This condition is **chronic bronchitis**. Chronic means long term. People with this condition find it difficult to move air into and out of their lungs as the bronchi are partly blocked.

Particles, bacteria and tar reach the alveoli. Phagocytic white blood cells digest a pathway through the lining of the alveoli to reach them. Eventually this weakens the walls of the alveoli so much that they break down and burst, reducing the surface area for gas exchange. This condition is **emphysema**, which leaves people gasping for breath as they cannot absorb enough oxygen or remove carbon dioxide efficiently. Many long-term smokers have both of these conditions. The cause of lung cancer is the carcinogens in tar. These substances promote changes in the DNA of cells lining the airways. The cells grow and divide out of control. This growth is very slow but eventually may form a small group of cells known as a **tumour**. If this is not discovered, it may grow to occupy a large area of the lung and block airways and blood vessels. Worse still, a part of the tumour may break off and spread to other organs.

From the graph above, you can see the sort of data that helped scientists establish the link between smoking and lung cancer. In the early part of the twentieth century, deaths from lung cancer were low, but these increased dramatically as the habit of cigarette smoking became more widespread. In contrast, deaths from other lung diseases, such as tuberculosis were falling, as a result of improved medical care and better housing.

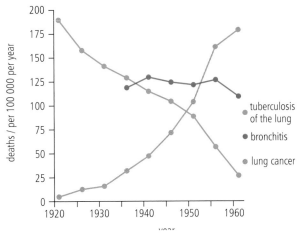

Deaths from lung diseases in England and Wales 1920–60.

A small cancerous tumour between alveoli in the lungs.

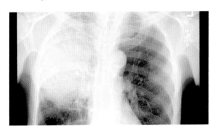

X-ray showing a cancer in the right lung. Compare with the X-ray on page 128.

Heart disease

Tobacco smoke increases the chances that fat will deposit in the walls of arteries. As nicotine increases the chance that blood will clot, there is an increased risk of developing coronary heart disease (see page 108). Blocked arteries reduce the supply of oxygen to the heart and this damages the heart muscle.

KEY POINTS

1 Tobacco smoke consists of nicotine, tar, carbon monoxide and smoke particles.

2 Smoking is associated with diseases such as bronchitis, emphysema, heart disease and lung cancer.

SUMMARY QUESTIONS

1 Define the following words:

   a nicotine   b tar   c carbon monoxide

2 Explain how cigarette smoking is linked to an increased risk of developing:

   a lung cancer   b bronchitis   c heart disease

3 a What is the function of the alveolus (air sac)?

   b Describe two ways in which a healthy alveolus is adapted to carry out its function.

   c Describe what happens to alveoli in people who develop emphysema.

   d Explain how emphysema affects the functioning of the lungs.

1 The best definition of the term *drug* is:

   A any substance taken into the body that influences chemical reactions in the body

   B any substance that affects the nervous system

   C substances that are taken to treat disease

   D substances that may become addictive

   (Paper 1) [1]

2 The drug in tobacco smoke is:

   A carbon dioxide

   B carbon monoxide

   C nicotine

   D tar

   (Paper 1) [1]

3 Which statement is *not* related to the taking of heroin?

   A The drug is a stimulant.

   B The drug is addictive.

   C Not taking the drug may lead to severe withdrawal symptoms.

   D The drug is a depressant.

   (Paper 1) [1]

4 Which component in tobacco smoke is the cause of lung cancer?

   A carbon dioxide

   B carbon monoxide

   C nicotine

   D tar

   (Paper 1) [1]

5 Which drug increases protein synthesis in muscle tissue?

   A adrenaline

   B antibiotics

   C insulin

   D testosterone

   (Paper 2) [1]

6 Which statement describes the effect of heroin on the nervous system?

   A blocks the activity of all sensory neurones

   B combines with receptors on nerves that transmit information about pain

   C prevents the release of neurotransmitters

   D stimulates neurones to send impulses about pain

   (Paper 2) [1]

7 Why is it necessary for drug companies to find and develop new types of antibiotics?

   A all viruses have become resistant to antibiotics

   B antibiotics have become drugs of misuse

   C antibiotics have been in use for over 60 years

   D many bacteria are resistant to some antibiotics

   (Paper 2) [1]

8 Tobacco smoke contains carbon monoxide. What is the effect of carbon monoxide on the body?

   A causes mutations in the epithelium of the bronchi

   B reduces the oxygen-carrying capacity of haemoglobin

   C stimulates goblet cells to make more mucus

   D stops the movement of cilia

   (Paper 2) [1]

9 Drugs may be classified as useful and harmful drugs.

   (a) Explain why antibiotics are classified as (i) drugs, and (ii) useful drugs. [4]

   (b) (i) Name one drug that is classified as a harmful drug. [1]

       (ii) Explain briefly why the drug you have named is classified as a harmful drug.

   (Paper 3) [2]

10 The graph shows the number of deaths in the UK in which alcohol was the major cause between 1991 and 2006. The death rates are shown as number of deaths per 100 000 men and per 100 000 women.

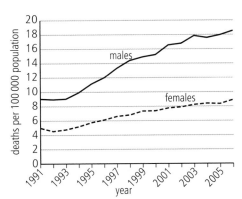

Source: Office of National Statistics. Crown Copyright material is reproduced with the permission of the Controller, Office of Public Sector Information (OPSI)

(a) Use the graph to describe the trends in deaths in which alcohol was the major cause. [3]

(b) Calculate the percentage increase in death rates for men between 1991 and 2006. [2]

(c) Cirrhosis of the liver is often caused by heavy consumption of alcohol. The number of cases of cirrhosis has increased in the UK, especially among young people. Explain how you would find out if the increase in the number of cases of cirrhosis is likely to be due to an increase in alcohol consumption. [2]

(d) Explain why driving a car under the influence of alcohol is dangerous. [3]

(Paper 3)

11 In many countries antibiotics are available only on prescription from a doctor.

(a) Explain why antibiotics are prescribed in these countries rather than freely available to buy. [2]

(b) Explain why antibiotics do not kill viruses. [3]

(c) Discuss the dangers of overusing antibiotics. [4]

(Paper 4)

12 The graph shows the number of cigarettes smoked by men each year in the UK from 1910 to 1980. The dotted line shows the number of deaths of men from lung cancer over the same period.

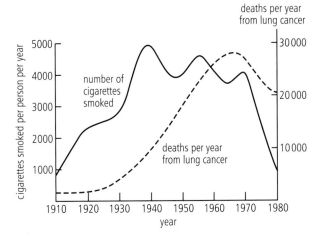

(a) State the year in which there was:

   (i) the highest consumption of cigarettes, and

   (ii) the highest number of deaths from lung cancer. [2]

(b) Describe what happened to:

   (i) cigarette consumption, and

   (ii) deaths from lung cancer over the period shown by the graph. [5]

(c) Discuss whether the graph provides any evidence to support the idea that smoking causes lung cancer. [3]

(Paper 4)

16 Reproduction

16.1

Asexual and sexual reproduction

Supplement

LEARNING OUTCOMES

- Define *asexual* and *sexual* *reproduction*
- Identify asexual reproduction from descriptions, diagrams and photographs
- Discuss the advantages and disadvantages of asexual and sexual reproduction

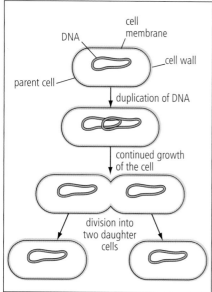

Figure 16.1.1 Binary fission in bacteria. Each new bacterial cell is genetically identical to each other and to the parent cell.

Reproduction means producing new living organisms. Animals and plants reproduce to make new individuals of the same species. There are two types of reproduction: **asexual** and **sexual reproduction**.

Asexual reproduction

In asexual reproduction there is only *one* parent. All the offspring are identical to the parent as they inherit exactly the same genetic information. They are **genetically identical** to the parent, which means there is little variation amongst the offspring. Any variation is due to the effect of the environment, for example the availability of nutrients and water determine how well organisms grow. Bacteria, fungi and plants reproduce asexually.

Bacteria are microscopic organisms made up of one cell. They do not have a nucleus, but a loop of DNA. When bacteria reproduce asexually the DNA loop is copied so that there is some for each new cell. You can find some information about how DNA is copied on page 218. The bacterial cell divides into two by making a new cell wall. This type of asexual reproduction is called **binary fission**, because the parent cell splits into two.

Pin mould is a fungus which grows on bread (see pages 6 and 7). It reproduces asexually by making **spores**. These are small and light, like specks of dust, and they float through the air. When a spore lands on a damp surface, it splits open and a thread grows out. This thread or **hypha**, grows over the surface of bread forming a dense network of threads called a **mycelium**. Eventually, short hyphae grow upwards and produce spore cases or **sporangia** at their tips. Inside each sporangium hundreds of new spores are formed asexually by division of the nuclei. Each nucleus gains a small quantity of cytoplasm and a protective spore case. When ready, the sporangia break open and the spores are dispersed.

Potatoes reproduce asexually by means of **stem tubers**, which are swollen underground stems that grow from the parent plant. Sucrose is transported in the phloem from the leaves into these underground stems that swell as they convert the sucrose into starch. The parent plant dies at the end of the growing season leaving the tubers in the ground to survive over the winter.

However, farmers dig them up and sell them. Some are kept as 'seed' potatoes to plant the next year. All the tubers produced from one parent plant are genetically identical. When they start regrowing, new shoots and roots emerge from the growing points of the potato known as 'eyes'.

Sexual reproduction

In sexual reproduction there are two parents. The parents have **sex organs**. The sex organs make sex cells or **gametes**. In animals the male gametes are **sperm cells**. In flowering plants the male gametes are nuclei inside **pollen grains**. In animals the female gametes are **egg cells**. In flowering plants the female gametes are inside structures called **ovules**. During sexual reproduction the gametes fuse together at **fertilisation**. The fertilised egg or **zygote** divides to form an **embryo** which may grow into a new individual plant.

At fertilisation, half the genetic material comes from the male gamete and half comes from the female gamete. The nuclei in the gametes contain one set of chromosomes. A zygote has a nucleus that contains two sets of chromosomes – one from each gamete.

Unlike asexual reproduction, the offspring produced are not genetically identical to the parents. Each zygote receives half its genes from its male parent and half from its female parent. This means that sexual reproduction brings about **variation** in the offspring.

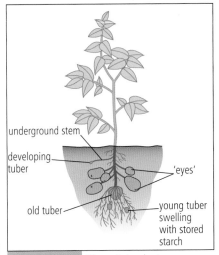

Figure 16.1.2 The potato plant.

Advantages and disadvantages of asexual reproduction in plants

An advantage of asexual reproduction is that it is fast. Organisms can reproduce rapidly to spread and colonise areas where the parent plants grow. Plants grow more rapidly from tubers than from seeds.

If left in the ground, potato tubers grow into new plants in the same place as the parent, but compete with each other for resources. Seeds produced in sexual reproduction are dispersed to new areas where there may not be as much competition. However, many seeds land in unsuitable places for growth so many are wasted.

There is very little, if any, variation as a result of asexual methods of reproduction. This can be a disadvantage if these individuals are affected by a certain disease. Since they are all genetically identical, none may have resistance to the disease and they may all be killed. Disease could spread very quickly through a crop monoculture, in which all the plants are genetically identical. If environmental conditions change, it is unlikely that there will be individuals with features adapted to the new conditions. Sexual reproduction gives rise to variation. In the wild, many individuals may not be as well adapted to the existing conditions as their parents, but may be well adapted if the environment changes. If there is plenty of variation, species are better able to survive and evolve.

Potato plants reproduce asexually by producing tubers, as well as sexually with flowers.

SUMMARY QUESTIONS

1 Explain each of the following:

   a gamete   b fertilisation

   c zygote   d embryo

2 Describe each of these examples of asexual reproduction:

   a binary fission in bacteria

   b spore production in fungi

   c tuber formation in potatoes

3 Make a table to compare the advantages and disadvantages of asexual reproduction and sexual reproduction in plants.

KEY POINTS

1 Asexual reproduction results in the production of genetically identical offspring from one parent.

2 Sexual reproduction involves the fusion of male and female nuclei to form a zygote, producing offspring that are genetically different from each other and their parents.

3 Binary fission in bacteria, spore production in fungi and tuber formation in potatoes are all examples of asexual reproduction.

4 Asexual reproduction may produce many individuals rapidly, but it does not result in variation of offspring as in sexual reproduction.

The spiky pollen grains are from insect-pollinated flowers and the smooth grains are from wind-pollinated flowers.

The Titan arum from Sumatra in Indonesia produces a smell of rotting flesh to attract flies – the flies jump in, slide down the sticky walls and are covered in pollen; after a few days the walls of the plant change, allowing the flies to climb out to visit another flower of the same species.

Flowering plants carry out sexual reproduction by producing flowers, which have male and female parts. The male parts make pollen grains to carry male gametes to the female parts. The transfer of pollen grains is **pollination**. Female gametes are deep inside the female part and pollen grains grow a tube to reach them. Each pollen tube delivers a male gamete so it can fuse with the female gamete to form a **zygote**. Fusion of male and female gametes is **fertilisation**.

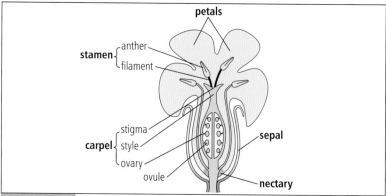

Figure 16.2.1 A half flower drawing of an insect-pollinated flower.

Flowers vary in structure depending upon their method of pollination.

Insect-pollinated flowers

Flower parts are often arranged in rings attached to the end of a swollen flower stalk. The sepals form the outer ring and the carpels form the inner ring.

- Sepals are leaf-like structures that protect the flower when it is a bud.
- Petals are brightly coloured and scented. Many have a nectary at the base, which makes sugary nectar. Visiting insects land on the petals to feed on nectar.
- Stamens are the male sex organs. Each one is made up of two parts: the anther, where the pollen is made, and the filament, a stalk which holds the anther. In this flower there are four stamens (eight in the full flower).
- Carpels are the female sex organs. Each carpel is made up of a stigma, a style and an ovary.
- The ovules are inside the ovary. Each ovule has a female gamete – the egg cell.

The pollen grains of insect-pollinated flowers are often spiky like those in the photograph, above left. They are also often sticky. The grains are adapted to attach to hairs on the surfaces of insect bodies. The grains become dislodged onto a stigma when an insect enters another flower of the same species. The flower in Figure 16.2.1 has a stigma in the centre of the flower which an insect will push past on its way towards the nectary.

Wind-pollinated flowers

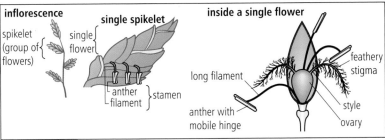

Figure 16.2.2 This species of rye grass is an example of a wind-pollinated flower.

Grasses and cereals are all pollinated by the wind. Rye grass, *Lolium perenne*, has many features typical of a wind-pollinated flower. The flowers are small, green and inconspicuous compared with insect-pollinated flowers. They have no scent and no nectar.

The anthers produce pollen grains that are small, smooth and light so that they are easily carried on the wind. These flowers make huge numbers of pollen grains as the chances of pollen grains landing on a stigma are very small. The anthers hang outside the flower so the wind can blow away pollen. The feathery stigmas are also positioned outside the flower. They act like a net, providing a large surface area for catching pollen grains that get blown into them. Wind-pollination can waste a lot of pollen, which is why the anthers produce so much. But the fact that plants such as grasses and cereals grow close to one another means that pollen is more likely to be transferred than if they were separated by large distances.

Two flowers of *Bromus*, showing large, loosely attached anthers and feathery stigmas.

Long anthers hanging out of rye grass flowers.

SUMMARY QUESTIONS

1 Copy and complete the sentences using these words (you may use some more than once):

> **carpels** **pollen** **stigma**
> **anther** **ovary** **stamens**

The male parts of a flower are called the _____ . Each is made up of an _____ and a filament. The _____ grains are made inside the _____ . The female parts of a flower are called the _____ . Each is made up of a _____ , style and _____ .

2 Make a table to compare the pollen of insect-pollinated flowers and wind-pollinated flowers.

3 Compare the structure of an insect-pollinated flower with a wind-pollinated flower.

Honey bee with pollen on its legs.

STUDY TIP

When you look at insect-pollinated flowers with a hand lens, identify how their structure helps them to be pollinated by insects. You should be able do this for a flower you have not seen before.

Pollination is the transfer of pollen grains from the anther to a stigma. Pollination may be carried out by insects or by the wind, and flowers are adapted to one of these methods. Insects and wind are two agents of pollination; there are others such as birds and bats.

When an anther is ripe, it splits open along its length and releases its pollen so that pollination can occur. Pollination is needed in order to bring the male gamete (inside a pollen grain) near to the female gamete so that **fertilisation** can occur.

Insect pollination

Look at the diagram of the flower below. It has a number of adaptations for insect pollination.

The flower is pollinated by bees, which land on the petals. Bees have long tongues and are able to reach the nectaries at the base of the petals.

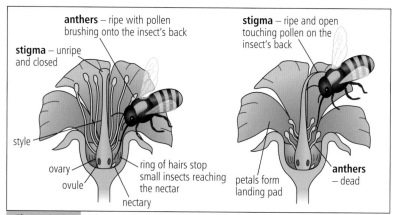

anthers – ripe with pollen brushing onto the insect's back

stigma – unripe and closed

style

ovary

ovule

ring of hairs stop small insects reaching the nectar

nectary

stigma – ripe and open touching pollen on the insect's back

petals form landing pad

anthers – dead

Figure 16.3.1 A bee transfers pollen from one flower to another of the same species.

The bees feed upon the sugary nectar made in the nectaries. The anthers are positioned in such a way that sticky pollen from them will brush against the bee's back as it pushes its head down to the base of the petals. When the bee enters another flower, it brushes some of the pollen against the ripe stigma and pollination is achieved.

Wind pollination

Look at the diagram of wind pollination on the next page. The anthers hang outside the flower so that they release their pollen when the wind blows them.

The stigmas are feathery and are also found outside the flower, where they act as a net to catch pollen grains in the air. Wind-pollinated flowers have light, smooth pollen grains which can easily be carried by the wind to another flower. The anthers produce many pollen grains so that some, by chance, will land on the stigma of a flower of the same species. Most of the pollen produced will be lost.

The structural features of insect- and wind-pollinated flowers are compared in the table.

feature	insect-pollinated flowers	wind-pollinated flowers
petals	present – colourful and scented to attract insects	absent or very small and difficult to see
nectaries	present – make nectar which is a sugary liquid food for pollinating insects	absent
stamens	present – usually with short filaments; anthers attached firmly to filaments; inside the flower for insects to rub against	long filaments so anthers hang outside the flower; anthers loosely attached to the filaments so pollen is easily blown away
pollen	small quantities of sticky, spiky pollen grains that stick easily to insects' bodies	large quantities of smooth, light pollen that can easily be carried by the wind
carpels	sticky, small stigmas usually inside the flower for insects to rub against	large, feathery stigmas to catch pollen grains in the air

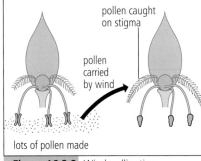

Figure 16.3.2 Wind pollination.

Hazel catkins release pollen to be carried by the wind.

Pollination and variation

Self-pollination occurs when pollen is transferred from the anther to a stigma of the same flower or to a different flower but on the same plant.

Cross-pollination occurs when pollen is transferred to a stigma of another plant of the same species. Cross-pollination is of a greater benefit to a species of plant, since it ensures exchange of genetic material between different plants and results in greater **variation** upon which **natural selection** can operate (see Unit 18).

Self-pollination results in much less variation, since genetic material is not exchanged with different plants but with the same one. However, self-pollination is an advantage if there are no pollinating insects, and for plants growing in isolation from others of the same species.

KEY POINTS

1 Pollination is the transfer of pollen grains from the anther of the male part of the flower to the stigma of the female part of the flower.

2 Insect-pollinated flowers have bright petals, scent and nectar. Wind-pollinated flowers are inconspicuous with feathery stigmas and anthers that hang outside the flower.

3 Self-pollination involves the transfer of pollen from the anthers to the stigma of the same flower or a different flower on the same plant. Cross-pollination involves the transfer of pollen from the anthers of one flower to the stigma of another flower on a different plant of the same species.

SUMMARY QUESTIONS

1 a Give two ways in which pollen from insect-pollinated flowers is different to the pollen from wind-pollinated flowers.

   b Write down three other differences between insect-pollinated and wind-pollinated flowers.

2 a What is the difference between self-pollination and cross-pollination?

   b State the advantages of cross-pollination over self-pollination.

LEARNING OUTCOMES

- State that fertilisation in flowering plants occurs when a pollen nucleus fuses with a nucleus in an ovule
- Describe the events that lead to fertilisation
- State the requirements for germination of seeds

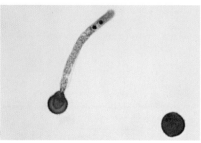

A pollen grain and a germinated pollen grain (left) with its pollen tube and male gamete nuclei.

STUDY TIP

Remember that the male gamete is inside the pollen grain and is delivered to the female gamete by the pollen tube. The pollen grain is _not_ the male gamete.

Scanning electron micrograph of pollen grains and pollen tubes of lily, *Lilium* sp.

Fertilisation

Pollination is complete when the pollen grains land on the stigma of the female part of a flower. If a pollen grain lands on a ripe stigma it starts to grow a pollen tube to take the male nucleus to the female nucleus. Fertilisation occurs when the male nucleus fuses with the female nucleus. This occurs inside the ovule.

Each pollen grain grows a **pollen tube** to take the male nucleus to the ovule. As it grows, the pollen tube gains nutrition from the tissues of the style and carries nutrition from the tissues of the style and carries the male gamete nucleus with it. The first pollen tube reaches the ovary. It enters the ovule through a small hole – the **micropyle**. The male gamete nucleus then fuses with the egg cell nucleus. This is fertilisation in which a zygote is formed by fusion of the two nuclei.

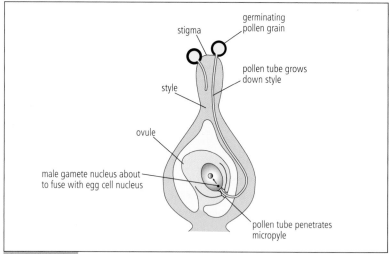

Figure 16.4.1 Fertilisation occurs within the ovule.

PRACTICAL

Growing pollen tubes

Find some flowers with anthers covered in pollen. Lily flowers are particularly suitable.

Cut some small squares of damp Visking (dialysis) tubing and transfer them to filter paper soaked in a nutrient solution rich in sucrose. Use a paintbrush to transfer some pollen grains from the anthers to the squares of dialysis tubing. Leave in a covered Petri dish for 24 to 48 hours. Remove the squares carefully with a pair of forceps, put on a microscope slide and observe.

Seed formation

After fertilisation, the zygote divides and grows into the embryo. The ovule forms the seed with the embryo inside it.

The ovary forms the fruit with the seeds inside it. Many of the parts of the flower are not needed when fertilisation has occured, so the sepals, petals and stamens wither and fall off. They have completed their functions.

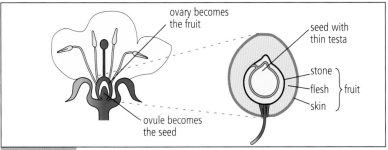

Figure 16.4.2 How a plum flower grows into a plum fruit.

Each fertilised ovule grows to form a seed. Each seed is made up of:

- the embryo
- an energy store
- a seed coat or testa.

The ovary forms the **fruit**.

Conditions for germination

Three conditions are needed for seeds to germinate:

- Water is needed for seeds to swell. This swelling breaks the seed coats of some seeds. Cells absorb water, develop vacuoles and expand. Cell expansion makes the radicle (embryonic root) grow out from the seed. Water is also needed to dissolve the soluble products made by enzymes from the food stores. Water also activates these enzymes.
- Oxygen is needed for aerobic respiration to provide the embryo with energy.
- A suitably warm temperature is needed so that enzymes can work efficiently.

16.5 The male reproductive system

The **testes** are the male sex organs. They produce the male gametes or **spermatozoa** – a word that is usually abbreviated to **sperm** or sperm cells. The testes also make the male hormone **testosterone**. This stimulates changes in a boy's body as he develops into an adult during puberty. This happens between about 10 and 16 years of age.

The testes are located inside a sac called the **scrotum** which hangs outside the body. This keeps the testes at a cooler temperature as sperm cells need a temperature cooler than 37 °C to develop properly and be stored. After puberty, the testes constantly produce sperm cells which are stored in small tubules just outside the testes where they mature. A much wider tube called the **sperm duct** connects these tubules to the urethra. The **prostate gland** and other glands secrete fluids in which the sperm cells can swim. The prostate secretes mucus and other glands secrete sugars which sperm cells use as a source of energy for their respiration. Sperm cells and the fluid together form **semen**, also known as seminal fluid.

The two sperm ducts join with the urethra, which runs down the centre of the **penis**. Urine and semen never pass down the urethra at the same time. A ring of muscle around the urethra contracts to prevent the loss of any urine from the bladder during sexual intercourse.

Male and female gametes

Gametes are sex cells. The male gamete is called the **sperm** and the female gamete is called the **ovum** or **egg**. The sperm cell is a specialised cell that is adapted for swimming by having a flagellum that lashes from side to side. Sperm cells carry genetic information from the male parent to the egg of the female parent. The nucleus contains the father's chromosomes. An egg is much bigger than a sperm because it provides the food store that supports the embryo after fertilisation.

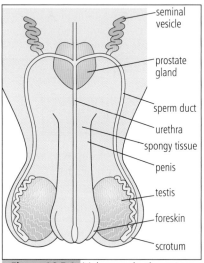

Figure 16.5.1 Male reproductive system: front view.

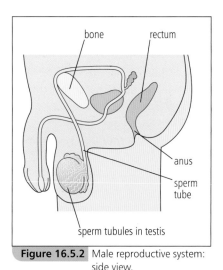

Figure 16.5.2 Male reproductive system: side view.

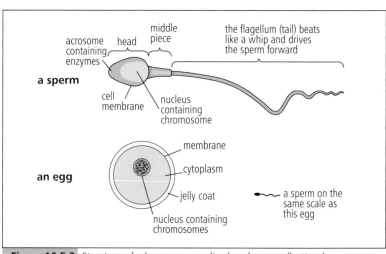

Figure 16.5.3 Structure of a human sperm (top) and an egg (bottom).

Its nucleus contains the mother's chromosomes. Gametes contain half the number of chromosomes as body cells. They are described as haploid cells. Body cells have the normal number of chromosomes and are described as diploid.

An egg is fertilised by a sperm cell. The head of the sperm contains enzymes to help it reach the surface of the egg cell. The cell membrane of the sperm cell fuses with the egg cell membrane. The sperm nucleus moves through the cytoplasm of the egg and the two nuclei fuse together to form a **zygote** (fertilised egg). After fertilisation, the jelly coat changes, so that no more sperm can enter.

Comparing male and female gametes

feature	sperm cell	egg cell
size	small	much larger than sperm cell
movement (mobility)	swims using its flagellum (or tail) that lashes from side to side	does not move itself – is moved along the oviduct by cilia and peristalsis
energy store	has very little – uses sugar in seminal fluid for respiration	protein and fat in cytoplasm – enough to last until implantation in uterus
number of chromosomes	23 (haploid number)	23 (haploid number)
number produced	millions constantly produced (after puberty) often throughout life	one a month after puberty until menopause, except when pregnant or taking a contraceptive pill

Each sperm cell has **mitochondria**, which releases energy to power swimming by the tail. The **acrosome** contains enzymes that digest a pathway through the jelly coat surrounding the egg.

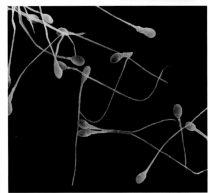

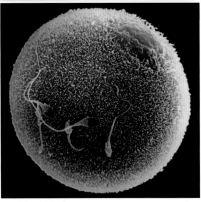

Human sperm (top) and sperm fertilising an egg.

SUMMARY QUESTIONS

1 Copy and complete the sentences using these words:

> **semen   urethra   testes   ejaculation   prostate
> sperm cells   penis   sperm duct   scrotum**

The male gametes are called _____ and are made in the male sex organs or _____ which are contained within a sac called the _____ . A tube called the _____ carries the sperm away from each testis. The _____ gland adds fluid to the sperm cells to form _____ . During an _____ the semen passes along the _____ and out of the erect _____ .

2 a   Give three differences between sperm cells and egg cells.

   b   Explain how sperm cells and egg cells are adapted for their functions.

KEY POINTS

1 The male reproductive organs consist of the testes, scrotum, sperm ducts, prostate gland, urethra and penis.

2 The sperm cells are much smaller than the female egg cell. They are produced in huge numbers and are able to swim using a tail.

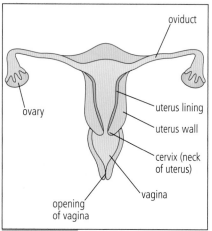

Figure 16.6.1 Female reproductive system: front view.

Ovaries are female sex organs that produce the female gametes called **ova** or **eggs**. The ovaries also make the female hormones **oestrogen** and **progesterone**, which starts to happen with girls between 10 and 15 years of age during puberty.

Oestrogen stimulates the development of the sex organs and secondary sexual characteristics in a girl's body as she starts to develop into an adult. Progesterone prepares the uterus so that it is ready to receive an embryo in the case of a pregnancy.

The ovaries are attached to the inside of the abdomen just below the kidneys. After puberty, an egg is released from an ovary about every 28 days. The ovaries tend to release an egg (or eggs) on alternate months. The egg passes out of the ovary and into the funnel-shaped openings of the oviduct in a process called **ovulation**. The egg moves slowly down the oviduct towards the **uterus** (womb).

If sperm cells are present in the oviduct the egg will be fertilised. If the egg is not fertilised it will die after about a day. If the egg is fertilised it will divide to form an embryo which may attach itself to the lining of the uterus where it develops into a fetus.

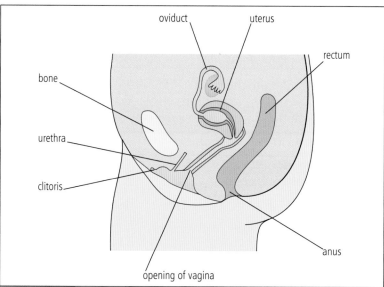

Figure 16.6.2 Female reproductive system: side view.

The lower end of the uterus has a ring of muscle called the **cervix**. It leads to a muscular tube called the **vagina** that opens to the outside of the body. Above the opening of the vagina is the opening of the **urethra** through which urine from the bladder passes out. Above the urethra is the sensitive **clitoris**. The outer opening of the vagina is called the **vulva**. The vagina is sometimes known as the birth canal, as it is through here that a baby passes at birth.

Sexual intercourse

During sexual intercourse males and females stimulate each other. Blood is pumped into special spongy tissue in the penis so that it becomes erect. The erect penis is placed into the vagina during sexual intercourse. Fluid made by the walls of the vagina lubricates movements of the penis within the vagina. This movement stimulates the penis and contractions begin in the sperm ducts to move sperm cells from the tubules around the testes towards the penis.

As they flow, secretions from the glands including the prostate gland are added to form semen or seminal fluid. Contractions of the urethra move the seminal fluid through the penis into the vagina. This is an **ejaculation** and at this time the man experiences feelings of pleasure called an **orgasm**. Repeated movements of the erect penis against the clitoris or against the vagina walls may also produce orgasms for the woman. Each ejaculation contains between 2 to 5 cm³ of semen with up to 500 million sperm cells.

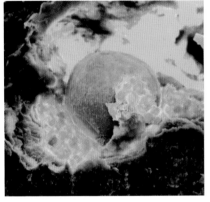

An egg cell leaves the ovary at ovulation. If sexual intercourse happens at around the time of ovulation there is a chance that fertilisation will occur.

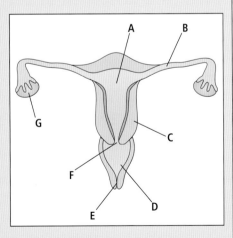

Human fertilisation

There are millions of sperm in a man's semen. If one of them is to meet an egg it must swim from the vagina to the oviduct – the muscular tube that links the ovary to the uterus.

After intercourse, sperm cells swim through the mucus in the cervix into the uterus and then all the way to the oviduct. Many sperm cells do not survive this difficult journey, which is why so many sperm cells are produced – to increase the chance of some of them reaching the oviduct.

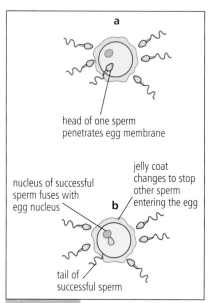

Figure 16.7.2 Events at fertilisation.

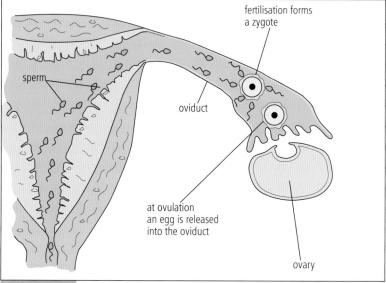

Figure 16.7.1 Ovulation and fertilisation.

If there is an egg in the oviduct, a sperm cell may succeed in penetrating it. Enzymes released by the acrosome on the head of the sperm digest a pathway through the jelly coat surrounding the egg.

After the sperm membrane has fused with the egg membrane the nucleus enters the egg cytoplasm and the flagellum is left outside. The sperm nucleus fuses with the egg nucleus to form the zygote nucleus which is diploid as it contains two sets of chromosomes – one from the mother and one from the father. The zygote nucleus now contains 46 chromosomes and these start being copied in readiness for the first cell division which occurs shortly after fertilisation.

This stops other sperm cells from entering so only *one* sperm is able to fertilise the egg. If there is no egg in the oviduct, no fertilisation can take place. However, the sperm can stay alive for 2 or 3 days. So if intercourse happened just before ovulation, the sperm can fertilise an egg if it is released during this time.

Implantation

Fertilisation takes place in the oviduct. After this, the fertilised egg or **zygote** begins to divide. It divides once to form a two-celled embryo. Then it continues to divide to give four cells and then eight, but after a while this cycle of divisions becomes less regular. Some cells continue to divide while others stop or slow down the rate at which they divide. After a few hours the embryo is a hollow ball of cells. It moves down the oviduct, pushed along by peristaltic contractions of the oviduct and the beating of the ciliated epithelial cells lining the oviduct.

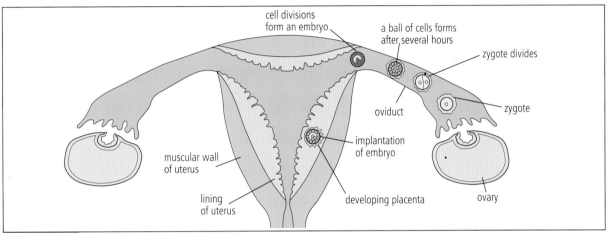

Figure 16.7.3 Implantation of the embryo.

It may take a number of days for the embryo to reach the uterus. The embryo embeds into the soft lining of the uterus. This is called **implantation**. The uterus lining has thickened in preparation and contains numerous blood vessels. The embryo obtains nutrients and oxygen from these blood vessels by diffusion. Carbon dioxide and chemical wastes diffuse out in the opposite direction.

SUMMARY QUESTIONS

1 Copy and complete the sentences using these words:
   zygote sexual ovaries implants uterus
   sperm cells vagina fertilisation weeks embryo

   Sperm cells enter the _____ of the female during _____ intercourse. The sperm cells swim through the cervix and up through the _____ and then enter the oviduct. An egg is made in one of the _____ about every four _____ . The egg passes into the oviduct where _____ may take place if _____ are present. The fertilised egg or _____ divides to form an _____ which passes down to the uterus where it _____ in the lining of the uterus.

2 a How many sperm cells are needed to fertilise an egg?

   b Why does a membrane form around the egg after fertilisation?

3 State the number of chromosomes in the nuclei of the following human cells:

   a an egg b a sperm cell c a zygote

KEY POINTS

1 Fertilisation occurs when the nucleus of a sperm cell fuses with the nucleus of an egg cell.

2 For fertilisation to occur, ovulation must have taken place and a sperm cell must have reached the oviduct.

3 The zygote divides to form a ball of cells called the embryo, which passes down the oviduct and implants in the lining of the uterus.

Pregnancy

LEARNING OUTCOMES

- Outline the growth and development of the fetus
- State the functions of the umbilical cord, placenta, amniotic sac and its fluid
- Describe the functions of the placenta and umbilical cord
- State that some toxins and pathogens can cross the placenta and affect the fetus

STUDY TIP

Once all the organs have formed and the embryo has features that are recognisably human, it is known as a fetus.

STUDY TIP

The placenta is the gas exchange surface for the fetus. It is also where dissolved food substances are absorbed. This is why it has the same features as the lung and the small intestine: a large surface area and a short distance for diffusion. Since the exchanges are between fetal and maternal blood there are <u>two</u> good blood supplies to maintain concentration gradients (see page 28).

Pregnancy

Pregnancy is the period of time between fertilisation and birth, which in humans is 9 months. This period of time is called the **gestation period**.

After the embryo has implanted into the lining of the uterus, it grows projections into the soft tissue to gain nutrients and oxygen to support its growth. These projections continue to grow into the **placenta**. An **umbilical cord** grows to attach the fetus to the placenta so it fills the **amniotic sac** which surrounds it.

- The placenta is the site of exchange of oxygen and nutrients for carbon dioxide and other wastes; they diffuse between fetal blood and maternal blood.
- The umbilical cord attaches the fetus to the placenta and contains blood vessels to transport blood to and from the placenta.
- The amnion sac makes **amniotic fluid** to surround and protect the fetus against mechanical damage, for example when the mother makes sudden movements.

The main stages in the development into a baby ready to be born at the end of pregnancy are summarised as follows.

- One month after fertilisation, a human embryo looks a bit like a fish embryo or tadpole. The embryo does not yet have arms or legs, but it is clear where these will develop. The heart has started to beat.
- Two months after fertilisation the embryo has a face, limbs, fingers and toes and looks human. It is now called a fetus. Most of the organs are formed.
- Three months after fertilisation the nerves and muscles of the fetus are developing rapidly.
- Five months after fertilisation, although only about 180 mm in length, the fetus has perfectly formed eyebrows, fingernails, fingerprints and body hair. Its movements may have been felt by the mother for the last month.
- Seven months after fertilisation development is almost complete.

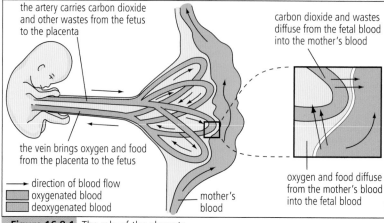

the artery carries carbon dioxide and other wastes from the fetus to the placenta

carbon dioxide and wastes diffuse from the fetal blood into the mother's blood

the vein brings oxygen and food from the placenta to the fetus

oxygen and food diffuse from the mother's blood into the fetal blood

→ direction of blood flow
oxygenated blood
deoxygenated blood

mother's blood

Figure 16.8.1 The role of the placenta.

The placenta is the gas exchange surface for the fetus. It is also where dissolved food substances are absorbed. This is why it has the same features as the lung and the small intestine: a large surface area and a short distance for diffusion. Since the exchanges are between fetal and maternal blood there are *two* good blood supplies to maintain concentration gradients (see page 28).

The fetus obtains all its food from its mother, including glucose, amino acids, fats, mineral ions, vitamins and water. The fetus produces some urea which diffuses into the mother's blood to be excreted through her kidneys.

The umbilical cord joins the fetus to the placenta. The cord contains an artery that transports deoxygenated blood from the fetus to the placenta and a vein that returns oxygenated blood to the fetus.

Amniotic fluid is formed from filtered blood and has a composition similar to plasma. Cells from the fetus' skin are shed into the amnion sac and when the organs have developed the fetus carries out some breathing movements taking the fluid into the lungs. The fetus urinates into the fluid and drinks it. The amniotic fluid does not provide any nutrients for the fetus.

The placenta acts as a barrier to toxins and pathogens. However, even though the womb is a sterile environment, the fetus is still susceptible to disease. Rubella (German measles) is caused by a virus that can get across the placenta and harm the fetus. This is why all young women should be vaccinated against rubella before they become pregnant. Young men should also be vaccinated since they can act as a reservoir of infection for the disease.

The HIV virus that causes AIDS may cross the placenta, so a baby may be born HIV positive if the mother is infected with the virus. However, with careful management during pregnancy a mother who is HIV+ can give birth to a child who does not carry the virus. Nicotine can also pass across the placenta (see Topic 16.9).

The fetus at 26 days.

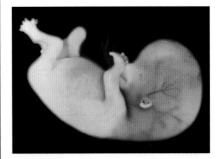
The fetus at 8 weeks.

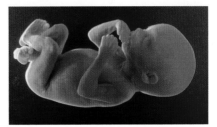
The fetus at 5 months.

SUMMARY QUESTIONS

1 a How is the fetus connected to the placenta?

  b List two substances that pass from the mother's blood to the fetus and two substances that pass in the opposite direction.

2 What is the *gestation period*? How long does it last in humans?

3 What are the functions of the amnion and amniotic fluid?

4 Describe two ways in which the structure of the placenta helps efficient diffusion between fetal blood and mother's blood.

KEY POINTS

1 The fetus grows and develops inside the uterus. It increases in size and develops all the major organs.

2 The umbilical cord attaches the fetus to the placenta, which is the site of exchange of substances between fetus and mother; the amnion makes amniotic fluid to protect the fetus from mechanical damage.

3 The placenta has a large surface area for gas exchange and diffusion of nutrients and waste between fetal blood and mother's blood.

4 Some toxins and pathogens cross the placenta into the fetus; the placenta acts as a barrier to many others.

16.9 Ante-natal care and birth

Supplement

LEARNING OUTCOMES

- Describe the ante-natal care of pregnant women
- Describe the processes involved in labour and birth
- Describe the advantages and disadvantages of breast milk and formula milk

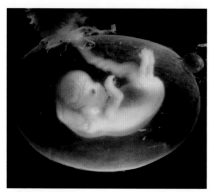

A fetus inside the amnion at 7 to 8 weeks of pregnancy.

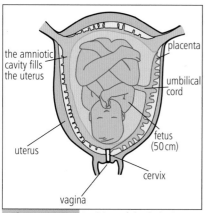

Figure 16.9.1 Position of the baby just before birth.

Ante-natal care

Ante-natal care, or care before birth, is vital for the health of the fetus while it is in the womb. The mother also needs to make sure that she provides the right nutrition and protection for the fetus during the pregnancy.

It is important that the mother has a balanced diet during this time so the fetus obtains all the nutrients it needs for growth and development. Particularly important are vitamins and minerals, such as iron and calcium. A pregnant woman will need to ensure that she gets adequate quantities of:

- **calcium**, as the bones of the fetus are growing
- **iron**, so that her body can make the extra red blood cells needed to carry oxygen to the fetus. Her growing fetus is also making red blood cells. They both need the extra iron in the diet to make more haemoglobin, the substance in red blood cells that carries oxygen
- **carbohydrate**, so that the pregnant mother has enough energy to move her heavier body around
- **protein**, to provide the amino acids that they both need to make new tissues. The mother makes muscle tissue in the uterus in preparation for birth. The fetus is also growing and developing and therefore making many new cells.

The mother should not smoke since nicotine and carbon monoxide cross the placenta and can result in **premature** or **underweight** babies. She should drink no alcohol, or very small quantities, as alcohol also crosses the placenta and can cause a variety of effects including birth defects and mental retardation. Drugs, such as heroin, cross the placenta so some babies are born with an addiction to heroin.

Birth

A normal pregnancy takes about 9 months. In some cases babies are born prematurely, perhaps because there has been a problem during pregnancy. An early breaking of the membrane (amnion) around the fetus is the most common reason.

A few weeks before birth, the fetus usually turns over inside the uterus. This positions its head above the cervix. Hormones released by the fetus and the build-up of pressure in the uterus stimulate hormonal changes in the mother. A hormone, oxytocin, is released from her pituitary gland that stimulates the muscles of the uterus to contract. The mother starts to feel small contractions of the uterus wall. This is the beginning of **labour**.

These contractions become stronger and more frequent. Stronger contractions slowly stretch the opening of the cervix and the amnion breaks, allowing the amniotic fluid to escape.

(Figure 16.9.1 labels: placenta; umbilical cord; fetus (50 cm); cervix; vagina; uterus; the amniotic cavity fills the uterus)

The muscles of the uterus wall now contract very strongly and start to push the baby towards the cervix. The cervix widens or dilates and the baby's head is pushed through the vagina. This part of the birth takes place quite quickly. As soon as it is born the baby breathes for the first time. It is important that the baby has airways clear of mucus and can breathe easily.

The umbilical cord is tied and cut just above the point where it attaches to the baby. The remains of the cord heal to form the baby's navel. After a few minutes the placenta comes away from the uterus wall. It is pushed out of the vagina as the **afterbirth**.

The process of labour with the muscular contractions of the uterus is painful for the mother. She can help the process by gentle exercises of the body before labour and by breathing in a special way during labour. Pain-killing drugs can also be given to ease the pain during birth. These may be given by an epidural. This involves inserting a catheter (tube) into the space between the spine and the spinal cord. The drugs prevent the transmission of impulses from pain receptors to the brain.

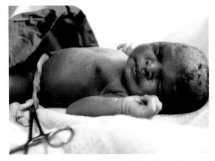
A newborn baby with its umbilical cord.

Breast-feeding

During her pregnancy the glands in the mother's breasts will enlarge. This is so that soon after birth she is able to produce milk.

Breast milk contains **antibodies** which pass to the baby and give it passive immunity to the diseases that the mother has had recently. **Passive immunity** (see Topic 10.3) means that the child has the mother's antibodies in the blood, but is unable to produce them itself. The antibodies remain in the child's blood while it is breast-fed and for a short time afterwards. Breast-feeding also enables a close bond to form between mother and baby that is beneficial to both.

Medical research has shown that breast-feeding is beneficial to both children and mothers. Children who are breast-fed are less likely to develop certain diseases such as childhood cancers and diabetes. Mothers are less likely to develop cancers of breast, womb and ovaries.

Some women are not able to breast-feed their babies or they find it difficult or embarrassing. They therefore rely on bottle-feeding. Formula milk may be more convenient, but since it comes in powdered form it needs to be mixed with water under sterile conditions. In some places women may not be able to do this, with the risk of babies getting infections from non-sterile milk or bottles. Formula milk also costs money, whereas breast milk does not. Of course, sometimes breast-feeding simply isn't an option at all!

KEY POINTS

1 During labour the muscles in the uterus wall contract.

2 At birth, the baby passes out of the vagina and takes its first breath. The umbilical cord is tied and cut. The placenta detaches from the uterus and passes out as the afterbirth.

3 Ante-natal care involves the mother having enough calcium, iron, carbohydrate and protein in her diet. She should avoid smoking, drugs and alcohol.

4 Breast milk has advantages over formula milk, notably that it contains antibodies that protect the baby from common infections.

SUMMARY QUESTIONS

1 Describe what happens to each of the following during the birth of a baby:
   a the muscles in the uterus wall   b the cervix
   c the amnion and amniotic fluid   d the placenta

2 Make a table to show the advantages and disadvantages of breast-feeding and bottle-feeding.

Supplement

Sex hormones

Puberty and adolescence

You are born with a complete set of sex organs. But they become active only later in life. Between the ages of about 10 and 14, the testes start to make sperm and follicles start to develop in the ovaries. This time of development in your life is called **puberty**.

Girls usually develop earlier than boys. The actual age of puberty varies from person to person. The changes that take place are all controlled by **hormones**. At the beginning of puberty the **pituitary gland** at the base of the brain starts to make hormones that stimulate the testes and the ovaries.

These make the sex organs active. The sex organs start to produce sex hormones which develop our **secondary sexual characteristics**.

Puberty in boys

The testes start making **testosterone**, which stimulates:

- the growth of the male sex organs
- the testes to make sperm cells
- growth of hair on the face
- the deepening of the voice
- development of muscles in the body.

Puberty in girls

The ovaries start making **oestrogen**, which stimulates:

- the growth of female sex organs
- the start of the first menstrual cycle and the first period
- growth of hair on parts of the body
- growth and development of breasts
- widening of the hips.

A person becomes an **adolescent** when puberty starts. Adolescence finishes when you stop growing at about 18 years of age. Adolescence can be an emotional time. Hormones can bring about mood changes and increased sexual urges.

Menstrual cycles are the changes that occur in the ovary and the uterus. A complete cycle lasts for about 28 days and is controlled by hormones which coordinate the activity of the ovary and the uterus. This coordination is important so that the lining of the uterus is ready to receive an embryo so that it can start its development.

Girls start to have periods between the ages of about 10 to 15 years old. During a period, the lining of the uterus breaks down and blood and cells pass out of the vagina. This is called **menstruation**. When this happens for the first time, it shows that a girl has had her first menstrual cycle.

A group of adolescents.

STUDY TIP

When you make your revision notes for Paper 4, include a table that lists the roles of all the hormones here and in Topic 16.11. Keep referring to your table when you read Topics 16.12 and 16.13 so you understand how they interact to control the menstrual cycle and how we use them in for contraception and treating infertility.

Stages in the menstrual cycle

Girls are born with a very large number of potential egg cells in their ovaries. They do not produce any more during their lifetime. Each potential egg is surrounded by a small group of cells and together they form a **follicle**. There are thousands of these follicles in each ovary.

At puberty, some of the follicles start to develop. This development involves the egg dividing by meiosis which reduces the number of chromosomes in the nucleus by half. The cytoplasm fills with stored food. Each month one or a few follicles start to develop. This is stimulated by hormones released by the pituitary gland (see page 164).

A follicle starts to develop in the ovary at the beginning of the cycle as menstruation finishes. As the follicle grows it enlarges and fills with fluid and moves towards the edge of the ovary. After 2 weeks, the follicle bursts, releasing the egg, some follicle cells and the fluid into the oviduct. This is called **ovulation**. The follicle cells left behind in the ovary form the **yellow body** which remains for the next 2 weeks. If implantation does not occur, it will then decrease in size. If implantation occurs it remains active during pregnancy.

While the egg is developing in the ovary the lining of the uterus starts to thicken. In the week after ovulation it has a thick lining of blood vessels and glands. If fertilisation occurs, the fertilised egg divides to form a ball of cells which **implants** into the thick uterus lining and the woman becomes pregnant.

If fertilisation does not occur, the egg dies and passes out of the vagina and the yellow body in the ovary breaks down. The thick lining of the uterus breaks down and is lost during menstruation. The cycle now begins again.

If a pregnancy does occur, the embryo releases a hormone that stimulates the yellow body to remain active which in turn stimulates the lining of the uterus to continue to thicken. This supplies the embryo with food and oxygen as it continues its development. It also ensures that menstruation will not occur.

Between the ages of 45 and 55 years a woman's periods stop as her menstrual cycles have stopped. This is called the **menopause**.

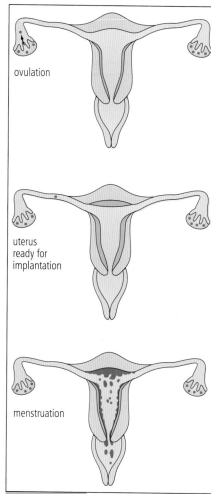

Figure 16.10.1 During the menstrual cycle the activities of the ovaries and the uterus are synchronised.

(labels on figure: ovulation; uterus ready for implantation; menstruation)

SUMMARY QUESTIONS

1 Explain the role of the testes and the ovaries in the development and regulation of secondary sexual characteristics in males and females at puberty.

2 a List the effects that testosterone has on the body of a boy at puberty.

  b List the effects that oestrogen has on the body of a girl at puberty.

3 Explain the meaning of the following terms:

  a menstruation   b ovulation   c implantation

4 Draw a timeline to show the changes that occur in the ovary and in the uterus during a menstrual cycle.

KEY POINTS

1 Testosterone stimulates the changes in secondary sexual characteristics that occur in males during puberty.

2 Oestrogen stimulates the changes in secondary sexual characteristics that occur in females during puberty.

3 During a menstrual cycle a follicle develops inside the ovary and the lining of the uterus thickens. If implantation does not occur, the lining of the uterus and vagina break down. This is menstruation.

LEARNING OUTCOMES

• Describe the sites of production of oestrogen and progesterone in the menstrual cycle and in pregnancy

• Explain the role of the hormones FSH, LH, oestrogen and progesterone in controlling the menstrual cycle

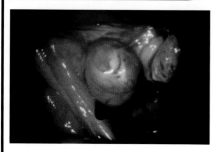

In the centre is a yellow body which enlarges in the second half of the menstrual cycle to produce and release progesterone.

There are four hormones that control the menstrual cycle.

The pituitary gland at the base of the brain secretes:

• **Follicle Stimulating Hormone (FSH)**
• **Luteinising Hormone (LH)**.

The ovary secretes:

• **oestrogen** and **progesterone**.

Each cycle starts with the secretion into the blood of FSH from the pituitary gland. This hormone is carried in the blood to the ovary where it stimulates the development of one or more follicles. Inside each follicle there is a cell that has the potential to grow into an egg. At the beginning of the cycle each of these cells starts to grow and divide.

FSH stimulates other cells in the follicle to secrete oestrogen into the bloodstream.

Oestrogen has three main effects.

• It stimulates the repair and thickening of the lining of the uterus.

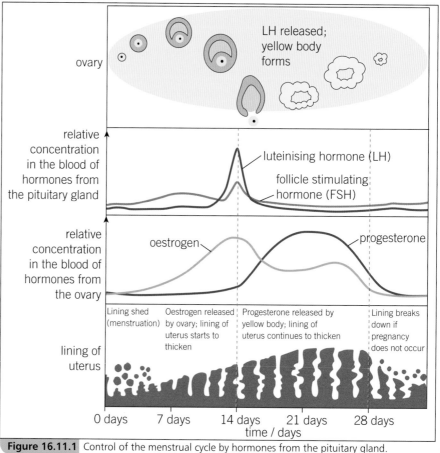

Figure 16.11.1 Control of the menstrual cycle by hormones from the pituitary gland.

- It stops the secretion of FSH from the pituitary gland.
- It stimulates the pituitary gland to secrete LH.

The lining of the uterus thickens in preparation to receive an embryo.

As you can see from the graphs in Figure 16.11.1, FSH and oestrogen are the important hormones during the first half of the menstrual cycle from day 0 to day 14. During this time the egg develops and increases in size within the follicle.

LH stimulates the follicle to burst and release its egg. This happens at around day 14 of the cycle.

The top part of Figure 16.11.1 shows that after ovulation the remains of the follicle become the yellow body which secretes progesterone.

In the second half of the cycle, progesterone together with oestrogen secreted by the ovary have these effects:

- they stimulate further growth of the lining of the uterus,
- maintain the lining in its thickened state and prevent it breaking down, and
- stop the secretion of the FSH and LH from the pituitary gland.

The graph shows that the concentration of LH decreases in the second half of the cycle. This reduces the stimulation of the yellow body so eventually it stops secreting progesterone and decreases in size.

As the concentration of progesterone decreases there is less stimulation for the lining of the uterus and menstruation occurs.

If intercourse occurs around the middle of the cycle and the egg is fertilised, then it may implant in the uterine lining so that pregnancy begins. If this happens, progesterone and oestrogen continue to be secreted by the yellow body so that the lining of the uterus remains thick. They also stop the menstruation occurring and stop the menstrual cycle starting again.

During the early part of pregnancy the embryo secretes a hormone that stimulates the yellow body to continue secreting progesterone. This is important as there is no LH from the pituitary gland. Some of this hormone is excreted in the urine and this is the basis of pregnancy tests. If a woman misses a period she may be pregnant. She can use a pregnancy testing kit to find out.

KEY POINTS

1 The pituitary hormone FSH stimulates the development of a follicle inside the ovary and the secretion of oestrogen.

2 Oestrogen stimulates the thickening of the uterus lining, inhibits FSH production and causes the pituitary to release LH.

3 LH stimulates ovulation and the formation of the yellow body from the remains of the follicle. The yellow body secretes progesterone which maintains the lining of the uterus.

DID YOU KNOW?

Nobody knows how many embryos do not implant in the lining of the uterus. One group of researchers estimated that it is about 20%, but no one is certain because there is no way of finding out whether an embryo has implanted or not.

During pregnancy, the yellow body continues to secrete progesterone and oestrogen, but gradually the placenta takes over this role. Progesterone from the placenta continues to maintain the lining of the uterus. This prevents menstruation occurring during pregnancy. This does happen sometimes and is known as a miscarriage.

SUMMARY QUESTIONS

1 Name four hormones involved in coordinating the menstrual cycle and state where each is secreted.

2 Describe the roles of these hormones in controlling the menstrual cycle.

3 Explain why it is important that the lining of the uterus is thick at the time of ovulation.

4 State what happens as a result of the decrease in concentration of oestrogen and progesterone towards the end of the menstrual cycle.

5 Explain why it is important that progesterone is secreted if pregnancy occurs, but FSH is not.

6 Suggest why it is difficult to find out what proportion of a woman's embryos never implant in the uterus.

Methods of birth control

- Describe the following methods of birth control:
  natural, chemical, barrier and surgical

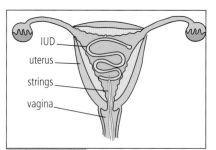

Figure 16.12.1 Intra-uterine device

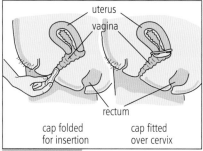

Figure 16.12.2 Cap

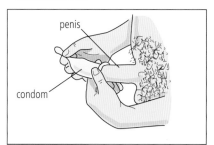

Figure 16.12.3 Condom

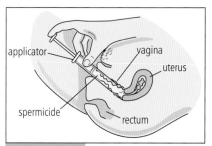

Figure 16.12.4 Spermicide

Birth control lets couples decide whether or not to have children and if so, how many. To prevent pregnancy, the method of birth control should either:

- stop the sperm from reaching the egg,
- stop the eggs from being made, or
- stop the fertilised egg from implanting and developing in the uterus.

Barrier methods

The **cap** or **diaphragm** is a rubber dome. The woman places it over her cervix before intercourse. It prevents the man's sperm from entering the uterus. It should be used with sperm-killing cream or foam.

The **condom** is a thin rubber tube. It is rolled over the man's erect penis before intercourse. It stops the sperm from entering the woman's body. It also gives protection against sexually transmitted diseases. It is the most common form of birth control and is easy to use.

A **femidom** is the female equivalent of a condom and is inserted into the vagina.

Chemical methods

The **IUD** (**Intra-Uterine Device**) is a small plastic device that is wrapped in copper or contains hormones. It is fitted inside the woman's uterus and it is thought to work by preventing sperm passing through the uterus. It may also prevent an embryo from implanting in the uterus should conception occur.

IUS (Intra-uterine system) is a small T-shaped plastic device that is placed in the uterus. It contains progesterone that is released slowly and causes the mucus in the cervix to become thick and sticky, making it unlikely that sperm will swim through.

'**The pill**' is an oral contraceptive as it is taken by mouth. There are different types of pill available. They contain the hormones oestrogen and progesterone in varying concentrations. There are also progesterone-only pills. These work by preventing the ovary releasing eggs. (They do this by inhibiting the production of FSH by the pituitary.) As a result, no eggs mature to be released by the ovaries and so pregnancy cannot occur. The woman has to take a pill every day.

For many couples the pill is a very reliable and convenient method of contraception. Drawbacks are that failure to take the pill regularly can result in a pregnancy. The side effects are sore breasts, weight gain, depression and painful periods – each pill has its own set of side effects. In a very small number of women, the pill can be the cause of heart and circulation problems.

A **contraceptive implant** is a small flexible rod that is placed just under the skin of the upper arm. It releases progesterone which prevents ovulation and works for up to 3 years.

Contraceptive injections also contain progesterone. There are different types that work between 8 and 13 weeks.

Spermicides are chemicals that kill sperm. They are sold as foam, cream or jelly. The woman puts the cream into her vagina before intercourse. Spermicides are not very effective on their own and should be used with another method such as the cap or the condom.

Natural methods

Some couples may not be able to use the methods described here or they may have religious or moral objections to using contraception. To avoid becoming pregnant they may abstain from having sex at times when fertilisation is most likely to happen. They may use the **rhythm method**. This relies on determining when ovulation is most likely to occur and abstaining from sex in the days just before and just after that date. A woman keeps a record of when she has her periods and then predicts each month when ovulation should occur. This is about 2 weeks after having her period. This is not a very reliable method, especially in women who do not have regular periods.

A better method is to follow changes in symptoms associated with ovulation. For example, checking for the increase in body temperature and changes in cervical mucus which becomes wetter and less sticky just before ovulation. This symptoms-based method is more effective. It is useful for people who do not have access to family planning clinics and the other methods of contraception described here.

Surgical methods

A sperm duct is also known as a vas deferens. The operation to cut them is a **vasectomy** and is a minor operation for a man. His sperm tubes are cut and tied. The operation is not usually reversible, so the man must be sure that he does not want any more children. The man can still ejaculate but there are no sperm in the semen, just fluid.

Sterilisation is a minor operation for a woman. Her oviducts are cut and blocked. The operation is not reversible, so the woman must be sure that she does not want any more children.

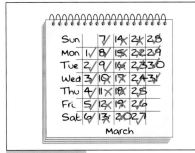

Figure 16.12.5 Calendar for rhythm method

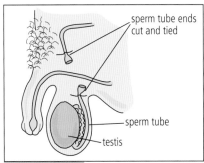

Figure 16.12.6 Vasectomy

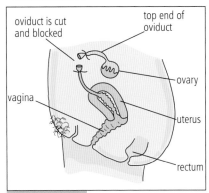

Figure 16.12.7 Sterilisation

SUMMARY QUESTIONS

1 a What is meant by the term *contraception*?

  b Under what circumstances would a couple choose to use birth control methods?

2 Explain how the following birth control methods work:

  a cap or diaphragm   b condom   c spermicides

  d vasectomy

3 a Explain how oral contraceptives ('the pill') act to prevent pregnancy.

  b What are the disadvantages associated with this form of birth control?

KEY POINTS

1 Barrier methods include the condom, femidom, IUD and IUS.

2 Chemical methods include the contraceptive pill, implants, injections and spermicides.

3 Methods of birth control can be natural (abstinence, rhythm method).

4 Surgical methods involve vasectomy and female sterilisation.

STUDY TIP

Before you read this section make sure you understand how the menstrual cycle is controlled and the function of FSH in stimulating the development of follicles in the ovary.

Fertility drugs

Some couples want to have children but cannot because the man or woman is infertile. Couples are regarded as being infertile if they have had regular, unprotected sexual intercourse for 12 months without the woman becoming pregnant.

Often the cause of infertility in women is that the ovaries do not release eggs. This may be due to a lack of FSH production by the pituitary. Treatment for this type of infertility may involve regular injections of a **fertility drug** containing FSH. The FSH stimulates the ovaries to release eggs.

Other treatments involve tablets that make the pituitary insensitive to oestrogen. Remember that oestrogen inhibits the production of FSH. So if the effect of oestrogen on the pituitary is blocked by the drug, then the pituitary continues to release FSH and ovulation occurs.

Unfortunately, fertility treatment does not always work for a variety of reasons that may not be easy for doctors to discover. On the other hand, it can work too well; too many eggs may be released, resulting in twins, triplets, quadruplets or even more!

Artificial insemination

Some couples cannot have children because the man is infertile. This may be because he does not produce enough sperm or there are problems with ejaculation. It is possible that semen may be collected from the man and placed via a fine plastic tube into the woman's uterus. The process is called **artificial insemination** (AI).

If the man does not produce enough sperm for this, or if the sperm are defective, then a **donor** may provide a sample of semen for AI. Artificial insemination by donor can lead to problems later in life because the child's father is not the *biological* father who probably remains unknown.

In vitro fertilisation

In vitro fertilisation (IVF) is another technique that can be used to treat infertility when other methods have been unsuccessful. *In vitro* means 'in glass'. Eggs and sperm are collected from the potential parents and mixed together in a laboratory dish where fertilisation occurs.

To increase the chances of success, the woman is often injected with FSH to stimulate the development of several follicles so that several eggs can be collected from her ovaries. A small incision is made in the body wall and a fine plastic tube inserted to collect the eggs from the follicles on the surface of the ovaries. The eggs are kept in a solution that contains nutrients and oxygen.

A couple with triplets resulting from fertility treatment.

Semen is collected from the man (or is used from a sperm donor) and is mixed with the eggs and left for up to 24 hours for fertilisation to occur. A technician checks that fertilisation has occurred and that the zygotes have started to divide into embryos. These are kept in the solution for a few days and then a doctor places them into the woman's uterus. If the procedure is successful, the embryo will develop into a fetus.

It is also possible to take the nucleus from a defective sperm cell and inject it into an egg. The zygote is then cultured for several days so it grows into an embryo and is then transferred into the woman in the hope that it implants in the uterus.

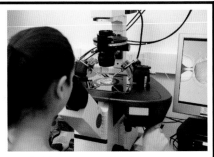

A technician at an IVF clinic injects an egg with a sperm nucleus.

The social implications of contraception and fertility treatments

The introduction of a safe contraceptive pill in 1960 made it much easier for couples to plan their families and decide when to have children. It also gave women more control over their own fertility. It also had a great effect on women's health, giving them the opportunity not to put their health at risk by pregnancy and birth. Death rates for mothers and infants at and around the time of birth have decreased considerably since the introduction of the pill. Methods of birth control have been so successful that birth rates are falling steeply in many parts of the world. In Thailand, for example, they have fallen from an average of 7 children per woman in the 1970s to just 1.6 in 2013.

The ability to plan families means that women can make choices about their education and careers. Contraception has also changed sexual behaviour and has led to greater promiscuity with a steep increase in sexually transmitted infections (see Topic 16.14).

IVF is an expensive procedure and is not available to all. Success rates for IVF are often about 30% for women under 35 years of age. In countries with a publically funded national health service, decisions have to be made about who receives treatment and how many repeat treatments they may have if the first proves unsuccessful. In many parts of the world childless couples are not able to afford this treatment. A problem is what to do with 'spare' embryos that are not implanted. Some people maintain that it is not acceptable to just dispose of them or use them for medical research.

> **STUDY TIP**
>
> The use of hormones in contraceptive pills is described in Topic 16.12.

> **KEY POINTS**
>
> 1 Infertile females can take fertility drugs containing FSH to stimulate the release of eggs and increase the chances of pregnancy.
>
> 2 If the male is infertile, couples may opt for artificial insemination where donor sperm are used to fertilise a woman's fertile eggs.
>
> 3 IVF is a technique in which fertilisation takes place inside a Petri dish.
>
> 4 Oral contraceptives allow women to decide whether they wish to become pregnant or not.

SUMMARY QUESTIONS

1 a Explain how fertility drugs can increase the chance of a pregnancy in infertile women.

  b Explain how artificial insemination can help bring about a pregnancy when the man is infertile.

2 Explain a why eggs to be used in IVF are kept in a solution of nutrients and oxygen, and b why embryos are kept for a few days before implanting them into the uterus.

3 Discuss what should happen to embryos from IVF treatments that are not implanted. Should they be destroyed, frozen for possible future use or used for research?

Supplement

Supplement

LEARNING OUTCOMES

- Define *sexually transmitted infection (STI)*
- State that HIV is an STI
- Describe the methods of transmission of HIV and ways in which its spread can be prevented
- State that HIV infection may lead to AIDS
- Describe how HIV affects the immune system

A sexually transmitted infection (STI) is a disease that is transmitted via body fluids during sexual contact. STIs include HIV, syphillis and gonorrhoea amongst the major infections that can be transmitted.

Human immunodeficiency virus (HIV)

The human immunodeficiency virus (HIV) is a human pathogen that was first identified in the early 1980s. Like all viruses it is composed of a few genes and a protein coat. It enters human cells and uses each cell to make more viruses which then enter even more cells. The main type of cell that it infects is a type of lymphocyte.

HIV infection may lead to the development of **Acquired Immunodeficiency Syndrome** or **AIDS**. If someone has been infected with HIV they are described as being HIV positive (HIV+). The virus weakens the body's immune system. It is easier for a person who is HIV+ to be infected by other diseases, such as tuberculosis (TB) and pneumonia. People who do not have HIV are described as HIV negative (HIV–).

STUDY TIP

AIDS is not a disease; it is a collection of diseases which result from a weakening of the immune system. This is what the word 'syndrome' means. Often the term HIV/AIDS is used.

HIV and lymphocytes

HIV attacks and destroys an important type of lymphocyte that coordinates the immune system. During an infection these T lymphocytes stimulate other lymphocytes to produce antibodies. During an HIV infection the number of T lymphocytes decreases and so fewer antibodies are produced every time there is another infection. By reducing the number of T lymphocytes, HIV weakens the body's ability to fight disease. Eventually the person infected with HIV will succumb to any number of other infections because of their weakened immune system.

Lymphocyte infected with HIV virus (yellow) (×5000).

A health worker treating a patient with AIDS in Cambodia.

AIDS (acquired immune deficiency syndrome) is the name given to a collection of diseases brought on by the weakening of the body's immune system.

The early symptoms of AIDS are very much like flu, with swollen glands and a high temperature. Later, symptoms might include weight loss, various types of cancer and a decrease in brain function. Not all HIV+ people develop AIDS. Some people remain HIV+, but without any symptoms at all.

How is HIV transmitted?

HIV is transmitted in the blood and semen. The virus can pass from one person to another during unprotected sexual intercourse.

Either partner may infect the other. The virus can also be passed via hypodermic needles contaminated with infected blood. In this way, HIV has spread very quickly amongst drug addicts who share needles.

Unborn babies are also at risk from HIV. This is because the virus can pass across the placenta to the fetus. Even more likely is its passage from the mother's blood to the baby's blood at birth when the two bloodstreams come into close contact. HIV can also be transmitted in breast milk. Blood used for transfusions is another way in which HIV has been transmitted.

How can HIV/AIDS be prevented?

There is no cure for HIV/AIDS and as yet no vaccine for HIV. Scientists are trying to develop drugs that will inactivate HIV, but the problem is that the drugs may damage the host lymphocytes. There are anti-viral drugs, e.g. **zidovudine (AZT)**, and some of these prevent the virus multiplying inside the body cells. Antibiotics do not work against viruses, but people with AIDS take them to treat the bacterial and fungal infections that they have.

Since HIV infection cannot be cured, it is important to prevent the transmission of the virus. **Contact tracing** is important in the control of the spread of HIV in the UK. Any person diagnosed with HIV positive is asked to identify people who they may have put at risk, either as a result of sexual intercourse or by sharing needles. These people are then offered an HIV test which reveals the presence of HIV antibodies in their blood. However, these antibodies will appear only several weeks after the person has been infected.

Other methods to reduce its spread include:

- the use of condoms to provide a physical barrier to the transmission of the virus during sexual intercourse

- setting up free needle exchange schemes for those people who inject drugs; this reduces the risk of transmission from the use of shared needles and syringes

- screening donated blood for HIV antibodies and eliminating contaminated blood being used for transfusion.

- education programmes to make people aware of the methods of the transmission of the HIV virus and how it can be prevented.

KEY POINTS

1 A sexually transmitted infection can be passed on via body fluids during sexual contact.

2 The spread of STIs can be controlled by the use of condoms, preventing the sharing of hypodermic needles, and education programmes.

3 HIV is a virus that infects cells in the immune system. It is transmitted in the semen or in the blood. With time, HIV infection may lead to a collection of diseases which is known as AIDS.

4 The spread of HIV can be reduced by men using condoms during sexual intercourse, by reducing the use of shared needles between drug users and by careful screening of donated blood used in transfusions.

5 HIV destroys lymphocytes leading to a weakening of the immune system. This makes people susceptible to many diseases, such as TB.

SUMMARY QUESTIONS

1 a What is the difference between AIDS and HIV?

  b What exactly is HIV?

  c Why can an HIV infection not be treated by antibiotics?

2 a State the ways in which HIV is transmitted.

  b State the measures that can be taken to reduce the spread of HIV infection.

3 Describe the effect of HIV on the immune system and explain how HIV infection leads to the development of AIDS.

Supplement

1 The drawing shows a half flower. Which row is the correct identification of the labels **1** to **4**?

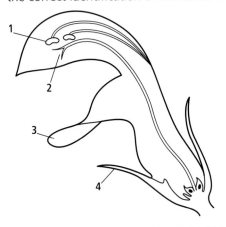

	1	**2**	**3**	**4**
A	anther	stigma	petal	sepal
B	petal	stigma	sepal	anther
C	sepal	petal	anther	stigma
D	stigma	anther	sepal	petal

(Paper 1) [1]

2 Which row shows the correct functions of an insect-pollinated flower?

A	anthers produce pollen grains	carpels contain ovules	sepals provide landing site for insects
B	sepals attract insects	anthers produce pollen grains	carpels contain ovules
C	stigmas receive pollen grains	ovaries contain ovules	anthers produce pollen grains
D	petals protect a flower in the bud	stigmas are site of pollination	anthers produce ovules

(Paper 1) [1]

3 The diagram shows the female reproductive system. Which row identifies the places where the events occur?

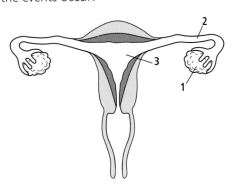

	1	**2**	**3**
A	fertilisation	implantation	ovulation
B	implantation	ovulation	fertilisation
C	ovulation	production of eggs	fertilisation
D	production of eggs	fertilisation	implantation

(Paper 1) [1]

4 Which organ secretes testosterone?

  A ovary

  B pituitary gland

  C placenta

  D testis

(Paper 1) [1]

5 The following four processes occur during sexual reproduction in flowering plants.

  1 Pollen grains lands on the stigma.

  2 The male gamete fuses with the female gamete.

  3 The male gamete travels down the pollen tube.

  4 The pollen tube grows out from the pollen grain.

In which order do these processes occur?

  A 3, 4, 1, 2

  B 4, 3, 2, 1

  C 1, 4, 3, 2

  D 4, 2, 1, 3

(Paper 2) [1]

6 Which hormone is secreted by the yellow body in the second half of the menstrual cycle?

A follicle stimulating hormone (FSH)

B luteinising hormone (LH)

C oestrogen

D progesterone

(Paper 2) [1]

7 The diagram below shows the changes that occur to the uterus during the menstrual cycle.

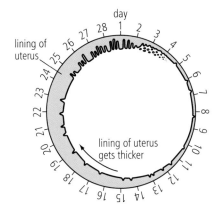

Which row shows the times (in days) when the events during the cycle occur?

	ovulation is likely to occur	menstruation occurs	uterus lining gets thicker
A	10–27	25–28	4–10
B	13–15	1–4	6–25
C	22–25	1–4	7–14
D	10–27	13–15	25–28

(Paper 2) [1]

8 Potato plants reproduce by means of flowers and stem tubers. Which results in the least variation among offspring in potato plants?

A cross-pollinating plants of different varieties

B cross-pollinating plants of the same variety

C production of stem tubers

D self-pollination

(Paper 2) [1]

9 The diagram shows a fetus in the womb just before birth.

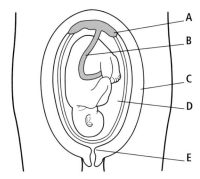

(a) Name body parts **A** to **E**. [5]

(b) Describe the changes that happen to the structures **A** to **E** during birth. You may use the letters rather than the names in your answer. [5]

(Paper 3)

10 (a) Define the term *sexual reproduction*. [3]

(b) Explain the difference between the following pairs of terms: **(i)** ovule and ovary, and **(ii)** pollination and fertilisation. [4]

(c) State three factors required for the germination of seeds. [3]

(Paper 3)

11 Clomiphene is a drug that is used to treat infertility in women. It works by causing an increase in production of FSH from the pituitary gland.

(a) Explain how a deficiency in FSH causes a woman to be infertile. [2]

(b) Outline the social issues involved with fertility treatment. [4]

(c) Describe the roles of oestrogen and progesterone in pregnancy. [3]

(Paper 4)

Chromosomes, genes and DNA

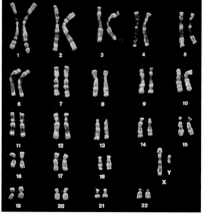

People who belong to the same family often resemble each other – sometimes quite closely; in the case of identical twins, very closely. Characteristics or traits such as physical appearance, personality and capabilities at such things as sport and music, often run in families. We say that someone has inherited features from their parents.

The transfer or transmission of these features from one generation to the next is called **inheritance**. The characteristics are controlled by genes which are like coded instructions. The study of inheritance is known as **genetics**.

In Unit 16 we looked at how new generations of flowering plants and humans are produced. Genes from one generation are transmitted to the next in the gametes. At fertilisation, the gametes fuse to form a zygote which contains the genetic information from both parents. A new individual grows from the zygote – half of its genetic information comes from its male parent and half from its female parent.

Chromosomes

We cannot see genes through a light microscope, but we can see **chromosomes**. Most cells have nuclei containing chromosomes. The first step in visualising the genetic information that we inherit is to look at these chromosomes.

Chromosomes are visible through a microscope during cell division. The photograph shows the chromosomes of a human white blood cell specially prepared so that they can be counted and analysed.

There are 46 chromosomes in the nucleus of a human cell. This is called the **diploid number**. The chromosomes are different shapes and sizes. They can be sorted into pairs based on their size and shape. In the photograph, a human male's chromosomes are sorted into 23 pairs. In each pair, one of the chromosomes has been inherited from the male parent and one from the female.

You can see that each of the chromosomes in the photo on the left is double stranded. They have two strands joined together usually at the centre, but not always. Figure 17.1.1 at the top of the next page shows the two strands of one chromosome in two different colours.

If we could unravel a chromosome, it would form an extremely long thread. That thread is made up of a long chain molecule called **DNA** which stands for **deoxyribonucleic acid**. The DNA is wound around molecules of protein. During cell division the DNA and protein are packed very tightly together and the chromosomes are coiled up. At other times they are uncoiled so the cell can use the information in the DNA.

What is a gene?

Genes are lengths of the DNA that you can see in the diagram below. Each chromosome has many genes along its length, although the actual number depends on its length. You can see that some of our chromosomes are much longer than others; these chromosomes have large numbers of genes.

Each gene is a unit of inheritance in that it codes for a specific protein. It is a chemical code that the cell interprets as an instruction to make a protein molecule. You know about several proteins – haemoglobin, amylase and lipase for example. There is a gene for each of these proteins and many more. Humans have between 20 000 to 25 000 different genes. The proteins produced by cells influence how the body works and what it looks like. Some features that genes control are influenced by the environment. Your height is determined by both genes you inherit and your environment – for example by your diet and how much exercise you take. Other genes are not affected by the environment. Your blood group is determined by a gene and the environment has no affect on this at all.

Figure 17.1.1 A double-stranded chromosome.

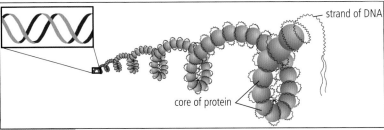
strand of DNA

core of protein

Figure 17.1.2 The arrangement of DNA in a chromosome.

SUMMARY QUESTIONS

1 Copy and complete the sentences using these words:

   **chromosomes   DNA   genes   genetic information
   gametes   nuclei   resemble**

   Offspring _____ their parents because of the _____ passed on to them in the _____ (sex cells) from which they developed.

   This information is contained in the thread-like structures called _____ that are inside the _____ of most cells. These threads are made of proteins and _____ . The information is found in small units called _____ .

2 a Define the terms *gene* and *chromosome*.

   b Name three human proteins.

   c Write a few sentences to explain the relationship between the following:

      nucleus, chromosome, gene, protein.

3 a What is meant by the *diploid number*?

   b State the diploid numbers of three non-human species.

Protein synthesis

- Explain that the sequence of bases in a gene code for the correct order of amino acids in a protein
- Explain how a protein is made

Each chromosome is made up of a long, super-coiled molecule of DNA. Each molecule is divided into thousands of shorter sections called **genes**. A gene is a small part of the DNA strand.

The length of DNA, making up a particular gene, carries the information needed to make a particular protein. Remember that inside every cell there are thousands of different chemical reactions taking place.

Enzymes control all these chemical reactions. You should remember that all enzymes are proteins. Because DNA codes for proteins, it determines which enzymes are produced in each cell and therefore which chemical reactions take place inside cells.

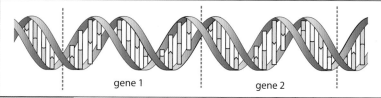

Figure 17.2.1 A gene is a short section of DNA.

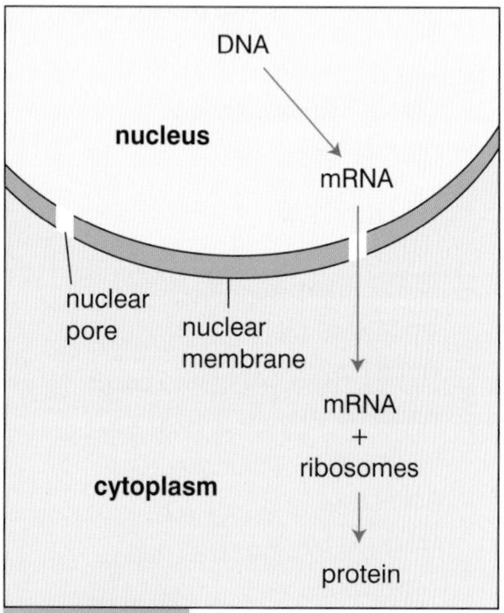

The structure of DNA (see Unit 4) contains bases. The information is found in the **sequence of bases** along the length of DNA. These determine the sequence of amino acids in the protein. DNA also controls the production of other important proteins, including antibodies and receptors for neurotransmitters.

DNA carries information to build proteins from amino acids. The information moves from the DNA to the site of protein synthesis in the ribosomes by a kind of 'messenger molecule'. This molecule is RNA and is given the name **messenger RNA** (mRNA). The role of mRNA is to carry a copy of the base sequence on DNA out of the nucleus to the ribosomes in the cytoplasm where protein synthesis occurs.

Figure 17.2.2 The vital role of mRNA carrying the code to the ribosomes.

The genetic code

There are **four** different bases found in DNA. We refer to them by their letters A, T, C and G. Each amino acid is coded for, by a sequence of *three* of these bases on DNA. There are about 20 different amino acids and each is coded for by a different base triplet on DNA. When mRNA copies the code on DNA it does so by the 'rule of base pairing', so for instance C will copy G, and G will copy C. However, mRNA does not have the base T – this is replaced by another base U (uracil). Therefore, A on DNA will code for U on mRNA.

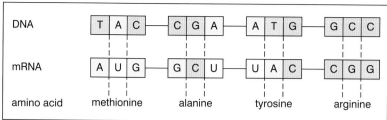

DNA	T A C	C G A	A T G	G C C
mRNA	A U G	G C U	U A C	C G G
amino acid	methionine	alanine	tyrosine	arginine

Figure 17.2.3 How DNA base triplets code for amino acids.

The coded information on mRNA is used to assemble amino acids in the correct sequence for each protein.

• The mRNA arrives at the ribosomes from the nucleus.

• The mRNA strand contains the base triplets for each particular amino acid in the protein.

• The mRNA then passes through the ribosomes and each ribosome assembles amino acids into protein molecules.

• The amino acids bond together forming a long chain – the protein.

• The specific order of amino acids is determined by the sequence of bases in the mRNA and each amino acid is joined to the next one by a peptide bond.

All body cells in an organism contain the same genes. However, in a particular cell many genes may not be expressed because the cell makes only the specific proteins it needs.

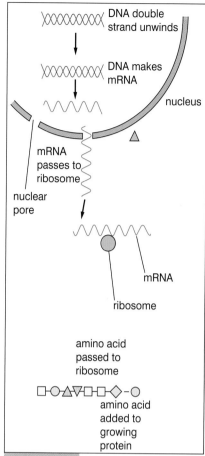

Figure 17.2.4 How DNA controls protein synthesis.

SUMMARY QUESTIONS

1 What is the function of each of the following in protein synthesis?

   a   DNA

   b   mRNA

   c   ribosomes

2 Look at the part of a molecule of messenger RNA here:

   AUGACGCAUGCAGUCCGA.

   a   How many base triplets are shown in this section of mRNA?

   b   What is the role of each of these base triplets in protein synthesis?

   c   Write down the DNA base triplets that coded for this mRNA.

3 Indicate if these statements are true or false.

   a   DNA is a protein

   b   DNA has a double helix

   c   DNA contains four bases: A, G, C and U

   d   The molecular shape of DNA is maintained by hydrogen bonds between complementary base pairs.

KEY POINTS

1 The sequence of bases on DNA codes for the correct sequence of amino acids in a protein.

2 The role of mRNA is to copy the DNA code and carry it to the ribosomes in the cytoplasm.

3 The ribosomes assemble amino acids into protein molecules.

LEARNING OUTCOMES

- Define *mitosis* and state its roles in growth, repair, replacement of cells and asexual reproduction
- Define *meiosis* and state its role in gamete production
- State that chromosomes are duplicated before mitosis
- Outline what happens to chromosomes during mitosis
- Describe the role of stem cells

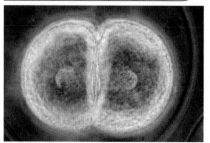

A zygote of a sea urchin has just divided in two by mitosis. You can see that each cell has a nucleus.

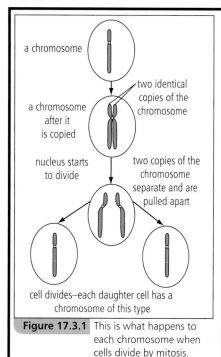

a chromosome

a chromosome after it is copied

two identical copies of the chromosome

nucleus starts to divide

two copies of the chromosome separate and are pulled apart

cell divides—each daughter cell has a chromosome of this type

Figure 17.3.1 This is what happens to each chromosome when cells divide by mitosis.

New cells are produced by the division of old ones. In the photograph below left, the fertilised egg of a sea urchin is in the process of dividing into two **daughter cells**. Before the cell divides, the nucleus must divide. As the nucleus is finishing its division into two, the cytoplasm divides. In animal cells a constriction occurs between the two nuclei to divide the cell into two. In plants a new cell wall is formed between the nuclei.

There are two ways in which nuclei divide.

- **Mitosis** is the type of nuclear division that occurs during growth and asexual reproduction. The daughter cells are genetically identical.
- **Meiosis** occurs in sex organs to form gametes. It is the nuclear division that gives rise to daughter cells that are not genetically identical.

The process of mitosis

Before a cell can divide, new copies of the genetic information in the DNA of the chromosomes must be made. This is a very reliable process in which exact copies are made. This copying process occurs *before* the nucleus divides. While it is going on, the DNA in the chromosomes is uncoiled and arranged very loosely in the nucleus. You cannot see chromosomes in the nuclei when this is going on.

The new copy of DNA of each chromosome is attached to the original copy. As mitosis begins, the DNA coils up so that each chromosome becomes thicker. They become visible under the microscope when a suitable staining technique is used.

During mitosis the two copies separate so each new cell gets a copy of each chromosome. The diagram shows what happens to one chromosome.

As a result of **mitosis**, each daughter cell has the same chromosome number as the original parent cell. As the chromosomes have been copied by a reliable system, they are genetically identical to each other and to the parent cell. If they were genetically different they would be rejected by the body's immune system.

The significance of mitosis

Mitosis produces cells that are an exact copy of the parent cell. These daughter cells have the same number of chromosomes and are genetically identical to the parent cell. Mitosis is important for the following processes:

- **growth** – in animals this happens all over the body, in plants it happens in special growing areas such as the tips of stems and roots
- **repair** of wounds – your skin cells divide by mitosis to repair damaged tissues and wounds
- **replacement** of cells that wear out and die, such as red blood cells which only live for a short time

- **asexual reproduction** – this occurs in fungi and in plants, but is rare in the animal kingdom. There are some examples of asexual reproduction on page 184.

Growth: As multicellular organisms grow, the number of cells making up their tissues increases. The new cells must be identical to the existing ones. Growth by mitosis takes place over the whole body in animals. In plants, group is confined to certain areas called **meristems**.

Repair and replacement of cells: Damaged cells must be replaced by identical new cells. Your skin cells and the cells lining your gut are constantly dying and being replaced by identical cells. Red blood cells which live only for a short time will wear away and die. They also need to be replaced by mitosis.

Asexual reproduction: Asexual reproduction results in offspring that are identical to the parent. Mitosis occurs when unicellular organisms reproduce, such as yeast and potato tubers (see Topic 16.1).

KEY POINTS

1 Mitosis produces daughter cells that are genetically identical and have the same number of chromosomes as the parent cell.

2 Stem cells are unspecialised cells produced by mitosis. They produce daughter cells that become specialised for specific functions.

Supplement

Stem cells

Supplement

Some of the cells that form from an embryo remain as unspecialised cells. They do not develop into any type of specialised cell. These cells keep dividing by mitosis to provide the body with a constant supply of new cells. These cells are needed to replace cells that get worn out and die. There is a layer of stem cells just under the outer layer of skin cells. These stem cells continually divide to make cells that become toughened for protection. They get rubbed off and need to be constantly replaced. When skin is wounded, these stem cells produce more cells to repair the damage. There are many stem cells in the bone marrow to produce new red blood cells and phagocytes as replacements.

Each time a stem cell divides, one daughter cell starts to become specialised for some specific function and the other daughter cell remains unspecialised to go through the process of mitosis again. The cells in meristems in plants have the same role as stem cells in animals.

SUMMARY QUESTIONS

1 a Which type of cell division produces cells that are identical to the parent cell?

  b Give two examples of when mitosis occurs **i** in plants and **ii** in animals.

2 Describe the role of stem cells.

Supplement

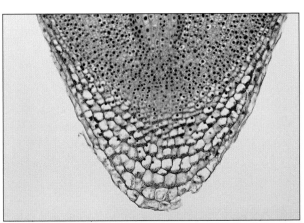

Mitosis takes place in the meristem behind the root tip.

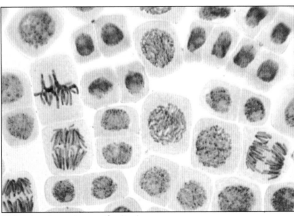

Cells at various stages of mitosis.

17.4 Meiosis

Meiosis

Meiosis is a type of cell division that is essential for sexual reproduction. It takes place in the reproductive organs and results in the formation of gametes with half the normal chromosome number. This is known as the **haploid number**. (As opposed to the full chromosome number, which is the **diploid number**).

If human sperm cells and egg cells both had 46 chromosomes the zygote formed at fertilisation would have 92 chromosomes. If this continued generation after generation the number of chromosomes in the nuclei would double every generation. This does not happen. Instead, the number remains at 46. This is because there is a different sort of nuclear division in the life-cycle involved in producing gametes. This type of nuclear division is a reduction division called **meiosis**.

Meiosis halves the number of chromosomes so egg cells and sperm cells have only 23 chromosomes each. The zygote has 46 chromosomes, 23 from the mother, in the egg, and 23 from the father, in the sperm. In sexual reproduction the number of chromosomes stays constant from generation to generation.

Meiosis occurs in sex organs in humans – the ovaries and testes. In flowering plants meiosis occurs in the anthers and in the ovules.

In meiosis the daughter cells are not identical. They are genetically different and this contributes to genetic variation that provides the raw material for selection and allows organisms to evolve in response to changing environments.

Figure 17.4.2 shows what happens to one pair of chromosomes during meiosis. As you can see there are **two divisions** of the cell resulting in four cells. Each of these cells has one chromosome, so the number has been halved to give a haploid gamete.

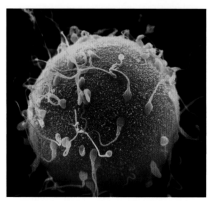

LEARNING OUTCOMES

- Define meiosis as a reduction division in which the chromosome number is halved, resulting in genetically different cells
- Explain how meiosis produces variation by forming new combinations of maternal and paternal chromosomes

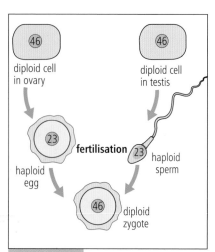

Human sperms and eggs contain 23 chromosomes as a result of meiosis.

Figure 17.4.1 The number of chromosomes in the gametes is called the haploid number. The number in a zygote is the diploid number.

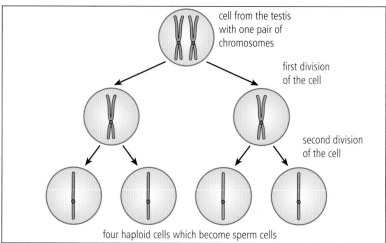

cell from the testis with one pair of chromosomes

first division of the cell

second division of the cell

four haploid cells which become sperm cells

Figure 17.4.2 This shows what happens to one pair of chromosomes during meiosis. There are two divisions of the cell to give four cells. Each of these has one chromosome so the number has been halved.

Figure 17.4.2 shows what happens to one pair of chromosomes during meiosis. In each pair, one chromosome is paternal in origin as it was inherited from the father, and one is maternal as it was inherited from the mother. When the 23 pairs of chromosomes separate in the first division of meiosis, the daughter cells gain different combinations of paternal and maternal chromosomes. This is one way in which the haploid cells produced in meiosis are genetically different from one another.

Differences between mitosis and meiosis

The differences between mitosis and meiosis are summarised in the table and in Figure 17.4.3.

Mitosis	Meiosis
One division	Two divisions
The number of chromosomes remains the same	The number of chromosomes is halved
The daughter cells are genetically identical	The daughter cells are genetically different from the parent cells
Two daughter cells are formed	Four daughter cells are formed

A fertilised human egg or zygote. The nuleus of the sperm is fusing with the nucleus of the egg to give a cell with 46 chromosomes.

STUDY TIP

Be careful when you write the words _mitosis_ and _meiosis_. They are easily confused and you should learn how to spell them correctly.

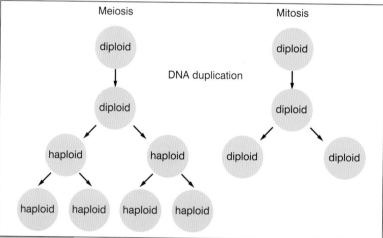

Figure 17.4.3 Meiosis and mitosis compared.

SUMMARY QUESTIONS

1 a How many pairs of chromosomes are there in the nucleus of a human body cell?

 b State the number of chromosomes in these human cells:

 i an egg cell   ii a sperm cell

 iii a zygote   iv a red blood cell

2 a State three ways in which meiosis differs from mitosis.

 b Explain how meiosis produces variation.

 c Explain the advantage of genetic variation among gametes.

KEY POINTS

1 Meiosis is a reduction division in which the chromosome number is halved from diploid to haploid resulting in cells that are genetically different.

2 Meiosis produces variation by forming new combinations of maternal and paternal chromosomes.

Symbols for pedigree diagrams:

square = male; circle = female

shaded circle or square = individual with the condition studied

unshaded circle or square = individual not showing the condition

individuals may be identified by generation and number, e.g. I – 2, II – 5 and III – 1.

Cat breeders know that if they breed together two cats which both have long hair, all the kittens will have long hair. If two short-haired cats are bred together then there may be some long-haired and some short-haired kittens in the litter, or there may be no long-haired kittens at all. Breeders can make pedigree diagrams to show the features that appear within a family of cats as in Figure 17.5.1. You will see that a pedigree diagram looks very like a family tree, but symbols are used to indicate the gender of individuals.

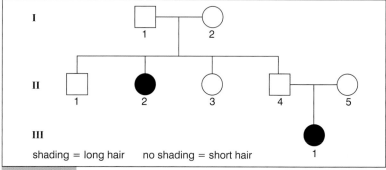

shading = long hair   no shading = short hair

Figure 17.5.1 A pedigree diagram showing the inheritance of hair length in a family of cats.

Genes and alleles

A gene is a length of DNA that is the code for making a protein molecule. Each gene is always located in the same place on one of the chromosomes. Each species has its own genes and every individual from the same species has the same genes. However, genes vary between individuals within each species. The different versions of each gene are known as **alleles**. There may be two alleles of a gene (see below), but usually there are many more.

In Figure 17.5.2, you can see three genes on one pair of chromosomes. You can also see the alleles **A/a**, **B/B** and **d/d**.

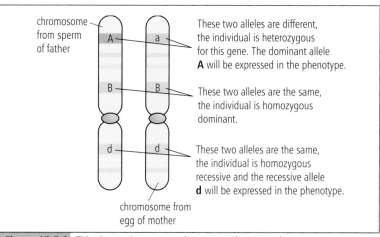

chromosome from sperm of father

These two alleles are different, the individual is heterozygous for this gene. The dominant allele **A** will be expressed in the phenotype.

These two alleles are the same, the individual is homozygous dominant.

These two alleles are the same, the individual is homozygous recessive and the recessive allele **d** will be expressed in the phenotype.

chromosome from egg of mother

Figure 17.5.2 This shows three genes that are on the same chromosome.

Sometimes an organism gains two copies of the same allele so there is only one type of instruction to use during growth: development or metabolism. But many organisms contain two different alleles. If an allele is expressed and affects the appearance of the individual, it is said to be a **dominant allele**. If it is not, then it is said to be a **recessive allele**.

In the 1850s, Gregor Mendel investigated inheritance in pea plants. He noticed that amongst garden pea plants one variety grew to be tall while another remained short, or dwarf. He used pea plants that always bred true – from generation to generation they were either tall plants or dwarf plants.

When he cross-bred these two varieties, all the next generation were tall. None of them were dwarf. But when this new generation of pea plants self-pollinated and produced seeds, the next generation consisted of many tall plants and some dwarf. He reasoned that dwarfness had been passed on although it had 'skipped' a generation. He did not know anything about genes and DNA, but he used his knowledge of algebra to give letters to represent what he called 'factors' that controlled height.

Using modern terminology there is a gene that controls height in pea plants. This gene has two alleles.

Let **T** represent the allele for tall, and let **t** represent the allele for dwarf. A capital letter is used for the dominant allele and a lower case letter for the recessive allele.

The **genotype** of an organism is the alleles it has. So the genotype of a pea plant could be **TT**, **Tt** or **tt**. If the two alleles are the same, **TT** or **tt**, the plant is **homozygous**. They can be **homozygous dominant** (**TT**) or **homozygous recessive** (**tt**). Individuals are **heterozygous** if they have two alleles that are different, for example **Tt**.

The **phenotype** is the way the alleles are expressed in an individual. It refers to all the aspects of an organism's biology except its genes. We will apply the term to one feature of an organism, for example height in pea plants. The genotypes **TT** and **Tt** result in plants with the tall phenotype, whereas the genotype **tt** results in dwarf plants.

STUDY TIP

Notice that we always show the dominant allele as a capital letter and the recessive allele as a small letter. Never use different letters for the alleles of the same gene.

Mendel studied the inheritance of seven different features in pea plants including the shape of the seeds – round and wrinkled.

KEY POINTS

1 Pedigree diagrams show how features are inherited in families.

2 An allele is a version of a gene. Most genes have many alleles.

3 The genotype is the genetic makeup of an organism; the phenotype is all the features of an organism other than its genotype.

4 A dominant allele is always expressed in the phenotype.

5 A recessive allele is expressed only when no dominant allele is present.

6 A homozygous individual has two identical alleles of a gene and a heterozygous individual has two different alleles of a gene.

SUMMARY QUESTIONS

1 Chromosomes occur in pairs in all cells except the gametes. Explain why they are not in pairs in gametes.

2 a What are alleles?

  b Explain what is meant by the terms *dominant allele* and *recessive allele*.

3 Explain each of the following genetic terms:

  a homozygous   b heterozygous

  c genotype   d phenotype

4 Explain how you would find out whether the allele for wrinkled seed shape is dominant or recessive.

Monohybrid inheritance

Monohybrid inheritance concerns the inheritance of a *single* characteristic, such as plant height or flower colour.

It involves the inheritance of the alleles of one gene. We will use the example of Mendel's peas.

There is one gene for height in pea plants. The gene for plant height has two alleles: tall, **T**, and dwarf, **t**.

There are three possible genotypes for plant height:

TT = homozygous tall

Tt = heterozygous tall

tt = homozygous dwarf.

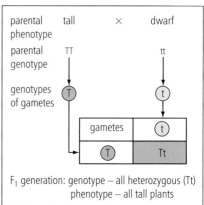

Figure 17.6.1

The parental cross

The genetic diagram above shows a cross between a homozygous tall plant, **TT**, and a homozygous dwarf plant, **tt**. These represent the pure breeding plants that Mendel used in his crosses. This is called the parental cross.

As a result of meiosis, all the gametes from the tall plant contain the dominant allele, **T**, and all the gametes from the dwarf plant contain the recessive allele, **t**. When fertilisation occurs, the new plants receive one dominant allele and one recessive allele so will be heterozygous and all have the genotype, **Tt**. Their phenotype will be tall, because the allele, **T**, is dominant to the recessive allele, **t**.

The first generation is known as the F_1 generation, and in this particular genetic cross, all the plants in the F_1 generation are heterozygous tall (**Tt**).

Crossing the F_1 generation (3:1 ratio)

The F_1 plants are allowed to self-pollinate (see genetic diagram, left). When meiosis occurs, half of the gametes of *each* plant will have the dominant allele, **T**, and half will have the recessive allele, **t**.

This time when fertilisation occurs, there are three possible combinations of alleles in the F_2, or second generation:

TT (homozygous tall)

Tt (heterozygous tall)

tt (homozygous dwarf).

Amongst this generation ¼ (or 25%) will be homozygous tall; ½ (or 50%) will be heterozygous and therefore tall; ¼ (or 25%) will be homozygous dwarf. Since the phenotypes of **TT** and **Tt** are the same, ¾ of the F_2 plants will be tall and ¼ will be dwarf. This can also be written as 3 tall : 1 dwarf.

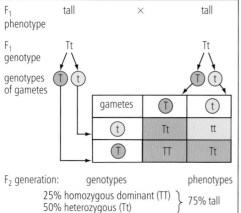

Figure 17.6.2 75% : 25% is a 3:1 ratio.

A test cross

If you look at the last two crosses, you will see that the recessive allele is not expressed in the F_1 generation but is expressed in the F_2 generation. This means that we do not know the genotype of a tall plant just by looking at it because plants with genotypes **TT** and **Tt** have the same phenotype. We do know the genotype of a dwarf plant because it can only be **tt**.

A test cross is used to determine the genotype of an individual to find out whether it is homozygous or heterozygous (see genetic diagrams below).

If our tall plant is homozygous (**TT**), then crossing it with a dwarf plant (**tt**) gives all tall plants (**Tt**). If, however, our tall plant is heterozygous (**Tt**), then crossing it with a dwarf plant (**tt**) will give half tall plants (**Tt**) and half dwarf plants (**tt**) in a ratio of 1:1.

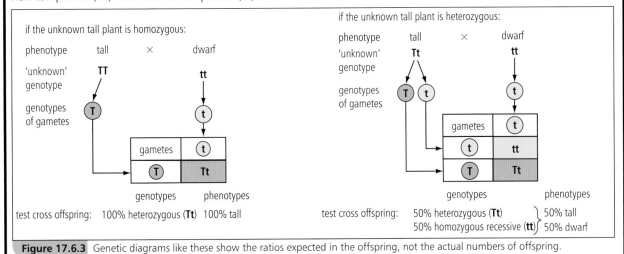

Figure 17.6.3 Genetic diagrams like these show the ratios expected in the offspring, not the actual numbers of offspring.

SUMMARY QUESTIONS

1 Complete these sentences by choosing the correct word from inside the brackets.

   a A pea plant with a tall (genotype/phenotype) could have the (genotype/phenotype) **TT** or **Tt**.

   b A tall pea plant with genotype **TT** is (homozygous/heterozygous) dominant.

   c A tall pea plant with genotype **Tt** is (homozygous/heterozygous).

   d A dwarf pea plant with genotype **tt** is (homozygous/heterozygous) recessive.

2 Fruit flies have a gene that determines the length of their wings.

   a Using the symbols **W** = long wing and **w** = short wing:

      i   State the genotype of a fruit fly which is heterozygous for this characteristic.

      ii Give the genotypes of its gametes.

   b A fruit fly heterozygous for wing length is mated with one which is homozygous for short wing. Use a genetic diagram to predict the proportions of each wing type in their offspring.

KEY POINTS

1 Monohybrid inheritance involves the inheritance of a single characteristic such as plant height or flower colour.

2 Crossing a homozygous tall plant with a homozygous dwarf plant produces all heterozygous tall plants.

3 Crossing two heterozygous tall plants gives a ratio of 3 tall plants to 1 dwarf plant (3:1 ratio).

4 Crossing a heterozygous tall plant with a homozygous dwarf plant gives a ratio of 1 tall plant to 1 dwarf plant (1:1 ratio).

Codominance

LEARNING OUTCOMES

- Explain the term *codominance*
- Describe the inheritance of ABO blood groups

Codominance

Up to now the examples we have looked at involve alleles that are either dominant or recessive. The recessive allele is not expressed in the phenotype of individuals with the heterozygous genotype. Sometimes, however, *both* alleles are expressed and neither is dominant.

So the phenotype is a mixture of the effects of each allele.

This condition is known as **codominance** and the alleles are called codominant alleles. The four o'clock plant can have red flowers or white flowers. Flower colour is determined by one gene. However, if you cross homozygous red-flowered plants with homozygous white-flowered plants, the offspring are all pink. The two parents produce heterozygous offspring (the F_1 generation) in which both alleles are expressed to give pink – an intermediate colour.

You can see in Figure 17.7.1 what happens when the pink plants are pollinated amongst themselves. There are red-, pink- and white-flowered plants in the F_2 generation in a ratio of 25% : 50% : 25% or 1:2:1. Notice that the ratio of the phenotypes is the same as the ratio of the genotypes when the alleles are codominant.

Notice also that each allele is represented as a capital letter, so red flower = C^R and white flower = C^W. This shows that neither allele is recessive and that they both exert an equal effect on the phenotype.

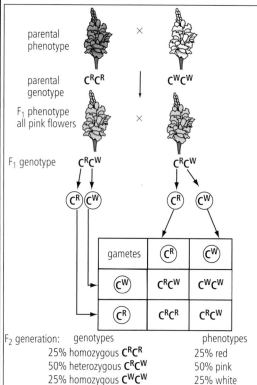

parental phenotype ×

parental genotype C^RC^R C^WC^W

F_1 phenotype all pink flowers ×

F_1 genotype C^RC^W C^RC^W

C^R C^W C^R C^W

gametes	C^R	C^W
C^W	C^RC^W	C^WC^W
C^R	C^RC^R	C^RC^W

F_2 generation: genotypes — phenotypes
25% homozygous C^RC^R — 25% red
50% heterozygous C^RC^W — 50% pink
25% homozygous C^WC^W — 25% white

Figure 17.7.1

Multiple alleles

So far, we have looked at examples where a gene has only two alternative alleles. There are examples where a gene may have three or more alleles.

The four groups of the human ABO blood group system are determined by one gene, **I**, with three different alleles: I^A, I^B and I^O.

The alleles I^A and I^B are codominant and code for slightly different molecules on the surface of red blood cells. The allele I^O is recessive to both I^A and I^B and does not code for one of these molecules.

Genotype	Blood group
I^AI^A	A
I^AI^O	A
I^BI^B	B
I^BI^O	B
I^AI^B	AB
I^OI^O	O

Look at the table. You can see the possible genotypes for each blood group. Remember any person can only have two alleles as we are all diploid.

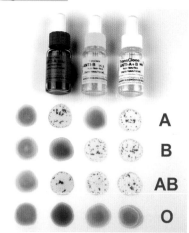

A

B

AB

O

Blood of the four different groups tested with anti-sera A (blue), anti-sera B (yellow) and anti-sera A and B (white). Red blood cells clump together if the anti-sera reacts with the A or B molecules on their surfaces.

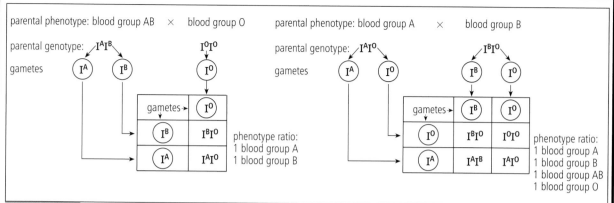

Figure 17.7.2

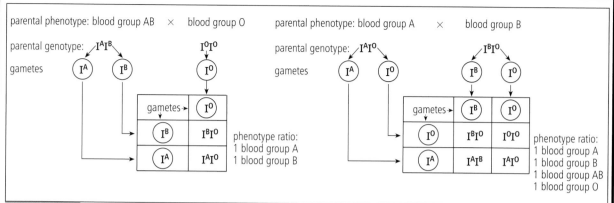

(The genetic diagrams show two parental crosses.)

Left cross:

parental phenotype: blood group AB × blood group O

parental genotype: $I^A I^B$ and $I^O I^O$

gametes: I^A, I^B and I^O

gametes →	I^O
I^B	$I^B I^O$
I^A	$I^A I^O$

phenotype ratio:
1 blood group A
1 blood group B

Right cross:

parental phenotype: blood group A × blood group B

parental genotype: $I^A I^O$ and $I^B I^O$

gametes: I^A, I^O and I^B, I^O

gametes →	I^B	I^O
I^O	$I^B I^O$	$I^O I^O$
I^A	$I^A I^B$	$I^A I^O$

phenotype ratio:
1 blood group A
1 blood group B
1 blood group AB
1 blood group O

Look at the genetic diagrams. Two parents have blood groups AB and O. The children will have blood groups A or B. None of them will have the same blood group as their parents.

Two parents have blood groups A and B. They could have children with all four blood groups – A, B, AB and O.

Flowers of the four o'clock plant.

SUMMARY QUESTIONS

1 There are red-flowered, white-flowered and pink-flowered varieties of the four o'clock plant (see photo on right). The results of a number of crosses between the varieties are shown in the table. Explain the results of each cross.

crosses	number of plants of each colour		
	red	pink	white
red × pink	126	131	
white × pink		92	88
red × white		115	
pink × pink	39	83	43

2 Use a genetic diagram to work out the possible blood groups of the offspring where the father is heterozygous for blood group A and the mother is blood group O.

3 The inheritance of coat colour in some cattle is an example of codominance. If a homozygous red cow is crossed with a homozygous white bull, all the calves have coats with a colour known as 'roan'.

The allele R^R causes red hairs to be produced and the allele R^W causes white hairs. Heterozygous cattle have coats that are a mixture of red and white hairs giving roan.

Use genetic diagrams to show the genotypes and phenotypes of the offspring in the following crosses:

a   a roan bull and a roan cow

b   a roan bull and a red cow

c   a roan bull and a white cow

STUDY TIP

The ABO blood group system is an example of codominance and also of discontinuous variation. You can expect to be asked about both topics in the same question.

KEY POINTS

1 Codominance occurs when both alleles are expressed in the phenotype, as neither is dominant over the other.

2 The inheritance of ABO blood groups involves three alleles: I^A, I^B and I^O.

Supplement

227

Supplement

LEARNING OUTCOMES

- Describe how sex is determined and inherited in humans
- Define a *sex-linked* characteristic
- Describe colour blindness as an example of sex linkage
- Use genetic diagrams to predict the results of crosses involving sex linkage

Sex determination

Your chromosomes determine which sex you are. In humans there are 46 chromosomes and these occur in 23 pairs. In females all 46 chromosomes can be matched together into pairs. This is done by taking images of the chromosomes and matching them for size and shape. In males there are two chromosomes that are not alike. These are known as the X and Y chromosomes. Females have two X chromosomes and males have an X chromosome and a Y chromosome.

In Figure 17.8.1 you can see how sex chromosomes in the gametes determine the sex of individuals. Remember that the number of chromosomes is halved in meiosis so each gamete can have only one of the two sex chromosomes.

All egg cells contain an X chromosome. Half of the sperm contain an X chromosome and half of the sperm contain a Y chromosome. At fertilisation, the egg may fuse with either a sperm with an X chromosome or a sperm with a Y chromosome. Since there are equal numbers of sperm with the X chromosome and sperm with the Y chromosome, there is an equal chance of the zygote inheriting XX or XY and the child being female or male.

Sex inheritance

As you can see in the electron micrograph opposite, the Y chromosome is very small and much smaller than the X. Therefore the Y chromosome has far fewer genes than the other chromosomes and far fewer than the X. One gene on the Y chromosome stimulates the development of testes in the embryo. If it is not present (or if it has mutated) then the body that develops is female.

Sex linkage

The genes that are located on the X chromosome are described as sex-linked even though they have nothing to do with determining gender or controlling sexual characteristics. Among the genes on the X chromosome are genes involved with controlling vision and blood clotting.

Males only have one copy of the genes that are on X chromosomes, if any of them are recessive, then the effect will be seen. Because women have two X chromosomes they are less affected by sex-linked recessive alleles and this is why sex-linked conditions are more common in boys than in girls. Two examples are discussed here.

Colour blindness

One of the genes on the X chromosome controls the ability to see red and green colours. The gene works in the receptors, known as cones, in the retina of the eye. There is an allele of this gene which does not

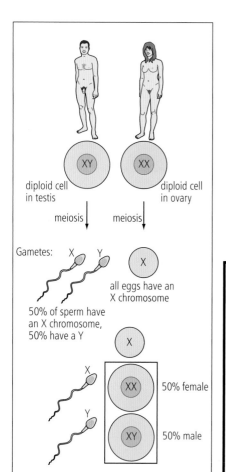

Figure 17.8.1 The diagram shows the inheritance only of the sex chromosomes, X and Y. Do not forget that each of the gametes (egg and sperm) contain 22 other chromosomes to give the haploid number of 23.

produce a protein necessary for colour vision. The allele is recessive, so any girl or woman who is heterozygous, **Rr**, has normal colour vision. Males only have one X chromosome so if they have inherited the allele **r** they will be colour blind.

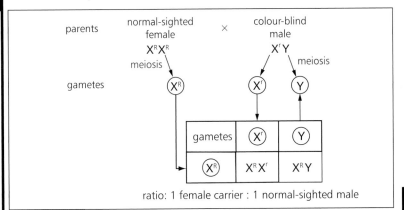

Figure 17.8.3

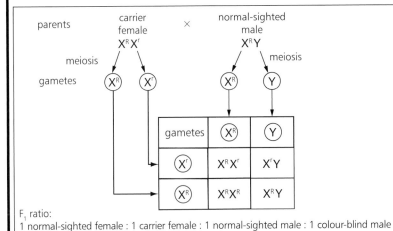

Figure 17.8.4

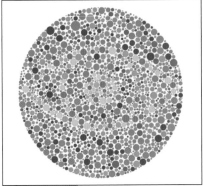

Figure 17.8.2 What can you see? If you cannot see something familiar then you may have red-green colour blindness.

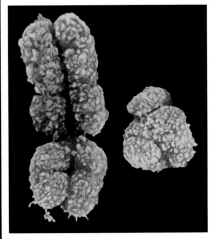

An X chromosome on the left and a Y chromosome on the right, photographed in an electron microscope.

Figure 17.8.3 shows what to expect if the mother is homozygous dominant and the father is colour blind. Notice that none of the children can expect to be colour blind, but the girls will inherit the allele for colour blindness on the X chromosome from their father. They have this allele, but it does not affect their phenotype. Any females who are heterozygous are **carriers** of colour blindness.

Figure 17.8.4 shows what to expect if the mother is a carrier and the father has normal colour vision. In this case, there is a 1 in 4 chance that one of the children will be colour blind. Note that there is a 1 in 2 chance that any boy will be colour blind.

SUMMARY QUESTIONS

1 Explain how sex is determined in humans.

2 Explain the following terms:
   a sex linkage   b carrier   c colour blindness

KEY POINTS

1 Sex is determined by the sex chromosomes, X and Y. Males are XY and females XX.

2 A sex-linked characteristic is one in which the gene responsible is located on a sex chromosome, usually the X chromosome.

3 Males have only one X chromosome so re-issue alleles are more likely to be expressed in males than in females.

4 Females are carriers of sex-linked disorders, such as red-green colour blindness.

1 Which explains the advantage of meiosis occurring in a life cycle?

  A meiosis allows sexual reproduction to occur

  B meiosis produces four times as many cells as mitosis

  C meiosis produces haploid gametes

  D meiosis produces sperm and eggs

(Paper 1)   *[1]*

2 A couple discover that they are both carriers of the condition albinism in which skin pigment is not made. What is the probability that their first child will have this condition?

  A 0%

  B 25%

  C 50%

  D 100%

(Paper 1)   *[1]*

3 A recessive allele:

  A causes a harmful feature

  B is never expressed in the phenotype

  C is not expressed when in a genotype with a dominant allele

  D produces the same phenotype when homozygous as when heterozygous

(Paper 1)   *[1]*

4 Which is meiosis involved with?

  A asexual reproduction

  B growth of stems and roots

  C production of egg cells

  D repair of tissues

(Paper 1)   *[1]*

5 A man has normal colour vision. His wife does not have colour blindness but her father is colour blind. What is the probability that their sons will be colour blind?

  A 25%

  B 50%

  C 75%

  D 100%

(Paper 2)   *[1]*

6 The Indian muntjac deer has a diploid number of 6. How many chromosomes are there in a skin cell and in a sperm cell of this species of deer?

	skin cell	sperm cell
A	6	6
B	12	6
C	6	3
D	3	6

(Paper 2)   *[1]*

7 Which feature is controlled only by a gene and *not* influenced by the environment?

  A blood group

  B concentration of glucose in the blood

  C heart disease

  D lung cancer

(Paper 2)   *[1]*

8 Which describes the genetic code?

  A the four different bases that are found in DNA

  B the mRNA molecules that leave the nucleus in a cell that is making proteins

  C the sequence of bases in a gene

  D the triplet of bases in DNA and RNA that code for the different amino acids in proteins

(Paper 2)   *[1]*

9 (a) Define the term *inheritance*.   *[2]*

  (b) Explain the difference between the following pairs of terms:

    (i) genotype and phenotype   *[2]*

    (ii) gene and allele   *[2]*

(Paper 3)

10 Brachydactyly is a rare genetic condition in which people with the condition have short fingers and toes. A pedigree diagram was drawn for this rare condition:

Key

○ female without the condition

● female with the condition

□ male without the condition

■ male with the condition

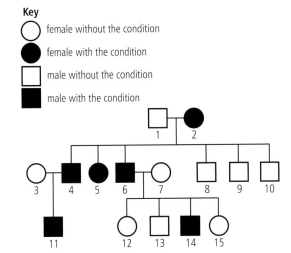

(a) (i) State whether brachydactyly is caused by a dominant or a recessive allele. *[1]*

(ii) Give evidence from the family tree that supports your answer to part **(i)**. Refer to the people shown in the tree by numbers. *[3]*

(b) What are the chances of parents 3 and 4 having another child with this condition? Draw a genetic diagram to explain your answer. *[3]*

(Paper 3)

11 In tabby cats the pattern of stripes in the coat is controlled by a gene. The allele for parallel stripes (**T**) is dominant to the allele (**t**) for a blotched pattern.

parallel   blotched

(a) Use genetic diagrams to predict the proportions of cats with different coat patterns in the next generation in the following crosses:

(i) homozygous dominant × homozygous recessive, *[4]*

(ii) heterozygous × homozygous recessive. *[4]*

(b) Explain why it is possible to state the genotype of a tabby cat with blotched stripes, but not one with parallel stripes. *[3]*

(c) Explain why male and female tabby cats with blotched stripes cannot produce a kitten with parallel stripes. *[2]*

(d) Use the example of inheritance in this question to explain the terms *gene* and *allele*. *[3]*

(Paper 3)

12 (a) Explain how you would use a test cross to find the genotype of an organism that could either be homozygous dominant or heterozygous. *[3]*

The four o'clock plant, *Mirabilis jalapa*, has a red-flowered variety and a white-flowered variety. When these are cross-bred the offspring have pink flowers.

(b) (i) Explain why this is so. *[3]*

(ii) Show, by means of a genetic diagram, the result of cross breeding the pink-flowered plants among themselves. *[5]*

(c) Plants with pink flowers were crossed with plants with red flowers. The seeds were collected and grown. When they flowered, 64 plants were of the pink variety and 78 of the red variety.

(i) Use a genetic diagram to explain these results. *[5]*

(ii) Explain why there were no white-flowered plants among the offspring in this cross. *[3]*

(Paper 4)

18.1   Variation

LEARNING OUTCOMES

- Distinguish between genetic variation and phenotypic variation
- State the differences between continuous and discontinuous variation
- Identify and explain examples of variation
- State that discontinuous variation is caused by genes alone

Although we show a lot of variation, we are all one species.

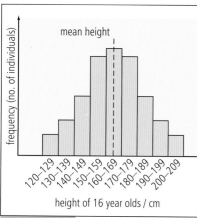

Figure 18.1.1 Frequency histograms are drawn to present data on continuous variation.

Types of variation

All humans have many features in common as we are all members of the same **species**. Now think about the similarities and differences between ourselves and monkeys and apes. We have many features in common, but there are significant differences which we use to categorise animals into different species. In Biology, the term variation means:

- differences between species – we use these when constructing keys
- differences within a species.

It is differences *within* a species that we are concerned with in this unit. Although we all belong to the same species, there are many differences between us – even between identical twins. Some people are taller or heavier, others have different coloured hair, skin and eyes. There is considerable variation between individuals of the same species.

Variation occurs in these features that we can see, but also in features that we cannot see but use specialist equipment to detect, such as the different types of haemoglobin that people have. These are all examples of **phenotypic variation**.

As we have seen in Unit 17, variation also exists at the genetic level. Individuals have different genotypes – different combinations of alleles of the genes that they have inherited. **Genetic variation** is the differences between the genotypes of individuals.

There are two types of phenotypic variation within a species:

- continuous variation
- discontinuous variation.

Continuous variation

If you measured the heights of all the pupils in your year group you would find a range of heights from the shortest to the tallest. If you divide the class into groups (170 cm to 179 cm, etc.) you can plot a frequency histogram. It is likely that the mean height will be in the middle of the range and correspond to the group with the largest number of people. This type of distribution is called a normal distribution.

This type of variation is called **continuous variation** because there is a continuous range of heights from shortest to tallest. There are many genes that contribute to your overall height. But other factors such as the quantity and quality of the food you eat and the amount of exercise you take influence your height. Also people go through their growth spurts at different ages and this will influence the results you get from your class.

Other features that you can measure also show this type of variation. Try measuring the length of your index finger or your hand span. You can collect data from the whole class and plot as frequency histograms.

Continuous variation also occurs in plants. You can measure the lengths of leaves or stems and you will get the same results – a range between two extremes with many intermediates in between. There are no clear divisions between the intermediates.

Discontinuous variation

Some people have attached ear lobes. Other people's ear lobes are free and not attached. This is an example of **discontinuous variation**. In this type of variation there is usually a small number of phenotypes (e.g. attached and not attached ear lobes) and no intermediates. If you collect data about a feature showing this type of variation, you plot it as a bar chart (see question 11 on page 243).

Discontinuous variation is caused by genes alone and the environment has no effect. The ABO blood group system is another example. Everyone has an ABO blood group – it is either A, B, AB or O. There are no intermediates. Some flowering plant species have flowers of distinctly different colours. This is another example.

Inheritance versus environment

Identical twins develop from the same embryo and are genetically identical. However, they are different in many ways. These differences will have come about because they developed slightly differently in the womb and have been exposed to different environmental influences. Identical twins separated at birth, or shortly afterwards, and brought up apart often show striking similarities. These are often to do with personality, which shows the influence that genes have on us.

The effect of environmental factors can be investigated with plants. The photograph shows the plant sometimes called the Mexican hat plant. It reproduces asexually to make lots of tiny plantlets all around its leaves. You can detach these and grow them in pots and keep them under different conditions. Any variation shown will be because of **environmental** factors.

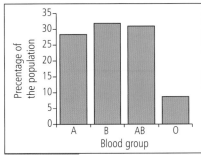

Figure 18.1.2 This bar chart shows the percentage of people with the four blood groups (A, B, AB and O) in Maharashtra, India.

These plantlets grow on the leaves of the Mexican hat plant. They are genetically identical.

STUDY TIP

When the words 'distinguish between' are used in a question, you need to give the differences between two things. Here you could be asked to distinguish between the two types of variation.

KEY POINTS

1 Variation is the difference between individuals of the same species.

2 Genetic variation is the differences in the genotypes of individuals; phenotypic variation is the differences in external and internal appearance.

3 Continuous variation is influenced by genes and the environment resulting in a range of phenotypes, e.g. height in humans.

4 Discontinuous variation results in a small number of phenotypes with no intermediates.

5 Discontinuous variation is cause by genes alone, e.g. human blood groups.

SUMMARY QUESTIONS

1 Distinguish between continuous and discontinuous variation and give three examples of each.

2 Describe how you would investigate the effect of an environmental factor on the growth of plants that all share the same genotype.

LEARNING OUTCOMES

- Define the term *mutation*
- State that mutation is the way in which new alleles are formed
- Describe the possible effects of radiation and certain chemicals on the rate of mutations
- Describe sickle cell anaemia as an example of mutation
- State that the heterozygous condition of sickle cell trait gives resistance to malaria

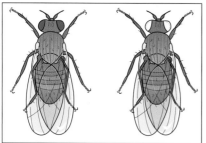

Figure 18.2.1 Red (normal) eye colour versus white (mutant).

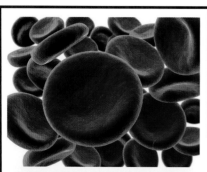

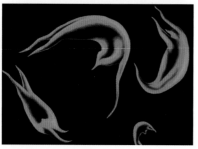

A **mutation** is a change in a gene or a chromosome that may cause a change in a phenotypic characteristic. A gene mutation is a change in the base sequence of DNA. Often this change can be harmful, but some mutations are beneficial and some have no effect at all.

Gene mutations

One of the first mutations studied was in the fruit fly. The normal eye colour of fruit flies is red, but a **mutant** form with white eyes was discovered. A gene controlling eye colour had changed. This was due to a change in the DNA so that the gene no longer coded for the production of the red pigment.

A similar example occurs in **albinos**. A gene controls production of the skin, hair and eye pigment **melanin**, which protects the skin and eyes from ultra-violet light. The gene can mutate to a form that does not produce melanin. It is a recessive allele and homozygous recessive individuals are albino. Gene mutations are the cause of many human genetic diseases. Sickle cell anaemia is an example.

Causes of mutations

Gene mutations are the only way in which new alleles are formed. They are caused by damage to DNA or by a failure in the copying process that occurs before nuclear division. They occur naturally at random, but the rate at which they occur is increased by exposure to ionising radiation and some chemicals. Ultra-violet radiation, X-rays and gamma rays are the most damaging. The greater the dose of radiation, the greater the chance of mutation. Benzpyrene in cigarette smoke is a cause of mutations.

If mutations occur in body cells they may affect the individual. For example they may lead to cancers, as happened after the atomic bombs were dropped in Japan in 1945 and the nuclear power station at Chernobyl in the Soviet Union exploded in 1986. It is only mutations that occur in gamete-forming cells, or gametes themselves, that are important in the long term. This is because these mutations can be inherited and provide new forms of variation.

Sickle cell anaemia – an inherited disease of the blood

The blood at the top shows normal red blood cells. The blood at the bottom is from a person with sickle cell anaemia (SCA). These red blood cells are in the shape of a sickle. (A sickle is a C-shaped tool for cutting plants like grasses and cereals.) These cells have abnormal haemoglobin which makes it difficult for the red blood cells to carry oxygen. SCA can prove fatal if the distorted red blood cells block blood vessels, or if the spleen destroys the abnormal sickle cells at a greater rate than red blood cells, so causing anaemia.

- SCA is a genetic disease resulting from a gene mutation. The normal allele, **Hb^A** (or **H^A**), codes for normal haemoglobin and the mutant allele, **Hb^S** (or **H^S**), codes for an abnormal form of haemoglobin. The two alleles are codominant so that people who are heterozygous have both forms of haemoglobin in their red blood cells.

Heterozygotes who carry one SCA allele have a much milder form of the disease called **sickle cell trait**. Clearly, the possession of two SCA alleles puts a person at a great selective disadvantage, since the abnormal form of haemoglobin is not very good at transporting oxygen which causes severe problems. This inherited disease is most common in West, Central and East Africa.

The SCA allele is common among people from areas where malaria is common. This is because people who are heterozygous have both the normal and abnormal forms of haemoglobin and do not usually have a problem with transport of oxygen. But they do have a resistance to malaria as the parasite that causes this disease cannot enter their red cells. Figure 18.2.2 shows that children of parents who are both heterozygous have a one in four chance of suffering from SCA. Note that alleles are shown as for codominance (see page 226), since both of them are expressed in the red blood cells of heterozygous people.

The link between SCA heterozygotes and resistance to malaria is an example of natural selection.

- There is a strong selection pressure against people who are homozygous for sickle cell anaemia, **H^{S}H^S**, because they could die from anaemia.
- There is a strong selection pressure against people homozygous for the normal allele, **H^{A}H^A**, because they could die from malaria.

But, in areas where malaria is common, there is a strong selective advantage for heterozygous individuals, **H^{A}H^S**, since they do not suffer badly from anaemia and they are protected from malaria.

So the sickle cell allele remains in the population in areas where malaria is an important **selective agent**.

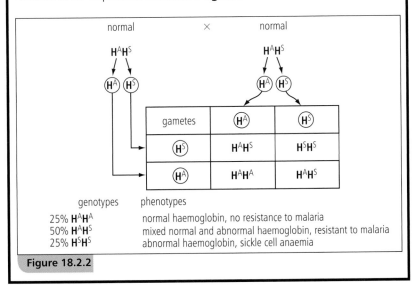

genotypes	phenotypes
25% H^{A}H^A	normal haemoglobin, no resistance to malaria
50% H^{A}H^S	mixed normal and abnormal haemoglobin, resistant to malaria
25% H^{S}H^S	abnormal haemoglobin, sickle cell anaemia

Figure 18.2.2

STUDY TIP

The allele for SCA arose in the first place by mutation, but has now increased because it gives the advantage of protection against a killer disease – malaria.

KEY POINTS

1 Mutation is a change in a gene caused by a change in the base sequence of DNA.

2 Ionising radiation and some chemicals increase the mutation rate.

3 Sickle cell anaemia is a genetic disease caused by an allele of the gene for haemoglobin.

4 Individuals with SCA in the heterozygous condition have resistance to malaria.

Adaptive features

LEARNING OUTCOMES

- Define the term *adaptive feature*
- Interpret images or other information about a species to describe its adaptive features
- Explain the adaptive features of the leaves, stems and roots of plants to pond and desert environments
- Define *fitness*

An **adaptive feature** is an inherited feature that helps an organism to survive and reproduce in its environment. The three photographs below show examples of adaptive features. The thistle has tufts on its seeds so that they are carried by the wind to disperse them widely. The mayfly nymph has a flattened, streamlined body and clings to the underside of rocks so it is not washed away. In winter, the Arctic fox grows white fur as a camouflage. Adaptive features are behavioural as well as structural; for example, penguins huddle together to keep warm.

Adaptive features like those in the photographs are likely to increase an individual's *fitness* which refers to its chances of surviving in its environment, reproducing and having offspring. These offspring are likely to show the same features.

The feathery tufts of thistle seeds are dispersed by the wind.

A mayfly larva has a flattened body so it does not get washed away in fast-flowing streams.

The Arctic fox has to endure extreme cold.

Pond plants

Hydrophytes are plants that grow submerged or partially submerged in water. Living in water has both its costs and its benefits.

Buoyed up by water and with no need for water transport, floating plants save energy since they produce little or no xylem tissue.

Roots, if present, are for anchorage and since there is no need for the roots to absorb water or mineral ions, there are no root hairs. The leaves and stems of hydrophytes have little or no cuticle, since there is no need to conserve water.

Water lilies, *Nymphaea alba.*

The problem for hydrophytes is that carbon dioxide, which is needed for photosynthesis, diffuses through water much more slowly than it does through air. The same applies for oxygen (needed for plant respiration) since it is not very soluble in water. Therefore many hydrophytes have an extensive system of air spaces in their stems and leaves through which gases diffuse quickly. These air spaces provide buoyancy to keep the plants close to the light and are a reservoir of oxygen and carbon dioxide.

Desert adaptations

All plants have to balance water uptake with water loss. It is important that they maintain the turgor in their cells, or they will wilt. Very high rates of transpiration can kill a plant if it cannot absorb enough water to prevent long-term wilting.

Xerophytes are plants are that are able to exist in conditions where water is scarce. Cacti are xerophytes that survive in hot, dry (arid) desert regions. Cacti reduce water loss and conserve water in the following ways:

- Their leaves are reduced to spines. This reduces the surface area of the leaf over which water can be lost.
- A thick, waxy cuticle covers the plant's surfaces and reduces transpiration.
- They have swollen stems containing water-storage tissue.
- They have a shallow, spreading root system to quickly absorb any water from rain and overnight condensation.
- Many cacti have a round, compact shape which reduces their surface area so there is less surface through which water can be lost.
- They have shiny surfaces which reflect heat and light.
- Their stomata are closed during the day to reduce water loss. They open their stomata at night to absorb the carbon dioxide which they store for use in photosynthesis during the day. Photosynthesis occurs in the outer layers of cells in their stems.

Cacti store water in their swollen stems.

Spines provide some shade and protection from herbivores.

SUMMARY QUESTIONS

1 Research how the following animals are adapted to extreme conditions:

  a a camel to desert conditions

  b an Arctic fox to arctic conditions.

2 Make a list of four ways in which desert plants are able to reduce water loss and conserve water.

3 Explain each of the following statements about pond plants.

  a They need little xylem in their stems and leaves.

  b Their roots do not have root hairs.

  c They have an extensive system of air spaces inside their stems and roots.

4 Freshwater aquatic plants are surrounded by a solution of a higher water potential than the cell sap in the cells of their stems and leaves.

  a Explain how this is an advantage to the maintenance of turgor in the cells.

  b What prevents these cells from bursting?

  c Ions in the water are at a very low concentration. State the mechanism by which these ions are taken up by plants.

KEY POINTS

1 An adaptive feature is inherited and helps an organism to survive and reproduce.

2 Fitness is the probability of an organism surviving in its environment, reproducing and having offspring.

3 Pond plants do not need much transport tissue since they are buoyed up in water. They have extensive air spaces to store carbon dioxide and oxygen which diffuse very slowly in water.

4 Desert plants such as cacti have thick cuticles, leaves reduced to spines, swollen stems to store water and extensive root systems. Often their stomata close during the day and absorb carbon dioxide at night.

LEARNING OUTCOMES

- Describe natural selection
- Describe evolution as the change in adaptive features over time
- Define the process of adaptation

Natural selection is the process by which organisms that are well adapted to their environment have a greater chance to breed and pass on their alleles to the next generation than those that are less well adapted. The process may be summarised as follows.

Variation within populations

Variation is the term used to describe all the differences that exist within populations of organisms. Gene mutation is the only way in which completely new genetic material is produced. Some mutations may give an advantage to the individual that expresses them. An example is having good camouflage to avoid being seen by a predator. But variation is also produced by sexual reproduction between two individuals.

During meiosis, the alleles of different genes are 'shuffled' to give new combinations in the gametes. At fertilisation, when the gametes fuse, alleles from the two different individuals are combined within the same nucleus. Gene mutation, meiosis and fertilisation give rise to variation between individuals in every generation.

Charles Darwin as a young man.

Over-population of offspring

These sockeye salmon return from the Pacific Ocean to rivers in British Colombia where they hatched in order to breed. The females produce millions of eggs, many of which are fertilised. We see the same in other species, for example a poppy plant releases over 15 000 seeds a year. Yet, in spite of this huge over-production of eggs and seeds, populations of animal and plant species remain fairly stable (see page 254).

Populations remain stable because, in the case of sockeye salmon, most of the eggs are eaten by predators as are many of the young that hatch from the eggs. Many young salmon die from disease and starvation. Similarly, many poppy seeds may be eaten by birds; others may land in places where they cannot grow as there is no soil or water. Production of large numbers of eggs and seeds ensures that enough offspring make it into the early stages of life to maintain the population from year to year.

Sockeye salmon.

Competition for resources

Among the organisms that survive the early stages of life there is competition for resources. Plants compete for space, light, water and nutrients. Animals compete for food, water, space (territories) and mates. This is referred to by biologists as the '**struggle for existence**' or '**struggle for survival**' (see page 255).

Vultures and a hyaena compete for food from a dead gnu.

Competition is fiercest between individuals of the same species as they have the same adaptive features to obtain their resources from the environment. Individuals of different species also compete, but often the competition is not as fierce. For example, one species may feed at night and the other during the day, or they have slightly different foods in their diets.

Reproduction

The individuals with features that adapt them to the conditions in their environment are those most likely to survive and reproduce. Individuals that are not so well adapted are likely to lose out in the competition for resources. They may die before they have a chance to reproduce or, if they do reproduce, have few offspring. The better adapted individuals have a greater chance to pass on their alleles to the next generation.

Natural selection

If the environment does not change, then natural selection maintains populations of organisms so they do not change much, as many are already well adapted to their environment. Any that are not well adapted are unlikely to survive to breed. When the environment changes, individuals with features that help them to survive in the new conditions are at an advantage over others. These individuals that are now better adapted, compete successfully, survive, breed and pass on their alleles. Natural selection will bring about a change to a species over time. Thus selection is the mechanism by which evolution occurs – an idea first proposed by Charles Darwin in 1859. Evolution is a process of adaptation whereby populations become more suited to their environment over generations.

Antibiotic resistance

Antibiotics are chemicals that kill bacteria or inhibit their growth. Soon after the introduction of antibiotics in the 1940s, some bacteria developed a resistance to their effects.

When bacteria are exposed to an antibiotic, such as penicillin, most are killed. However, there may be some individual bacteria that have a mutation giving them resistance to the antibiotic. The bacteria may be able to produce an enzyme that breaks down the antibiotic. These individual bacteria survive and now have more resources available as all their competitors – the non-resistant bacteria – have died. The resistant bacteria survive to reproduce and pass on the gene for resistance to their offspring. This is an example of natural selection. An example of this is the strain of the bacterium *Staphylococcus aureus* that is resistant to the antibiotic methicillin (MRSA).

Humans have been responsible for the change in the environment by introducing antibiotics, but we have not consciously chosen the bacteria which are resistant to antibiotics. There is a constant search for new antibiotics as bacteria develop resistance to existing ones.

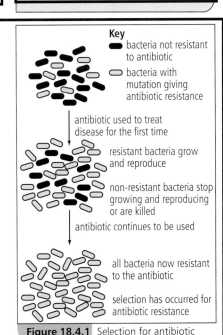

Figure 18.4.1 Selection for antibiotic resistance.

LEARNING OUTCOMES

- Describe the role of selective breeding in producing varieties of animals and plants with increased economic importance
- Describe examples of artificial selection
- Outline how selective breeding is carried out
- State the differences between artificial and natural selection

Selective breeding

People have grown crops and kept animals for at least ten thousand years. Over that time they have kept seed from one year to another and have bred their livestock. Domesticated plants and animals have changed considerably over that time due to this **artificial selection**. These changes have happened in this way:

1 Humans choose a desirable feature or features of an animal or plant to improve, e.g. fast growth, short stems (so less straw), docile (not fierce) nature, high yield, disease resistance, resistance to drought.

2 Animals or plants showing these features are bred to produce the next generation.

3 The offspring are checked to find those that show an improvement in the desired feature or features. These are kept for breeding the next generation.

4 This process of **selective breeding** continues for many generations.

Commercial farmers want to increase the yields of their crops and animals. They want to improve features of economic importance to maximise their profits. Plant breeders have increased the yield of grain in cereal crops such as rice, wheat and maize. Animal breeders have increased the milk yield in dairy cattle, and meat quantity and quality in beef cattle. Breeders of dogs and growers of ornamental plants are looking to improve other non-economic features, such as appearance.

This is called *artificial* selection as it is humans who are the selective agent, not the environment as it is with *natural* selection. Humans decide which feature to improve and which individuals survive to breed and pass on their genes.

Merino sheep have been selectively bred for their wool. They have been particularly successful since being introduced to Australia and New Zealand from Europe.

The danger of selective breeding is that there may be too much inbreeding between closely related individuals. This may result in harmful recessive alleles being passed on to the descendants and a reduction in variation. Many breeds of dog suffer from the effects of inbreeding, e.g. failure of the hip joint to develop properly leading to lameness.

Ankole longhorn (top) and Zebu are breeds of cattle well adapted to conditions in Africa such as shortages of water.

Merino sheep.

(a) These ewes from a flock of sheep have thick wool. They are mated with this ram who also has a thick coat

(b) to give these offspring – some of whom have thick wool

(c) but these sheep have thicker wool and are selected to breed the next generation

Figure 18.5.1 Artificial selection in sheep.

Differences between artificial selection (selective breeding) and natural selection

Some differences between artificial selection and natural selection (the evolutionary process) are shown in the table.

Artificial selection	Natural selection
Selection due to human influences	Selection due to environmental factors
Produces varieties of organisms very different from native generations	Produces greater biodiversity
Does not result in new species	May lead to new species
Inbreeding is common, leading to loss of vigour in the offspring	Outbreeding is common, leading to hybrid vigour
A relatively fast process	A slow process, taking many years
Proportion of heterozygous individuals in the population is reduced	Proportion of heterozygous individuals in the population remains high

There are two main methods of carrying out selective breeding:

Outbreeding involves the breeding of unrelated animals or plants. This may be used to combine the good characteristics of separate individuals, for example crossing a crop plant with high yield, with another crop plant that is resistant to disease, can produce offspring with both a high yield and is resistant to disease. Outbreeding often results in tougher individuals with a better chance of survival. This is called **hybrid vigour**.

Inbreeding involves breeding close relatives in an attempt to retain desirable characteristics. However, there can be harmful effects as a result of inbreeding. These can include a loss of vigour, with the population weakened by a lack of gene diversity and reduced fertility. There is also a greater susceptibility to disease as a result.

Hybrid maize plants display hybrid vigour.

DID YOU KNOW?

Selective breeding takes many years. It may take 10 years or more for plant breeders to develop a commercial variety that combines desirable features, such as disease resistance and high yield that come by crossing existing commercial varieties with low-yielding wild relatives.

KEY POINTS

1 Selective breeding (artificial selection) involves humans finding organisms with desirable features, crossing them and selecting the best from the next generation.

2 Selective breeding has produced new varieties of animals and plants with increased economic importance, e.g. high-yielding crops, cattle that produce more milk or better meat and sheep that produce more wool.

3 Differences between artificial and natural selection are summarised in the table.

SUMMARY QUESTIONS

1 State the useful features that are improved by breeders in:
   a race horses
   b sheep
   c cereal crops, e.g. wheat, rice or maize

2 Describe the procedure that a farmer would take to improve the milk yield of a herd of cows.

3 a List four differences between natural selection and artificial selection.

   b What is meant by i outbreeding and ii inbreeding?

   c What is meant by the term hybrid vigour?

1 Merino sheep have much thicker wool than wild sheep. Merino sheep are the result of:

  A mutation of a gene

  B natural selection

  C phenotypic variation

  D selective breeding

(Paper 1) [1]

2 Which is an example of discontinuous variation in humans?

  A blood group

  B body mass

  C foot length

  D hand span

(Paper 1) [1]

3 On very rare occasions a white mouse may appear in a wild population of mice that all have brown hair. This is the result of:

  A adaptation

  B artificial selection

  C competition

  D mutation

(Paper 1) [1]

4 An island was invaded by a species of bird that preyed on butterflies. In the population of butterflies, only those individuals that produced a toxic substance as a protection against predation survived. This is an example of:

  A artificial selection

  B competition

  C immunity

  D natural selection

(Paper 1) [1]

5 Individuals of a species of flightless sand-burrowing beetle range in colour from pale brown to almost black. The beetles have the same colour as the sand in which they live. The reason for this is likely to be:

  A Beetles are well camouflaged so that predators cannot see them.

  B Beetles change colour to match their surroundings.

  C Beetles are subject to artificial selection.

  D Dark beetles absorb more heat.

(Paper 2) [1]

6 Which row correctly identifies the causes of continuous and discontinuous variation?

	continuous variation	discontinuous variation
A	environmental factors only	genes only
B	genes and environmental factors	genes only
C	genes only	environmental factors only
D	genes only	genes and environmental factors

(Paper 2) [1]

7 Sickle cell anaemia is a genetic disease that is most common in areas where people are at risk from malaria. Which is an explanation of this similarity in distribution of the two diseases?

  A both diseases are associated with tropical climates

  B malaria causes sickle cell anaemia

  C people with sickle cell anaemia are more likely to catch malaria

  D people with the allele for sickle cell have a resistance to malaria

(Paper 2) [1]

8 Biologists use the term *fitness* to mean:

  A the ability to achieve great feats of endurance

  B the ability to leave very many offspring

  C the possession of an attractive phenotype

  D the probability that individuals survive and reproduce in the environment

(Paper 2) [1]

9 Match each of the following terms with the correct definition on the right:

mutation	outward appearance of an organism
variation	genetic constitution of an organism
selection	range of forms found in a species
phenotype	change in DNA
genotype	organisms best adapted to their environment survive and breed

(Paper 3) [5]

10 The distance between the outstretched thumb and little finger is the handspan. A group of students measured their handspans. The data is shown in the table.

handspan / mm	number of students	frequency / %
160–69	7	4.7
170–79	12	8.0
180–89	25	16.7
190–99	45	30.0
200–09	34	
210–19	19	12.7
220–29	8	5.3
Total	150	100.0

(a) (i) Calculate the percentage of students who had a handspan measurement between 200 and 209 mm. *[2]*

(ii) State the range in the handspan measurements. *[1]*

(b) Draw a frequency histogram of the results shown in the table. *[5]*

(c) (i) State the type of variation shown in your graph. *[1]*

(ii) Explain your answer to part **(i)**. *[2]*

(Paper 3)

11 Surveys were carried out in Pakistan and New Zealand to find the percentage of the population in the different blood groups of the ABO blood group system.

The results are shown in the table.

country	percentage of people in the sample in each blood group			
	A	B	AB	O
Pakistan	31.0	36.2	7.7	25.1
New Zealand	35.5	8.8	2.6	53.1

(a) Draw a bar chart to show the data. *[5]*

(b) (i) What type of variation is shown by the data? *[1]*

(ii) Explain your answer to part **(i)**. *[2]*

(c) A couple are blood group B. They have a child who is blood group O. Use a genetic diagram to explain how this is possible. *[5]*

(d) Before a kidney transplant operation may be carried out, the blood groups of the donor and recipient must be checked. Explain why it is important to do this. *[3]*

(Paper 3)

12 Xerophytes are plants that grow in dry places. They have many adaptive features.

(a) Define the term *adaptive feature*. *[2]*

(b) Describe three features of xerophytes that are adaptations for growing in dry habitats. *[3]*

(c) An area becomes very dry as a result of climate change. Explain how a species of plant may change over time to become adapted to this environmental change. *[6]*

(Paper 4)

13 Gonorrhoea is a bacterial disease. Antibiotic sensitivity tests are carried out on samples of bacteria taken from people with gonorrhoea to ensure that they are treated with an appropriate antibiotic.

(a) Describe how bacteria become resistant to antibiotics. *[3]*

(b) Explain how antibiotic resistance is transmitted from one bacterial cell to others. *[2]*

(c) Explain why it is important to carry out antibiotic sensitivity tests before prescribing a course of antibiotics. *[2]*

(d) Explain why antibiotics should not be used to treat viral diseases. *[3]*

(e) Antibiotic resistance is a serious medical problem worldwide. Suggest ways in which this problem may be overcome. *[3]*

(Paper 4)

19.1   Energy flow

LEARNING OUTCOMES

- State that the Sun is the main source of energy for biological systems
- State that a food chain shows the transfer of energy from one organism to another
- Define the terms *producers*, *consumers* and *decomposers*
- Construct simple food chains
- Define *trophic level*

Figure 19.1.1 A food chain.

producer

primary consumer

secondary consumer

Sunlight is the source of energy for all living things in this ecosystem.

In this unit we are looking at the flow of energy through living organisms and the recycling of nutrients to organisms in biological systems.

There is an important difference between these two processes. The Sun is a principal source of energy input into biological systems. Light energy is absorbed by photosynthetic organisms and made available to all other organisms as chemical energy. **Energy transfer** occurs when animals feed on plants and when animals feed on other animals. Eventually the energy is transferred to the environment as heat and is 'lost'. Nutrients on the other hand, consist of the chemical elements that make up living organisms. They are taken up by plants, passed on to animals and eventually recycled back into biological systems. This is known as **nutrient cycling**, since they are constantly reused.

Communities

Living organisms can be placed into groups based on similar features. However, when looking at a community of living things, it is often more useful to group them according to the way that they feed. Living organisms can be divided up into those that can make their own food and those that cannot.

Producers make their own organic nutrients from simple raw materials, such as carbon dioxide and water. Green plants use light as a source of energy to make sugars from carbon dioxide and water by photosynthesis. Some bacteria are also producers and they are either photosynthetic or obtain their energy from simple chemical reactions. Some of the bacteria involved in recycling nitrogen obtain their energy from reactions involving compounds of nitrogen (see page 253).

Producers make energy available for all the other members of the community in the form of energy-rich carbon compounds, such as carbohydrates, fats and proteins. Sunlight is the ultimate source of energy for almost all food chains. The major exceptions are the vent communities in the deep ocean that rely on chemical energy rather than light energy.

Consumers obtain their energy by eating other organisms, either plants or animals or both. All animals are consumers. They cannot make their own food, so they have to eat (consume) it. Herbivores are primary consumers as they eat the producers. Carnivores are secondary consumers and they eat herbivores. Tertiary consumers are carnivores that eat other carnivores, and are sometimes called top carnivores.

Decomposers are fungi and bacteria that gain their energy from waste organic material (see pages 6 and 7 for details of how they feed).

Each of these feeding groups is a **trophic level**. (The word *trophic* comes from a Greek word meaning 'to feed'.) You can put each organism in an ecosystem into a trophic level as you will see in these food chains.

Food chains

Food chains show what living organisms feed on in a community. They show the flow of food and energy from one organism to the next. Look at the food chain in Figure 19.1.2 from an East African ecosystem. The arrows show the direction in which energy in food is transferred from one organism to the next. This food chain tells us that the gazelle ingests grass and that the lion ingests the gazelle. Notice that the food chain always begins with a producer, often a green plant. This can include parts of a plant, such as seeds, fruits or even dead leaves. Food chains not only show the flow of food, they also show the flow of energy. Producers gain their energy from the Sun, so an *energy chain* would look like this:

Sun ⟶ producers ⟶ primary consumers ⟶ secondary consumers

But we do not include the Sun in food chains, as it is not a 'food'. Food chains start with producers.

Energy flows to decomposers, but they are rarely included in these simple food chains.

Food webs

In most communities, animals will eat more than one type of living organism. A food web gives a more complete picture of the feeding relationships in an ecosystem.

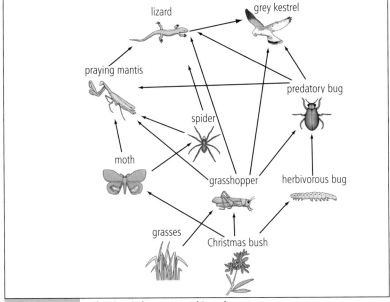

Figure 19.1.2 A food web for a West African forest ecosystem.

The over-harvesting of food species by humans in an unsustainable manner can result in a decrease in their numbers and extinction, e.g. overfishing. Also, the introduction of foreign (alien) species can displace native organisms due to predation and competition for resources, e.g food. Both of these result in disruption of food chains and webs, due to the removal of a species or a decrease in their numbers.

Pyramids of numbers and biomass

Pyramids of numbers

Food chains and food webs show feeding relationships in a community, but they do not show *how many* living organisms are involved. For instance, grasshoppers feed on many grasses and other plants and many grasshoppers and lizards are eaten by grey kestrels. Food chains and food webs do not show this.

Look at Figure 19.2.1. There are far more plants than caterpillars because each caterpillar eats leaves from many plants. There are far fewer owls than shrews because each owl eats many shrews.

Look at the numbers in this food chain:

plants 600 ⟶ caterpillars 100 ⟶ mice 10 ⟶ owl 1

This information can be shown in a **pyramid of numbers**. The area of each box in the pyramid shows roughly how many living organisms there are at each trophic level.

The producers are the first level, herbivores are the second level and carnivores are the third. Notice that there are fewer primary consumers than producers. There are also fewer secondary consumers than primary consumers. But secondary consumers are bigger than primary consumers as they have to eat them!

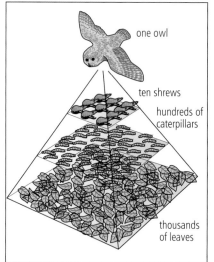

Figure 19.2.1 | This shows the producers and consumers that support an owl for several days.

one owl

ten shrews

hundreds of caterpillars

thousands of leaves

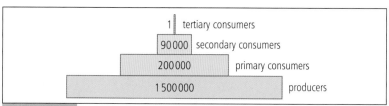

Figure 19.2.2 | Pyramid of numbers for a grassland community in 0.1 hectare.

1 | tertiary consumers
90 000 | secondary consumers
200 000 | primary consumers
1 500 000 | producers

A problem with pyramids of numbers is that they do not take into account the **size** of organisms at each trophic level.

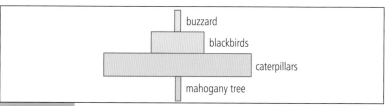

Figure 19.2.4 | A pyramid of numbers for the organisms living in a mahogany tree.

buzzard

blackbirds

caterpillars

mahogany tree

For instance, a mahogany tree and a grass plant each count as one organism. But one mahogany tree can support many more herbivores than one grass plant. As a result some pyramids of numbers can have unusual shapes.

In Figure 19.2.5 the tertiary consumers are parasites. Many of them feed on a single ladybird so this pyramid is inverted and it looks top heavy.

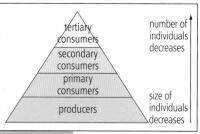

tertiary consumers
secondary consumers
primary consumers
producers

number of individuals decreases

size of individuals decreases

Figure 19.2.3 | Pyramids of numbers and biomass show these trends.

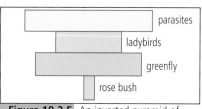

parasites

ladybirds

greenfly

rose bush

Figure 19.2.5 | An inverted pyramid of numbers.

Pyramids of biomass

One way to overcome the problem of size is to measure **biomass** instead of numbers. Biomass is the mass of living material. So a biomass pyramid shows the actual weight or mass of living things at each trophic level.

To draw a biomass pyramid, the data first needs to be collected. A sample of the organisms from each trophic level is taken and weighed. The average mass for the sample is calculated. The average mass is then multiplied by the estimated number of organisms present in the community.

The dry mass is determined for the plant matter by drying it in an oven until it reaches a constant mass. It is usually possible to estimate the dry mass of animals: approximately 65% of their body mass is water and this does not change as it does in plants. Dry mass tells us how much useful biological material is present as carbohydrates, proteins and fats as food for the organisms in the community. The data is usually expressed in grams per square metre.

The dry mass recorded is taken at one instant in time. Biomass pyramids do not take into account how fast an organism grows. For instance, grass often grows at a fast rate, but because it is grazed by animals, such as grasshoppers and cattle, its biomass at any instant in time will be low. This means it is better to calculate the biomass produced over a period of time, e.g. over a year or a growing season.

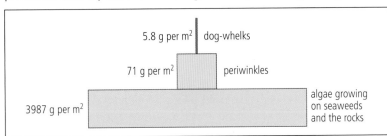

Figure 19.2.6 A pyramid of biomass for a rocky seashore community.

Phytoplankton consists of tiny photosynthetic organisms that float in water – both freshwater (lakes and ponds) and in the sea. The plankton grow quickly at times of the year when there are nutrients and the sea is warm with plenty of light. They live for only a few days as they are grazed by animal plankton. So their biomass at any one time is small. But over say a year, their biomass is huge. This biomass pyramid records only a few days' growth and so is inverted:

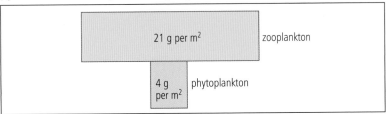

Figure 19.2.7 Biomass pyramid for the English Channel.

Biomass can vary with the seasons. In temperate latitudes, deciduous trees shed their leaves in winter. Their biomass is greater in summer than it is in winter as they will have leaves, flowers, fruits and seeds.

KEY POINTS

1 Pyramids of numbers show how many individuals there are at each trophic level but give no indication of their size.

2 Pyramids of biomass indicate the mass of living material at each trophic level, but give no indication of the rate of growth.

SUMMARY QUESTIONS

1 Draw two pyramids of numbers with different shapes. Explain their shapes.

2 Draw a pyramid of biomass for an ecosystem and explain its shape. Make sure that you include the unit of measurement for biomass on your pyramid.

3 What criticisms have been made about pyramids of numbers and pyramids of biomass?

STUDY TIP

Pyramids of energy show energy transfer in an ecosystem. Another way is a flow chart as shown opposite. This shows the energy losses to respiration and decomposers. You should be able to interpret flow charts that show energy flow.

STUDY TIP

As you will know from Physics, energy transfer processes are never 100% efficient. Much energy is 'lost' as heat. Respiration is no exception. Birds and mammals retain this heat to maintain a constant body temperature (see page 168).

The efficiency of energy transfer

The efficiency of energy transfer between trophic levels can be seen by using **energy pyramids**. These show the energy transferred from one trophic level to the next. The energy pyramid in Figure 19.3.1 shows that 87 000 kJ per m² per year is passed to the tadpoles from the plant plankton. The tadpoles pass on 14 000 kJ per m² per year to the small fish, and so on.

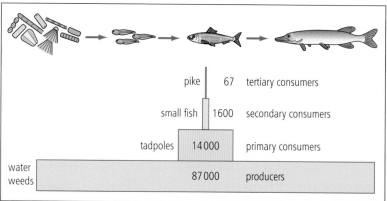

Figure 19.3.1 An energy pyramid for a lake. Figures are in kJ per m² per year.

The tadpoles obtain 87 000 kJ per m² per year from the water plants but only pass on 14 000 kJ per m² per year of this to the small fish. The remaining 73 000 kJ per m² per year has been lost from the food chain. The tadpoles will have used a lot of energy in swimming and they will have passed out some in waste – urine and faeces. The only energy that they do pass on to the small fish is the energy in their biomass – their muscles and other organs that the fish eats every time it swallows a tadpole.

Energy is always 'lost' in this way as it passes from one trophic level to the next. Of the 87 000 kJ per m² per year of energy in the producers only 67 kJ per m² per year of it becomes energy in the flesh of the top carnivore, the pike. Since only some of the energy is passed from one trophic level to the next an energy pyramid is never inverted. Its shape is not affected by the size of the organisms or how many of them there are since it simply looks at the energy that is passed on.

Energy losses

Energy flow through ecosystems is relatively inefficient. At best, plants absorb only 2 to 5% of the light energy that strikes their leaves. Much of this energy is not trapped in photosynthesis because it is reflected from the surface of the leaves, passes straight through them, or is green light which they cannot absorb.

Plants use much of the energy they trap in photosynthesis in their own respiration. As an energy transfer process, respiration is about 40%

efficient so about 60% of the energy is lost as heat to the plant and then to the atmosphere.

Much of the plant's biomass is lost to decomposers in the form of dead leaves, twigs, fruits, etc. About 10% of the energy trapped by the plant is available to secondary consumers. Much of what they eat is not very edible or easy to digest. So primary consumers make available only about 10% of what they have eaten to secondary consumers.

However, the animal flesh in primary consumers, for example grasshoppers and gazelles, is easier to digest and carnivores may obtain slightly more than 10% of the energy.

At each trophic level, energy is lost as heat to the environment and consumers lose energy in their faeces and urine to decomposers which means that there is less for the next trophic level. Food chains on land rarely have more than four trophic levels.

When farmers grow crops, such as rice, for human consumption, all of the energy in the grain is available to us as primary consumers. Livestock farmers grow or buy plant food to feed their animals. They may buy cereals such as barley or maize or they may grow fodder crops, such as grass, alfalfa and clover, for grazing. Animals, such as pigs, cattle and chickens, use up much of the energy in their food. This means that there is not as much energy available to us.

A vegetarian diet can support far more people. If we cut down the number of links in the food chain more individuals at the end of the food chain can be fed. This is because we are cutting down the 90% 'wastage' of energy that occurs between each trophic

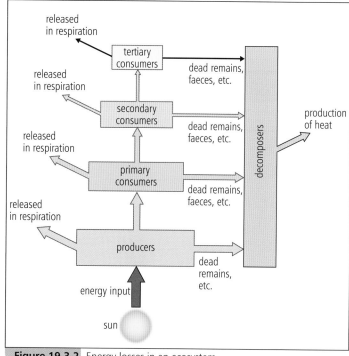

Figure 19.3.2 Energy losses in an ecosystem.

level, i.e. the energy that is uneaten, undigested or used in respiration at each level.

People in many countries tend to have diets that mainly consist of food of plant origin with some meat or fish. In Western developed countries many people have a varied diet which includes a higher proportion of meat from poultry, fish, lamb, beef and pork.

The human population is increasing. If farmers are to provide enough food for everyone, our diet may have to change to one that includes more plant foods and less food of animal origin.

SUMMARY QUESTIONS

1 a What is the ultimate form of energy input in a food chain?

 b Give two ways in which energy is 'lost' from food chains.

2 Explain why food chains usually have fewer than five trophic levels.

3 a Explain why it is efficient, in terms of energy transfer, to supply green plants as human food.

 b Explain why a diet rich in meat products is relatively inefficient in terms of energy loss.

KEY POINTS

1 Energy losses occur between trophic levels as a result of respiration, and in waste materials produced.

2 More energy is available in foods for human consumption if we feed as primary consumers rather than as secondary consumers.

3 A vegetarian diet can support far more people than one that includes meat products.

Nutrient cycles

Nutrient cycling

So far we have been concerned with food and energy. There is a constant flow of energy into ecosystems from the Sun, and energy flows through the different trophic levels and is then lost to the atmosphere. This topic and the next are about some of the elements that comprise living organisms and how they are cycled in nature.

Most living matter (95%) is made up of six elements: carbon, hydrogen, oxygen, nitrogen, phosphorus and sulfur. Living things must have a constant supply of these elements if they are to make proteins, carbohydrates, fats and other organic materials. Unlike energy, there is a finite source of these nutrients. Life has existed on Earth for millions of years. For it to continue, these elements must be recycled or they will become locked up in dead bodies and life will become extinct.

Bacteria and **fungi** feed on dead and decaying matter. They also feed on the waste matter of animals (urine and faeces). They feed by breaking down carbohydrates, fats and proteins into small molecules that they absorb. Many of these are respired. To do this they require water and a warm temperature as the processes of digestion and respiration are catalysed by enzymes. They also need oxygen for aerobic respiration.

Decomposers release carbon as carbon dioxide when they respire. They also break down amino acids releasing ammonia into their surroundings. In this way they recycle carbon and nitrogen – two of the most important elements for organisms. Simple compounds are absorbed by plants and converted into complex compounds.

These nutrients are then passed on to animals when they eat the plants. We say that **nutrients** are **cycled**.

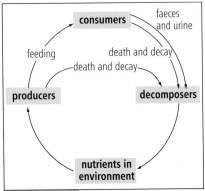

Figure 19.4.1 The role of decomposers in ecosystems.

The carbon cycle

Carbon is used to make carbohydrates, proteins, fats, DNA and other important biological molecules. The carbon comes from carbon dioxide in the air. Plants absorb carbon dioxide from the atmosphere and use it in photosynthesis to make food. Animals get the carbon compounds by eating plants.

Carbon dioxide gets into the air in the following ways:

- Plants and animals use some of their food for respiration, releasing carbon dioxide (as well as water and energy).
- Decomposers use dead plants and animals for food. They also use some of the decaying material for respiration, releasing carbon dioxide (plus water and energy).
- Fossil fuels like oil, peat, coal, and natural gas contain carbon. Carbon dioxide is one of the gases released into the air during combustion when these fuels are burnt.

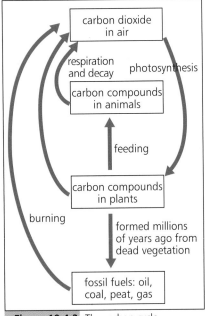

Figure 19.4.2 The carbon cycle.

These processes release carbon dioxide into the air and this balances the uptake by photosynthesis so the concentration of carbon dioxide in the atmosphere remains constant. Sometimes, dead plants and animals do not decompose. Their dead bodies become fossilised to form coal, oil and gas – fossil fuels.

The effects of combustion of fossil fuels and deforestation on the concentration of carbon dioxide in the atmosphere are discussed in Topics 21.2 and 21.5.

The water cycle

The water cycle demonstrates how important water is to life on Earth. Follow the arrows on the diagram to see how the world's water moves between living organisms and their environment.

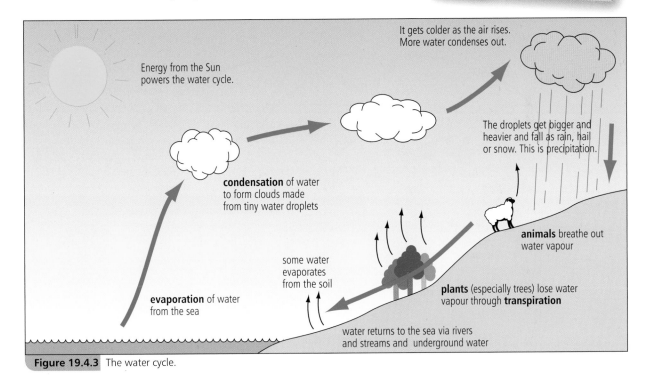

Figure 19.4.3 The water cycle.

Nitrogen

Nitrogen is an element required for many biologically important molecules. Amino acids, proteins, DNA and chlorophyll contain nitrogen. Approximately 80% of the air is nitrogen gas (N_2) but it is inert and very few organisms can make use of it. Plants, animals and most microorganisms cannot use it in this form. It has to be available to them as fixed nitrogen in the form of compounds such as amino acids and proteins, and as ions such as nitrate and ammonium ions.

The diagram below shows the nitrogen cycle which summarises all the changes that happen to nitrogen as it is recycled.

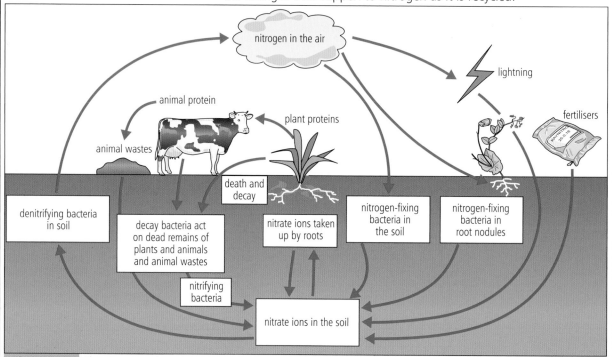

Figure 19.5.1 The nitrogen cycle.

Most of the nitrogen that plants obtain they absorb in the form of nitrate ions (NO_3^-). They use this to make amino acids in their leaves. The amino acids are used to form proteins, such as enzymes. Animals eat plants and so obtain their nitrogen in the form of plant protein. Any excess amino acids are broken down by animals to ammonia by the process of deamination. Animals that live on land, such as the cow in the cycle above, convert ammonia to urea and excrete it in their urine.

Decomposers (bacteria and fungi) break down dead remains and animal wastes releasing ammonium ions (NH_4^+) into the soil. Bacteria also break down urea in urine to ammonium ions. This process is sometimes called ammonification. Deamination results in the production of the excretory product urea, which passes out in animal urine (see Topic 13.1).

Nitrifying bacteria in the soil change ammonium ions into nitrate ions. These bacteria gain their source of energy from this oxidation reaction instead of absorbing light or feeding as a decomposer.

If you are tracing the cycle in the diagram you will realise we are back where we started with absorption of nitrate ions by plants.

Two other groups of microorganisms help to recycle nitrogen.

Nitrogen-fixing bacteria are found in the soil. These can convert nitrogen gas from the air into compounds of nitrogen that they use themselves. When they die and are decomposed, this fixed nitrogen is available to plants.

Nitrogen-fixing bacteria are also found in the roots of legume plants like peas, beans, alfalfa and clover and many tropical trees, such as flame trees. The bacteria are inside swellings on the roots called **root nodules**. These bacteria change nitrogen gas into ammonia that the legume plants can use to make amino acids. In return the legume plants provide a suitable environment for the bacteria (in the nodules) and also provide all the sugars that they need. Nitrogen fixation requires a lot of energy.

Denitrifying bacteria live in water-logged soil. They change nitrate ions to nitrogen gas. It is thought that they balance the uptake of nitrogen gas by nitrogen-fixing bacteria.

Lightning causes nitrogen and oxygen to react together at high temperatures. Nitrogen oxides are formed in the reactions. These are washed into the soil by rain where they form nitrate ions.

Nitrate ions are lost from the soil before plants can absorb them as they can be washed out of the soil by rainwater. This is called **leaching**.

Farmers add fertilisers containing nitrogen, e.g. ammonium nitrate. These are manufactured from ammonia made in the Haber process and add extra sources of nitrate ions for crop plants. The fertilisers replace the nitrate absorbed by the plants and removed at harvest. Some farmers do not use inorganic fertilisers, but instead use natural fertilisers such as manure. This provides dead material that enters the nitrogen cycle at the point where the decomposers act.

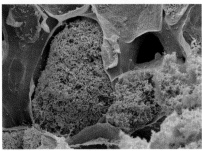

The top photo shows some nitrogen-fixing root nodules. In the bottom photo you can see the cells of a root nodule broken open to reveal the thousands of nitrogen-fixing bacteria that live inside.

KEY POINTS

1 Nitrogen is in the atmosphere as a gas (N_2) but is not available to most organisms as it is not reactive.

2 Some bacteria fix nitrogen (N_2) converting it to ammonia. Some of these live inside root nodules of legumes. They are called nitrogen-fixing bacteria.

3 Bacteria decompose protein and urea to ammonia. Nitrifying bacteria convert ammonia to nitrate ions.

4 Denitrifying bacteria convert nitrate ions to nitrogen gas (N_2).

5 Plants absorb nitrate ions and convert them to amino acids which they use to make proteins. Animals obtain the amino acids they need by eating and digesting plant or animal protein.

SUMMARY QUESTIONS

1 List four substances present in living organisms that contain the element nitrogen.

2 Describe the roles of microorganisms (bacteria and fungi) in the following processes:

  a decomposition   b nitrification

  c nitrogen fixation   d denitrification

3 Explain how nitrogen in the protein of dead animals and plants is made available to plants in the form of nitrate ions.

4 Explain how farmers can increase the quantity of nitrate available to crop plants.

Supplement

LEARNING OUTCOMES

- Define the term *population*
- State the factors affecting the rate of growth of a population
- Identify the different phases of growth in a population growth curve
- State that food supply, predation and disease are the main factors that affect the rate of population growth
- Define the terms *community* and *ecosystem*
- Describe and explain the different phases of the population growth curve

Wildebeest in East Africa.

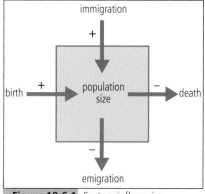

Figure 19.6.1 Factors influencing population size.

Population growth

A **population** is a group of individuals of the same species living in the same habitat at the same time. All the individuals in a population may interbreed. Sometimes it is difficult to tell the geographical limits of a population. This is easiest where the organisms live in a restricted area, such as on an island, like Komodo dragons on islands in Indonesia, or on a mountain, like *Protea kilimanjaro*, or they are highly visible, like wildebeest in the game parks of East Africa.

Some species are very widely distributed and it is difficult to divide them into separate populations. However, there are usually barriers to reproduction, such as rivers and mountains, that divide them into separate populations. Some individuals migrate across these barriers, so there is interbreeding between populations.

The table shows some animal and plant examples.

population	habitat	place
creole wrasse	coral reef	reefs around Tobago in the Caribbean
wildebeest	grassland	East Africa
Komodo dragon	dry grassland and forest	Komodo Island in Indonesia
Nepenthes rajah – carnivorous plant	mountain forest	Mount Kinabalu in Malaysia
Protea kilimanjaro	shrubland	Mount Kilimanjaro in Kenya

A **community** includes all the populations of different species in an ecosystem. An **ecosystem** is a unit containing the community of organisms and their environment, interacting together, for example in a decomposing log or a lake. An ecosystem is made up of the **biotic** component (the community) and the **abiotic** component (physical factors, such as light, water, pH and temperature).

Population size

Figure 19.6.1 shows the factors that determine the size of a population. The number of births adds to the population and the number of deaths decreases it. Individuals may enter or leave a population from neighbouring populations.

Animals migrate so there is mixing between populations. Flowering plants are fixed so you would think that migration is not possible. However, plants use seed dispersal to colonise new areas and in this way individuals leave one population of plants to join another. The main factors that affect the rate of population growth are food supply, predation and disease. Figure 19.6.2 shows the different stages of population growth. An organism that enters a new habitat may show the lag, exponential and stationary phases if there is plenty of food, no predators and no disease.

- **Competition for resources**. Plants compete for light, space, water and soil nutrients. Animals compete for food, space (territory) and mates. Competition for food is a common limiting factor for species of herbivores, such as wildebeest in East Africa.
- **Predation**. Predators often take young, sick individuals or less well-adapted individuals. Predators may limit the growth of a population of prey animals, but it is more often the case that the numbers of prey animals limit the population size of the predator.
- **Disease**. Disease is an important control factor when populations increase. Pathogens are transmitted between individuals more easily when organisms live close together.

The growth of a population: bacteria

The growth of a bacterial population is a good way to show how the size of a population changes. A small number of bacteria are placed into a flask with a warm nutrient solution which is aerated so the bacteria may respire aerobically. Bacteria grow and divide under these conditions. Some species of bacteria divide every 20 minutes if conditions are ideal. One cell divides to give two cells, two cells divide to give four, and so on. You can work out how many cells there will be after 4 hours and you will quickly reach a very large number.

Phases in bacterial growth

If you look at the graph in Figure 19.6.2 you will see four main phases in the growth of a population of bacteria.

- The **lag phase**, when doubling of the numbers has little effect as the numbers are so small. Bacteria take up water and nutrients, and make new cytoplasm, DNA and enzymes.
- The **exponential** (or **log**) **phase** when the population is increasing rapidly. The population increases by doubling and there are no limiting factors, such as food or water. During this phase of rapid growth there are no factors to limit population growth. But after a while, factors such as lack of food and build-up of wastes limit population growth, which slows down and stops.
- The **stationary phase**, when bacterial cells are dying at the same rate at which they are being produced. This may be because of shortage of food or because waste products are building up.
- The **death phase**, when more cells are dying than are being produced, so the population declines. Causes of death may be lack of food, shortage of oxygen or a build-up of toxic waste products. The curve shown in the graph is known as the **sigmoid growth curve**.

Some species show this population growth in the wild. When resources become available, species that can reproduce rapidly, like algae, grow exponentially. When the resources are used up, most of them die. Most species do not do this – their populations remain fairly stable over time.

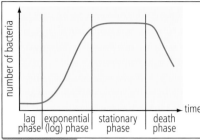

Figure 19.6.2 Growth of a bacterial population.

KEY POINTS

1 A population is a group of individuals of the same species, living in the same habitat at the same time.

2 The growth and final size of a population is influenced by the availability of food, by predation and by disease.

3 There are four phases in the sigmoid population growth curve: lag, exponential (log), stationary and death phases. These are influenced by limiting factors, such as food availability.

SUMMARY QUESTIONS

1 a Define *population*.

b Give three examples of populations in natural ecosystems.

c State the ways in which populations may increase and decrease.

2 Explain how each of the following can act to limit the growth of a population, giving an example in each case:

a food supply
b predation
c disease

3 With reference to Figure 19.6.2, describe and explain each of the following phases:

a lag b exponential (log)
c stationary d death

STUDY TIP

Ask yourself why the human population growth curve has a long lag phase and a short exponential phase. Look at the factors that control species – competition for food, predation and disease – and decide how the effects of these have changed. You may be asked how the control of human populations differs from that of animal species.

STUDY TIP

This is a good place to revise the section on birth control on page 206. You should be aware of how birth control fits in to the broader suject of human populations.

Human population growth

At present, the number of people on the planet is growing at an alarming rate. In 2011, there were estimated to be 7 billion people in the world. For thousands of years there was only a slow increase in the human population. Lack of food and shelter together with the effects of malnutrition, diseases and wars meant that mortality rates were high and people did not live long.

Six thousand years ago the world's population was about 0.2 billion people. People lived in small, scattered communities and much of the world was unaffected by human activities. However, people were beginning to make an impact on their environment by clearing forests, growing crops and keeping livestock. They relied less on hunting and collecting food from the wild, although these were still important and remain so in the case of the fishing industry.

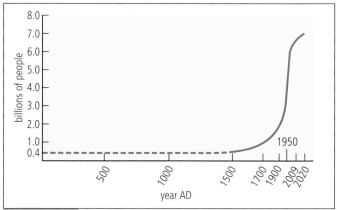

Figure 19.7.1 Notice the long lag phase and short exponential phase with no sign of a stationary phase.

The human population has increased exponentially over the past 300 years for the following reasons:

• Improved agriculture means that most people are better fed. Better nutrition means they are better defended against diseases, especially those caused by malnutrition.

• Public health has improved. There are clean water supplies, drainage of waste and sewage treatment. This has decreased epidemics of waterborne diseases such as cholera.

• Medical care has improved. Since the 1940s doctors have used antibiotics; people are also vaccinated against many infectious diseases, such as diphtheria and polio.

The effects of these improvements have been:

• decrease in the infant and child mortality rates,

• decrease in death rates from starvation and malnutrition,

• increase in life expectancy – people are now living much longer. In Europe and North America the average life expectancy is 76 years for men and 80 years for women. In India, average life expectancy is 68 years.

In spite of these improvements there are huge inequalities in food, housing, water and health services. Many of the world's human population live in poverty with high infant and child mortality rates, and life expectancies that are half those aforementioned.

Controlling human population growth

There are still controls on population increase. Famines, epidemics, floods, wars, tsunamis, earthquakes and other disasters lead to loss of life. But the overall trend is a rapid increase in the world's population. It is estimated that it could double in the next 30 years.

In Europe and North America the population is virtually stable. In other places populations continue to grow. The Earth has limited resources and limited space and the population will have to stabilise and even decrease at some time in the future. The most obvious way to do this is to reduce the birth rate. But many groups of people have strongly held religious or moral views on birth control and family size. This means they may often have large families. China has had a 'one-child' policy since 1979 and it is claimed this has restricted population growth in that country. No other country has such a policy.

The base of the population pyramids below show the percentage of children below the age of 5 years. The oldest people appear at the top. The rate of increase in populations is most influenced by the proportion of child-bearing women in the population.

Increasing human population size brings greater demands on resources and problems for the environment, such as the supply of food, housing, transport and energy needs. There is also the problem of how to dispose of our rubbish (see Unit 21).

KEY POINTS

1 Improved agriculture, better public health and medical care have contributed to an exponential increase in the human population.

2 Fewer children are dying from disease and lack of food, and people are now living much longer.

3 The increase in the human population has social implications, such as demands on food supplies, housing, energy needs and space to dispose of our rubbish.

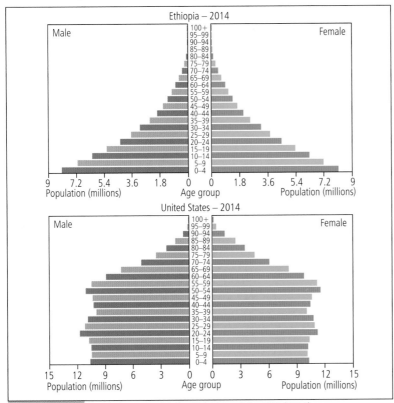

Figure 19.7.2 Population pyramids for Ethiopia and the USA (2014).

SUMMARY QUESTIONS

1 Study the population pyramids on the left.

   a State which country has the greater proportion of i young people, ii old people.

   b Suggest reasons for the differences between the populations of the two countries.

   c Suggest what you think will happen in each country to i the numbers of women of child-bearing age in 10 years' time, and ii the birth rate in 10 years' time.

   d Outline the likely effects that your answers to part c will have on the two countries in the future.

2 Explain why infant and child mortality rates have decreased and why people are living much longer.

1 The diagram shows a pyramid of numbers for an ecosystem. Which level indicates the tertiary consumers?

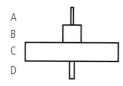

(Paper 1) [1]

2 Fungi and bacteria are important in recycling carbon and nitrogen in ecosystems because they are:

A decomposers

B herbivores

C parasites

D producers

(Paper 1) [1]

3 In the carbon cycle, which two processes add carbon dioxide to the atmosphere?

A combustion and respiration

B decomposition and fossilisation

C feeding and fossilisation

D photosynthesis and respiration

(Paper 1) [1]

4 Here is a simple food chain:

grass → grasshopper → bird

The arrows in a food chain represent:

A the decrease in numbers of the organisms

B the flow of energy between the organisms

C the increase in complexity of the organisms

D the increase in size of the organisms

(Paper 1) [1]

5 From which food chain is *most* energy lost?

A maize → beef cattle → humans

B soya beans → humans

C grass → dairy cattle → humans

D phytoplankton → mollusc larvae → small fish → tuna → humans

(Paper 2) [1]

6 Some species of bacteria live inside the root nodules of legumes. Which of the following do these bacteria provide to the legumes?

A compounds of nitrogen

B protection from infection by fungi

C sugars, such as sucrose

D water and ions from the soil

(Paper 2) [1]

7 What is the best explanation for the change in population size shown in the graph?

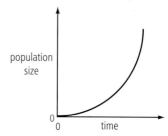

A birth rate and death rate are the same

B emigration

C an outbreak of disease

D no limiting factors

(Paper 2) [1]

8 The diagram shows a pyramid of biomass.

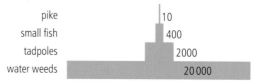

Which is a correct statement about the pyramid shown in the diagram?

A Biomass decreases less between primary consumers and secondary consumers than between producers and primary consumers.

B The biomass of each individual producer is very large.

C The biomass of the primary consumers is more than 20% of the biomass of the producers.

D There are very many producers.

(Paper 2) [1]

9 (a) Use this information to draw a food chain:

Flower beetles eat the pollen and nectar of black cherry trees. Kookaburras are birds that eat flower beetles. [2]

(b) State the role of each organism in the food chain you have drawn. [3]

(c) Define the term *population*. [3]

(d) State three factors that influence the rate of growth of a population. [3]

(e) Outline the reasons for the large increase in the human population over the past 250 years. [4]

(Paper 3)

10 The table shows the total biomass formed in one square metre of grassland during one year.

The figures were obtained by measuring the dry mass and not the fresh mass of the living material.

organisms	dry mass / g m² year⁻¹
green plants	480
herbivores	0.9
carnivores	0.1

(a) Calculate the percentage decrease in the biomass between green plants and herbivores. Show your working. [2]

(b) Explain why the biomass formed during the year decreases between trophic levels. [3]

(c) Explain the advantage of measuring dry mass rather than fresh mass when investigating biomass production in different ecosystems. [3]

(Paper 4)

11 The diagram shows the passage of energy along a food chain in a grassland ecosystem. The light energy that strikes the leaves of the grass each year is 400 000 kJ m².

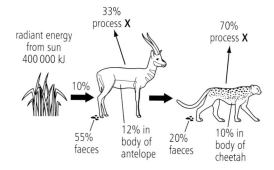

(a) The plant absorbs only about 1% of the energy that strikes the plant and uses it in photosynthesis to make simple sugars.

(i) State two compounds that are produced in plants from the simple sugars. [2]

(ii) State three ways in which the leaves of plants are adapted to trap as much light as possible. [3]

(b) Only about 10% of the energy trapped by the plant in photosynthesis is passed to herbivores like antelopes.

(i) State two ways in which the plant uses the energy that is not available to the herbivores. [2]

(ii) Calculate how much energy reaches the antelopes each year. Express your answer as kJ m⁻². Show your working. [3]

(c) Name process **X**. [1]

(d) Suggest why the antelope loses more of its energy intake as faeces compared with the cheetah. [2]

(e) This food chain has three trophic levels in it. Explain why there are no more. [4]

(Paper 4)

12 The following are biologically important substances that contain the element nitrogen:

urea, ammonia (NH₃), amino acids, nitrite ions, nitrate ions

(a) Choose one of the substances from the list to match each of the statements **(i)** to **(v)**. You may use each one once, more than once or not at all.

(i) the main nitrogen-containing excretory product of animals

(ii) absorbed by plants

(iii) a product of deamination

(iv) synthesised into proteins

(v) the end product of nitrification [5]

(b) Nitrogen gas (N₂) in the atmosphere is inert. Unlike carbon dioxide and oxygen, very few organisms are able to use it. Describe how nitrogen gas (N₂) present in the atmosphere is made available to be used by organisms. [4]

(Paper 4)

259

20 Biotechnology and genetic engineering

20.1

Microorganisms and biotechnology

LEARNING OUTCOMES

- State how bacteria are useful in biotechnology and genetic engineering
- Describe the role of yeast in breadmaking
- Describe the role of anaerobic respiration in yeast in the production of ethanol for biofuels

Some of the products of traditional biotechnology.

Many people have concerns about the welfare and health of chickens kept in these conditions.

The use of microorganisms and biotechnology

Biotechnology is the use of microorganisms, such as bacteria and fungi, to make useful products or to carry out services for us, such as making wastes harmless.

The use of both bacteria and fungi are in various processes is widespread.

- Bacteria reproduce very quickly with a generation time (the time taken for numbers to double) that is often as little as 30 minutes.
- Unlike animals or plants, microbes can convert raw materials into the finished product very quickly, i.e. in hours rather than months or weeks.
- The use of bacteria means that food production can be independent of climate.
- Bacteria can produce complex proteins (like enzymes and antibiotics) that pass out into the surrounding medium and can be harvested. Enzymes made by microorganisms are used in the food industry.

Examples of enzymes that are produced by bacteria for use in the food industry are:

- amylase for breaking down starch in the production of glucose syrup
- pectinase for extracting juice from fruit (see page 263)
- sucrase for breaking down sucrose in making confectionery
- protease for making meat more tender

Supplement

With the aid of genetic engineering, scientists can quickly alter microbes and so modify products. In contrast, breeding new varieties of plants and animals can take a long time.

In addition to the single loop of DNA, bacterial cells also contain small circles of DNA called **plasmids**. Plasmids are easy to work with, since they can replicate very quickly. Plasmids can be cut open by enzymes and a gene from another organism can be spliced into them. In nutrient media the bacteria multiply rapidly, making many copies of the plasmid and the inserted gene (see page 267).

The use of bacteria and fungi in the manufacture of complex chemicals does not raise the same ethical concerns over their manipulation and growth as would be the case in the use of other living organisms, for example keeping chickens in battery farms.

Biofuels and breadmaking

Many microbes can respire successfully without oxygen. Yeasts can respire with or without oxygen.

When yeast respires without oxygen (anaerobically) it is called fermentation.

Glucose $\longrightarrow$ alcohol + carbon dioxide + energy released

Biofuels

Biofuels are fuels made from biological material. Sugar-rich products from sugar cane and maize can be fermented anaerobically with yeast to produce ethanol. Some countries have no oil supplies of their own and have developed a biofuel substitute. Brazil, for example, uses fermenters to produce ethanol. The raw material often used is sugar cane juice which contains a lot of carbohydrates. Glucose from maize starch is another raw material. This is obtained by treating the starch with carbohydrase enzymes.

Ethanol is a good substitute for petrol because it has a high energy content and when burned does not produce toxic gases. Using ethanol as a fuel means that there is no overall increase in carbon dioxide in the atmosphere. As a result this fuel is **carbon neutral**.

Car engines need to be modified to be able to use pure ethanol. **Oil seed rape** is a crop that provides oil that is converted into fuel called **biodiesel**.

Bread-making

Baking is another example of using yeast to help produce a food. Bread is made from **dough**, a mixture of flour, water, salt, sugar and yeast. This mixture is kept at a warm temperature. The yeast starts to ferment the sugar, producing carbon dioxide gas and alcohol.

Bubbles of carbon dioxide are trapped inside the dough and make it rise. When the bread is baked in a hot oven the bubbles of the gas make the bread 'light' in texture. The heat causes the alcohol to evaporate, leaving behind the traditional taste of bread.

Some biofuels are made from waste material. This biobus runs on fuel made from used cooking oil and other waste products from the food industry.

Yeast produced carbon dioxide which caused the bread dough to rise and give these loaves their light texture.

Supplement

LEARNING OUTCOMES

- Investigate the use of pectinase in the production of fruit juice
- Investigate the use of enzymes in biological washing powders
- Explain the use of lactase in the production of lactose-free milk

BIOLOGICAL

Washing Powder

✓ Contains protease and lipase
✓ Works best at low temperatures (40°C)
✓ More effective than ordinary detergent

✗ Avoid using boiling water
✗ Do not use with silk

Figure 20.2.1 Biological washing powder.

Granules of biological washing powder. Some of the granules are partially opened. Inside these granules are enzymes which act as cleansing agents.

Enzymes have become very important in industry.

Biological washing powders

Biological washing powders may contain one or more of the following types of enzymes:

- **proteases** – break down protein stains, e.g. blood, grass and egg
- **lipases** – break down fats in grease stains, e.g. butter, lipstick and mayonnaise
- **amylases** – break down starch, e.g. food stains containing starch
- **cellulases** – break down cellulose fibres on the outside of cotton fabrics to remove the dirt attached to them

The enzymes listed above have been modified so that they withstand the high temperatures and alkaline conditions required for some washing powders. During a washing cycle, the enzymes break down stains; the products of the reactions dissolve or are suspended in water and are removed when the washing machine empties. For example, proteins are broken down into amino acids and starch and cellulose are broken down into glucose.

PRACTICAL

Comparing the action of biological and non-biological washing powder

1 Decide what sort of stain you are going to have, e.g. egg, soya sauce, mustard, etc.

2 Decide what sort of fabric you are going to use, e.g. cotton, linen or wool. Also, decide what size of fabric to use.

3 Add the same amount of stain to the fabric and let it dry before you put it into the beakers to 'wash'.

4 Weight out 5 g of each powder and dissolve each in 500 cm³ of water at 30 °C in a beaker.

5 Add a piece of the same fabric with the same stain to each beaker.

6 Decide how often you are going to stir the 'washing' over a 10-minute period.

7 Decide how to compare the cleanliness of each fabric after washing. You could produce a scale of cleanliness.

8 Repeat your tests to see if you get the same results.

Safety: Wear gloves and do not let the washing powders come into contact with your skin to avoid a possible allergic reaction.

Extracting fruit juice

Pectinases are enzymes that break down pectins, which are molecules that act like a 'glue' in plant cell walls. Pectinases are used for extracting fruit juices and for softening vegetables.

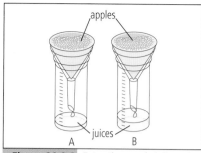

Figure 20.2.2 Extracting apple juice using pectinase.

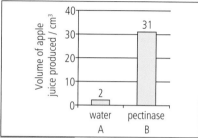

Figure 20.2.3 Sample results from the practical to extract juice from apples.

The use of lactase in making lactose-free milk

Some people do not produce enough lactase and therefore have problems digesting and lactose, the sugar found in milk. They may get diarrhoea, wind and stomach cramps, if they consume milk or milk products. Lactose is a complex sugar, similar to sucrose that can be broken down into simple sugars by the enzyme lactase.

$$\text{lactose} \xrightarrow{\text{lactase}} \text{glucose} + \text{galactose}$$
(complex sugar)   (2 simple sugars)

You can investigate the effect of lactase in breaking down lactose by using test strips that detect the concentration of glucose. These are the test strips that are used to detect glucose in urine. You cannot use Benedict's test as all three sugars in the equation above are reducing sugars and give a positive result.

SUMMARY QUESTIONS

1 Describe what each of the following enzymes do.
   a amylase b protease
   c lipase   d pectinase

2 Explain the advantages of using enzymes in biological washing powders.

3 Explain:
   a why water was added to beaker A,
   b the results in the graph above,
   c how pectinase releases juice from apple tissue.

4 a How is lactose-free milk produced?
   b Why is there a demand for lactose-free milk?
   c Outline an experiment to discover which is the best temperature to produce lactose-free milk.

KEY POINTS

1 Biological washing powders contain enzymes that work at low wash temperatures and in alkaline conditions.

2 Pectinase is used to extract fruit juice by breaking down the pectin in the cell walls in fruits such as apple.

3 Lactase breaks down lactose to produce lactose-free milk.

Supplement

263

Fermenters

LEARNING OUTCOMES

- Describe the role of *Penicillium* in the production of the antibiotic penicillin
- Explain how fermenters are used in penicillin production

You may recognise the word fermentation as meaning respiration without oxygen (see page 140). However, in an industrial context it refers to any process that uses microorganisms to produce a useful product or to carry out a useful process for us. The organisms involved may respire with or without oxygen.

Industrial production of penicillin

One important use of large-scale industrial fermentation is the production of antibiotics such as penicillin. The fermenter is inoculated with a culture of a suitable fungus, in this case *Penicillium chrysogenum*. This then proceeds to grow under the conditions maintained inside the fermenter. These conditions should include:

- Adding nutrients, such as sugar or starch, as a source of carbon for respiration
- Ammonia or urea is added as a source of nitrogen to make proteins
- Vitamin B complex is added for respiration
- Maintaining a constant temperature of about 30 °C
- Maintaining a constant pH of about 6.5

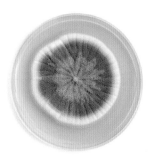

The fungus *Penicillium* growing on agar in a Petri dish. Fungi are cultured like this to find out whether they are suitable as sources of antibiotics.

Batch culture

It takes about 30 hours for penicillin production to start. Penicillin is secreted into the surrounding liquid by the fungus.

STUDY TIP

Fermenters provide suitable conditions for growth and metabolism of microorganisms, such as bacteria and fungi. If the temperature is too high, enzymes in the microbes denature and the organisms die.

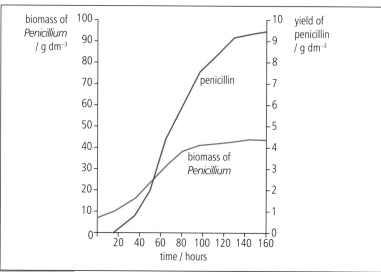

Figure 20.3.1 This graph shows the growth of the fungus *Penicillium* and the production of penicillin during a fermentation that lasts between 140 and 160 hours.

Look at the graph above. You should see that there is a delay in production. This is because penicillin is a **secondary metabolite**, a substance which is not necessary for the growth of the fungus.

Penicillin production starts after the exponential growth phase in response to one or more limiting factors. After about 6 days, the mixture in the fermenter is filtered, the penicillin is extracted using a solvent and is then purified into a crystalline salt. This type of production is known as **batch culture**. After the 6 day period, the fermenter is emptied, cleaned and sterilised ready for the next batch.

Industrial fermentation

Industrial fermenters are large tanks (see right) that can hold as much as 500 000 dm³ of fermenting mixture. Conditions inside these huge tanks are carefully controlled. We will use the production of the antibiotic **penicillin** to show how these fermenters work.

- The **fermentation vessel** is made of stainless steel and is filled with a medium containing the required nutrients. These include sugars and ammonium salts. To this some of the fungus *Penicillium* is added.

- The fungus grows well in the conditions inside the fermenter. Sugars provide energy for respiration and ammonium salts are used by the fungus to make proteins and nucleic acids (DNA and RNA). After a few days the fungus starts to produce penicillin, which accumulates in the fermenter.

- A **stirrer** keeps the microorganisms suspended so they always have access to nutrients and oxygen. Stirring also helps to maintain an even temperature throughout the fermenter.

- An **air supply** provides oxygen for the aerobic respiration of the fungus. (No oxygen is supplied as the yeasts respire without oxygen to make alcohol. See page 261.)

- A **water-cooled jacket** removes the heat produced by fermentation to give a constant temperature of 24°C.

- **Probes** monitor the temperature and make sure the pH is constant at 6.5 by adding alkalis if necessary.

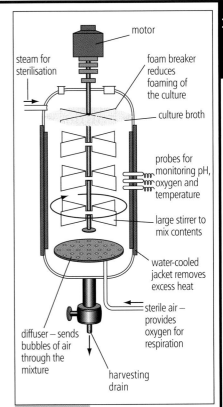

Figure 20.3.2 Large industrial fermenter.

Industrial fermenters.

SUMMARY QUESTIONS

1 Explain the function of each of the following in an industrial fermenter:

   a a stirrer b an air supply c a water-cooled jacket

   d temperature and pH probes

2 Explain the role of *Penicillium* in the production of penicillin.

3 State the conditions necessary for the production of penicillin in a fermenter.

4 Explain how and why these factors change during a fermentation process:

   a temperature b oxygen concentration c pH

5 What is meant by:

   a batch culture b secondary metabolite?

KEY POINTS

1 Microorganisms reproduce quickly the ideal conditions found inside large, industrial fermenters producing useful products such as penicillin.

2 The ideal conditions for fermentation include adding the correct nutrients and maintaining a constant pH and temperature.

Genetic engineering

LEARNING OUTCOMES

- Define the term *genetic engineering*
- State that genetic engineering can be used in the production of human medicines, herbicide resistant, insect resistant plants, and providing additional vitamins in crop plants
- Outline how genetic engineering is used to make insulin
- Discuss the advantages and disadvantages of genetically modifying crops

Herbicide-resistant soya can be sprayed with herbicides to reduce the competition from weeds

Assessing disease resistance in GM rice under glasshouse conditions.

Genetic engineering involves changing the genetic material of an organism by removing, changing or inserting individual genes. A gene is a section of DNA that codes for the production of a specific protein.

Genetic engineering is the transfer of a gene from the DNA of one species to the DNA of another species. This is a process that can never be achieved by artificial selection, because only in rare cases would one species breed with another. Genetic engineering allows the transfer of genes between totally unrelated species, for example from humans to bacteria and from bacteria to plants.

Examples of genetic engineering

Human medicines

Many human proteins, including insulin, human growth hormones and blood clotting agents, are produced by bacteria that have been genetically modified with the appropriate human genes. These proteins are mass produced and used as medicines.

Herbicide resistance in crop plants

A genetically modified variety of oilseed rape has been developed in which a gene is transferred into it from a soil bacterium. This gene makes oilseed rape resistant to the herbicide **glufosinate**. When applied to fields of oilseed rape this herbicide would not only kill the weeds, but also the crop itself. But the GM oilseed rape is not affected by the herbicide and so continues to grow while only the weeds are destroyed.

Insect resistance in crop plants

Maize and cotton are both important cash crops, but both can be attacked by insects which reduce their yields. The major pests of maize are root worm larvae and stem-boring caterpillars. Caterpillars and cotton boll weevils are pests of cotton plants.

At the beginning of the 20th century a soil bacterium was found that produced a toxin which kills certain caterpillars. In the 1990s the gene for this toxin was isolated and transferred to maize and cotton plants. The genetically modified plants produce their own toxin which kills insect larvae that feed on it. This has less environmental impact than spraying pesticides, because only the insects feeding on the crop plants are killed.

Additional vitamins in crop plants

Genetically modified varieties of rice have been developed by transferring genes from maize and a soil bacterium, that enables people to make vitamin A. The resultant rice is pale yellow in colour giving it its name, **Golden Rice**™. If, or when, this rice becomes available to grow on a large scale it will improve the diets of the people who eat it. They will be able to make vitamin A. This will reduce the high mortality rates as a result of poor immune systems.

Industrial production of insulin

Genetic engineering makes it possible to make insulin quickly and cheaply on a large scale. The human gene that codes for the production of insulin is identified. **Restriction enzymes** act as chemical scissors to cut the human insulin-making gene from the rest of the DNA.

A circular piece of DNA called a **plasmid** is removed from a bacterium. The same restriction enzymes are then used to cut open the plasmid. The two ends of the DNA of the insulin-making gene are an exact match with the two DNA ends and the plasmid. These are called '**sticky ends.**'

Another enzyme called ligase is used to attach the sticky ends of the insulin-making gene to the sticky ends and plasmid. The plasmid is now known as a **recombinant plasmid** and is inserted back into the bacteria. The bacteria now has the gene code for insulin. The bacteria multiply very rapidly inside an industrial fermenter and produce insulin.

Advantages and disadvantages of genetically modified crops

The benefits:

- **Solving global hunger** – genetic modification could feed more people as the crops that are produced are able to tolerate extreme climate conditions and soils. Food production could be increased in marginal areas.
- **Environmentally friendly** – GM crops, like soya, maize and rice, can be resistant to insects, weeds and diseases so there would be less use of pesticides. Also, genes that improve nitrogen uptake would mean less need for chemical fertilisers and lessen the environmental threat that they cause.
- **Consumer benefits** – GM crops have already been produced with an improved flavour and better keeping qualities. They are easier to produce and require fewer additives.

The concerns:

- **Environmental safety** – there are concerns that new GM plants will become successful weeds. Pollen from GM crops that are resistant to weedkillers may be transferred to other plants by insects or the wind.
- **Food safety** – new gene combinations may have effects that are so far unknown. They may result in harmful substances being produced.
- **Biodiversity** – increasing use of herbicides and plant breeding will reduce the number of plant varieties and wild relatives.

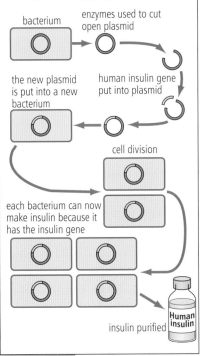

Figure 20.4.1 Human insulin is made by genetically engineered bacteria.

SUMMARY QUESTIONS

1 Explain what is meant by *genetic engineering*.

2 Describe three different ways in which crops have been genetically engineered to improve them.

3 What is meant by a plasmid?

4 List the stages involved in producing bacteria that make human insulin.

5 Discuss the benefits of GM crops and the concerns people have about their safety.

KEY POINTS

1 Genetic engineering involves taking a gene from one species and putting it into another species.

2 The human insulin gene can be transferred into a bacterial plasmid. The bacteria grow to produce lots of human insulin.

1 In order to transfer a feature from one species to another it is necessary to transfer:

   A a chromosome   **C** a genotype

   B a gene   **D** a nucleus

(Paper 1)   *[1]*

2 Yeast is used in the production of some biofuels. This is because:

   A yeast cells are a good source of energy

   B yeast cells produce oils that are added to petrol (gasoline)

   C yeast cells respire to produce carbon dioxide

   D yeast cells respire without oxygen to produce alcohol

(Paper 1)   *[1]*

3 A student is planning a laboratory investigation to compare the effectiveness of three washing powders at removing stains. Which combination of controlled factors influences the activity of the enzymes during the washing process?

   A size of container, frequency of stirring, type of washing powder

   B type of fabric, length of washing time, size of fabric

   C type of stain, size of fabric, final appearance of fabric

   D washing temperature, volume of water, pH of washing mixture

(Paper 1)   *[1]*

4 Which substance causes bread dough to rise?

   A carbon dioxide   **C** lactic acid

   B ethanol   **D** oxygen

(Paper 1)   *[1]*

5 Bacteria are used in many industrial processes. Which is not a reason for this?

   A bacteria do not have complex requirements from their environment

   B bacteria have many enzymes so can carry out complex sets of reactions to make useful products

   C bacteria reproduce slowly under optimum conditions

   D there are fewer safety concerns with modifying the genes of bacteria compared with modifying the genes of crop plants and livestock

(Paper 2)   *[1]*

6 Which is the *best* reason for controlling temperature in a fermenter that contains bacteria?

   A at high temperatures enzymes denature

   B at high temperatures the rate of fermentation increases too much

   C at low temperatures bacteria reproduce too fast

   D at low temperatures there is no diffusion of nutrients into bacteria

(Paper 2)   *[1]*

7 The following processes occur in the genetic modification of bacteria to produce a human protein:

   1 a gene is isolated from a human cell

   2 a plasmid is cut open by restriction enzymes

   3 bacteria divide by binary fission

   4 bacteria are grown in a fermenter to produce human protein

   5 recombinant plasmids are taken up by bacteria

   6 bacteria make multiple copies of the recombinant plasmid

   7 plasmids and human genes combine to form recombinant plasmids

Which sequence is the correct order of events?

   A 2, 1, 5, 7, 6, 3, 4

   B 2, 1, 7, 6, 4, 5, 3

   C 1, 2, 3, 7, 6, 5, 4

   D 1, 2, 7, 5, 6, 3, 4

(Paper 2)   *[1]*

8 An advantage of growing herbicide-resistant soya is that:

   A less herbicide needs to be sprayed on the crop

   B there is less competition between weeds and soya plants

   C there is less risk of disease attacking the soya plants

   D the soya plants grow faster

(Paper 2)   *[1]*

9 A student investigated the effect of pectinase on the extraction of apple juice. Several apples were cut up and crushed to make a pulp. The pulp was divided into two samples, A and B. Sample A was mixed with pectinase powder. Sample B was used as a control.

(a) Suggest:

   (i) what should be added to sample B as it is being used as a control, *[1]*

   (ii) three variables that the student should keep constant. *[3]*

(b) The student obtained 35 cm³ of juice from sample A and 15 cm³ from sample B. She noticed that juice from sample A looked more appealing than the juice from sample B.

   (i) Explain why there was much more juice from sample A. *[3]*

   (ii) Suggest why the student thought that the juice from sample A was more appealing. *[2]*

(c) State three ways in which the student could improve the investigation to find out how effective pectinase is at increasing the yield of fruit juice. *[3]*

(Paper 3)

10 Many organisms can be genetically engineered so that they can be used in biotechnological processes, but bacteria are one of the most common.

(a) Define the term *genetic engineering*. *[2]*

(b) State the advantages of using bacteria in industrial processes. *[2]*

(c) State three examples of genetic engineering. *[3]*

(Paper 3)

11 Look at the diagram of the fermenter on page 265. *Penicillium* is grown in fermenters to produce an antibiotic.

(a) Name the antibiotic produced by *Penicillium*. *[1]*

(b) State five conditions that are maintained inside the fermenter. *[5]*

(c) For each condition, describe how it is maintained and explain why it must be maintained inside the fermenter. *[5 + 5]*

(d) Describe how the antibiotic is obtained from the fermenter. *[2]*

(Paper 4)

12 Human growth hormone (HGH) promotes growth and is given to young people as a treatment for reduced growth. It is a protein made of 188 amino acids.

(a) (i) State how many bases in DNA code for this protein. *[1]*

   (ii) Explain your answer. *[2]*

Genetically engineered bacteria make human growth hormone and a similar hormone found in cattle. HGH is used to treat children who do not grow properly. There may be a genetic reason for this or they may have received radiation treatment for leukaemia or have had a brain tumour.

(b) Outline how bacteria are genetically modified to produce a human protein, such as HGH. *[5]*

(c) Suggest the advantages of using genetic engineering to produce HGH. *[3]*

(Paper 4)

13 People, especially children, who do not have a varied diet may suffer from blindness caused by a lack of vitamin A. Lack of this vitamin in their diet also harms their immune system.

(a) Explain the consequences of an ineffective immune system. *[4]*

Rice is the main foodstuff in many parts of the world. Researchers have used genetic engineering to produce a strain of rice that has seeds rich in the chemical beta carotene which the human body converts into vitamin A.

(b) Suggest why rice had to be genetically engineered and not simply produced by selective breeding. *[2]*

(c) State three other reasons for genetically modifying crop plants. *[3]*

(d) Discuss the arguments for and against genetically-modified food. *[3]*

(Paper 4)

21.1   Food supply

The waste from this Mongolian farmer's cows pollutes the surroundings of his farm.

Intensive fish farming in Thailand.

Increased food production

Modern technology has resulted in advances in agriculture over the past 60 years and improved food supplies in many parts of the world.

Arable farms use large **agricultural machines** to work very large fields. Examples are tractors and ploughs for preparing land for sowing seeds and combine harvesters for harvesting crops. **Chemical fertilisers** encourage the growth of crop plants, increasing the yield of the crop. **Pesticides** kill pests like insects that feed on crops. **Herbicides** kill weeds that compete with crop plants for water, light and nutrients. **Selective breeding** has increased yield and made crops more resistant to drought and diseases. **Genetic engineering** has transferred features, such as herbicide resistance, to crop plants from unrelated species.

Intensive farming is using modern technology to achieve high yields of crop plants and livestock. This involves growing crop plants over large areas at high densities and keeping livestock in large numbers, supplementing their food supply and often restricting their movement by keeping them indoors.

In many intensive systems animals are reared indoors in large numbers.

Intensive methods have negative effects on the environment. Animals, especially cattle, generate lots of methane which is a greenhouse gas (see Topic 21.5). Urine and faeces, often known as slurry, pollutes lakes, waterways and the sea where it can cause eutrophication (see Topic 21.4).

Fish, such as salmon, trout, sea bass and tilapia, are kept in large cages and have their food and growing conditions carefully controlled. The waste food from fish farms can have serious effects on the surrounding waters. High densities of fish mean that parasites and pathogens can spread easily. This means using pesticides and antibiotics to treat the fish, and with this come problems of resistance to these chemicals.

A **monoculture** is growing the same crop on large areas, year after year. This has the advantage of allowing farmers to concentrate on growing large quantities of specific crops, such as wheat, barley, maize, soya and rice.

There are negative effects of monoculture on the environment.

Herbicides and pesticides kill plant and insect species that are harmless and even some that may help the crop, such as parasitic wasps that lay their eggs inside pest species. Continuous use of these chemicals act as selective agents, resulting in the evolution of resistance among weeds, pests and plant pathogens (see Topic 18.4).

The large scale use of chemical fertilisers can reduce the structure of the soil so that over time it no longer supports the biodiversity of soil organisms that supports good crop growth. The addition of organic matter, such as manure, helps to maintain good soil structure but is not always carried out.

Machinery like this combine harvester allows farmers to grow crops in very large fields.

World food supply

As the world's population has increased, the ability of the world's farmers to produce enough food has also increased. There is enough food in the world to feed everyone, but every year many people die from starvation and malnutrition.

Shortage of food is most often caused by politics, economics, wars and poverty. Many poorer countries produce enough food to feed their populations or have enough land to do so, but have to sell food or grow non-food cash-crops to gain foreign currency.

Drought, flooding, pests and diseases may cause crops to fail, but famine is most often caused by a failure to supply food to those at risk of starvation. Often there are problems transporting food to places where there is famine because of political pressures or war.

Social, environmental and economic implications

Food production requires large areas of land. Often this has meant that groups of people have been displaced, sometimes forcibly, from places where their ancestors have lived for years.

Land clearance for plantations and for ranching has resulted in environmental damage that is described in Topic 21.2. As farming has become increasingly mechanised, there is not the demand for a large labour force so rural unemployment has driven the movement of people to cities.

Intensive agriculture requires large inputs of energy from burning fossil fuels. It also requires much investment by large corporations who own farms and plantations, often taking profits so that money does not reach local people.

Plantations of oil palm cover huge areas of land in South-East Asia. Here a lorry from a plantation in Indonesia carries away the fruits of these trees to be processed into palm oil.

KEY POINTS

1 The use of machinery, selective breeding and agricultural chemicals has resulted in an increase in food production.

2 Intensive crop and livestock production have many negative impacts on the environment and on people.

3 Famine can result from political problems, lack of food transport, drought, flooding and an ever increasing human population.

4 Intensive agriculture has social, environmental and economic implications.

SUMMARY QUESTIONS

1 How have the following increased food production?
  a improved agricultural machinery b artificial selection
  c use of fertilisers d use of pesticides

2 Describe the negative impacts of intensive livestock production and monoculture of crops.

3 Describe the reasons for food shortages in the world.

Habitat destruction

Open cast coal mining destroys large areas of land.

A sea otter eating a crab. Sea otters are protected and their numbers increased in the latter half of the 20th century, but now they are being preyed upon by killer whales that may have less prey to hunt because of overfishing.

The pressure of an increasing world population has resulted in significant losses of natural habitats. There are many reasons why the natural environment has been and continues to be destroyed. Land is deliberately cleared for a variety of reasons, but we have many indirect effects on natural ecosystems, such as pollution. There is no ecosystem on Earth that is unaffected by human activity.

Reasons for habitat destruction

The natural habitats of plants, animals and microorganisms are destroyed for a number of different reasons:

- The clearance of land for crop production (see Topic 21.1) and for the production of biofuels (see Topic 20.1).
- Space to build units for intensive livestock production and clearance of land for cattle rearing.
- Digging mines and quarries for the extraction of coal and mineral ores, such as iron ore and bauxite from which aluminium is produced.
- More people require more housing and the associated infrastructure: roads, industries, shops, etc.
- The human population produces huge quantities of waste. Space is needed for long-term storage of this waste and for recycling it (see Topic 21.8).
- Marine pollution. Major pollutants of the sea include fertilisers, which are thought to cause algal blooms in the sea, oil from oil wells and ships, and industrial chemicals.

At high risk of destruction by pollution are coral reefs which have very high biodiversity. In the Caribbean it is estimated that over 23% of coral reefs have been destroyed and many are considered at high risk.

Human influences on food chains and webs

If we haven't cleared areas of the natural environment completely or destroyed them by pollution, we have removed organisms either directly or indirectly:

- Killing of large predators – humans have killed many of these because they were dangerous (see also Topic 21.9).
- Killing of large herbivores for food – there were many different species of these in huge numbers. The population of American bison decreased from an estimated 60 million in 1492 to 750 in 1890.
- Overfishing has had significant effects.

Removing species results in unbalanced food webs and unbalanced ecosystems that are further at risk of damage by other factors.

Top predators control the populations of primary consumers. Their removal means that populations of herbivorous animals increases and so there is increased competition for the plant food that they eat. The effect can be overgrazing and habitat destruction.

The Pacific sea otter, *Enhydra lutris*, is a secondary consumer that feeds on invertebrates, such as sea urchins and crabs. In the 19th century the animals were hunted for their fur and there was a striking change to the whole of the food web as urchins exploded in numbers and ate many of the giant seaweeds that provided a habitat and food for many other species. The loss of one species, the sea otter, led to the catastrophic loss of many other animal species from the food web, such as many small invertebrates, fish, octopus and scallops.

Deforestation

Humans have been clearing forests for over 10 000 years to grow food and provide land for settlements and provide transport links. There are now very few forests in temperate regions that have not been cleared at some time in the past. Tropical forests in South-East Asia and South America have been cut down over the past 100 years. Many of these forests are rainforests rich in biodiversity.

Deforestation has resulted in a number of environmental problems:

- the extinction of species
- the loss of soil
- an increase in flooding
- the increase of carbon dioxide in the atmosphere.

Supplement

Soils in tropical rainforests are very thin and when the vegetation is removed the soil is easily washed away causing soil erosion, formation of gullies and loss of plant nutrients. The land is rapidly degraded after all the trees are cut down.

Local weather patterns change with more frequent and severe storms. Flooding happens more frequently as water runs off the land much quicker and is not absorbed by plants and transpired into the atmosphere. Forests act as 'stores' of water, their leaves slow down the rate of evaporation from the soil and decrease the rate at which water reaches the soil.

Carbon dioxide is added to the atmosphere because vegetation is burned. There is increased decomposition that also releases more carbon dioxide which is not absorbed by plants.

Rainforests have no effect on maintaining the correct balance of carbon dioxide and oxygen in our atmosphere. They produce about as much carbon dioxide as they use in photosynthesis and there is such a huge reserve of oxygen in the atmosphere that the quantity produced by all plants on Earth makes little difference to the overall oxygen concentration of the atmosphere. There is no evidence that the rainforests are the 'lungs' of the earth. There are however plenty of other reasons to stop cutting down forests. Destruction of rainforests means the loss of many habitats and the extinction of many species. There are many species in rainforests that have yet to be identified, studied and classified.

KEY POINTS

1 Habitats have been destroyed by clearance of land for farming, extraction of resources, such as fossil fuels and metal ores, housing and industry.

2 Pollution has destroyed many marine habitats, such as coral reefs.

3 Humans have negative impacts on habitats by removing species from food chains and webs.

4 Deforestation leads to extinction of species, loss of soil, flooding and increase of carbon dioxide in the atmosphere.

SUMMARY QUESTIONS

1 Outline some of the effects that an increasing human population has had on the environment.

2 It is estimated that between 1880 and 1980 about 40% of all tropical rainforest was destroyed.

 a Give three reasons for this large-scale deforestation.

 b Outline some of the effects that deforestation has on the environment.

3 Explain the negative effects of deforestation.

Supplement

Pollution

LEARNING OUTCOMES

- State the sources and effects of pollution of land and water
- Discuss the effects of non-biodegradable plastics in the environment, in both aquatic and terrestrial ecosystems

Anything released into the environment as a result of human activity that has the potential to cause harm is a **pollutant**. Pollutants are chemical substances, such as fertilisers, oil and carbon monoxide; biological material such as animal and human wastes; heat released from power stations; and noise from industrial and domestic sources.

Pollution is the release of substances from human activities that are harmful to the environment. We expect natural ecosystems to absorb and breakdown these substances. In Topics 21.3 to 21.5 we will look at some examples of pollution of land, sea and air.

We throw away mountains of rubbish every day. In many places this rubbish is taken to rubbish dumps, becoming a visual pollutant as well as a danger to the environment.

pollutant	sources	undesirable effects of pollutant on the environment
carbon dioxide	burning fossil fuels	enhances greenhouse effect (see page 278)
methane	cattle; paddy fields for growing rice; coal and oil extraction	enhances greenhouse effect (see page 278)
sewage	human and livestock waste contains urea, ammonia, protein, carbohydrates, fats and pathogens	reduces oxygen concentration in rivers; destruction of freshwater communities (see page 276)
industrial chemical waste – includes a variety of toxic substances such as lead, mercury and cyanide	factories	can be fatal to wildlife and humans; can accumulate in organisms
fertilisers (mostly N and P)	arable agriculture	eutrophication in fresh water (rivers, lakes)
herbicides for killing weeds	arable agriculture	spray drift kills harmless plants; persist in the environment
pesticides for killing pests and diseases	arable and livestock agriculture	accumulates in food chains; kills harmless (non-pest) species
solid waste (biodegradable and non-biodegradable rubbish)	domestic and industrial waste	buried in ground (landfill) or left on rubbish tips; health hazard; leakage of toxic liquids; release of methane
nuclear fall-out (beta and gamma radiation)	atomic bomb; accidents at nuclear power stations; nuclear tests	death with high exposure; cancers in humans; mutations in non-human species

Land pollution

Rubbish is not only a visual pollutant, but it contains many harmful chemicals that leach from rubbish dumps into the ground. Examples are heavy metals that enter water courses and the drinking water supply.

Agricultural chemicals, such as herbicides and pesticides, if not applied correctly can affect areas locally and also far away. Insecticides such as DDT and dieldrin have been detected in the body fat of Antarctic mammals and birds, many thousands of miles away from places where they would have been used to control crop pests and vectors of disease, such as mosquitoes. Herbicides, especially when used in forests, destroy the habitats of animals and when sprayed from aircraft can drift onto areas of natural vegetation and kill plants.

Supplement

Waste plastics

Some materials we throw away are broken down in the environment by decomposers – these are called **biodegradable**. Some are not and these are called **non-biodegradable**. Plastics are used for packaging because they last a long time and do not decay easily. But this becomes a problem when it comes to disposing of plastics. Most plastic waste goes into landfill sites or rubbish dumps where they take up a lot of space. Biodegradable plastics are designed to break down more quickly than conventional non-biodegradable plastics once they are dumped. For example, some plastics now have starch incorporated into their structures which is digested by bacteria in the soil.

We can recycle many thermoplastics that we use by melting them and remoulding them into new shapes. Burning plastics in incinerators reduces the volume of waste, but many plastics produce toxic gases as they burn.

Non-biodegradable plastics pose threats to aquatic life too. Sea turtles are threatened if they swallow plastic bags mistaking them for jellyfish. Fish and other aquatic animals can get entangled in discarded plastic nets and other forms of plastic waste. In 2014, the United Arab Emirates banned non-biodegradable plastic products to try to reduce pollution.

Most plastic bottles have a recycling symbol with a category number. These bottles are in category 1 as they are made of polyethylene terephthalate (PET or PETE) which can be recycled to make products such as packaging materials and carpets.

STUDY TIP

The table on the previous page summarises what you need to know for Paper 3. If you are taking Paper 4, it is an introduction to the rest of this unit.

KEY POINTS

1 Pollution is the harm done to the environment by the release of substances produced by human activities.

2 Pollutants affect the air, ecosystems on land and aquatic ecosystems like rivers, lakes and the sea.

3 Nuclear fall-out comes from the testing and use of nuclear weapons and leakages from nuclear reactors.

4 Some plastics are biodegradable; others are non-biodegradable but some can be recycled.

Supplement

SUMMARY QUESTIONS

1 List three pollutants of each of the following:
   a the atmosphere
   b aquatic ecosystems
   c the land

2 a State the main sources of radiation.
   b Describe the undesirable effects of nuclear fall-out.

3 Explain the problems caused to the environment by non-biodegradable plastics.

Supplement

Water pollution

LEARNING OUTCOMES

- Describe water pollution by sewage and toxic chemicals
- Describe the pollution of rivers and lakes due to sewage and the over-use of fertilisers
- Describe the negative impact of female contraception hormones in water courses
- Describe the effects of herbicides and pesticides on the environment

Pollution of rivers and seas

Domestic and industrial pollutants are often discharged straight into rivers and into the sea.

Rivers empty toxic wastes into the sea, resulting in the following:

- Fertilisers and sewage encourage the growth of algae that release toxins.
- Pesticides are concentrated in the tissues of some molluscs (shellfish).
- Radioactive chemicals are found in higher concentrations around coastal nuclear power stations.
- Toxic metals, such as mercury, copper and lead, are found in tissues of marine organisms.

Sewage is the biggest single pollutant. It encourages the growth of algae and bacteria which use up lots of oxygen, killing fish and small invertebrates. If raw untreated sewage is dumped in a river it encourages the growth of bacteria that feed on the organic matter. The shortage of oxygen means that many freshwater organisms cannot survive so die or migrate away.

As water travels downstream, the water gradually improves as suspended wastes settle out and are decomposed by bacteria. If no more sewage is dumped in the river, the community recovers as the oxygen concentration of the water increases. In many countries sewage is treated so that raw sewage is not deposited into rivers (see page 284).

Fertilisers drain from the land into rivers and lakes. We have seen how using fertilisers can increase the yield of crops. But farmers need to know the best type of fertiliser to use for their particular soil and crop and how much of the fertiliser to add. Problems can occur if a farmer uses too much fertiliser or if the fertiliser is added at the wrong time, e.g. before a period of heavy rain.

The result of either of these is that fertilisers can cause water pollution resulting in **eutrophication**, which means that the water is enriched with plant nutrients. This is the sequence of events:

- Fertiliser can be washed through the soil into rivers and streams – this is called **leaching**. The rivers may flow into a lake.
- Once in the water, this stimulates population growth of algae. Nutrients in the fertilisers are usually in low concentration in the water and so limit the growth of the algae.
- Animals that eat the algae do not multiply fast enough to control their growth.
- Algae cover the surface layers of water, reducing the light reaching plants at the bottom of the lake.
- These plants eventually die and rot on the river bed.
- Algae also die as there is competition for resources and many are shaded by the algae on the surface.

In many parts of the world raw sewage drains into rivers.

- Decomposers, such as bacteria, feed on dead plants and algae.
- Bacteria respire aerobically, multiply rapidly and use up a lot of dissolved oxygen.
- The concentration of oxygen decreases and this kills fish and invertebrates that cannot respire properly.

The same chain of events can happen if sewage gets into waterways. Bacteria multiply quickly, use up oxygen in respiration, which can result in the death of fish.

The problems of eutrophication are caused by nitrate and phosphate. These ions are in short supply in most natural aquatic ecosystems, so they act as a limiting factor to the growth of plants. When they are added in the form of fertilisers, plant growth increases.

Farmers are encouraged to reduce the fertilisers they apply and to make sure that they are applied at a time when crops will take them up. Phosphate tends to remain in soils and much of the phosphate in water comes from domestic sources. Removing phosphate from detergents has helped to reduce this type of pollution. Similar events happen in the sea – nitrate in the sea causes growth of algae, many of which produce toxins and kill other marine life.

The effects of contraceptive hormones

Many women take contraceptive pills containing oestrogen (see Topics 16.12 and 16.13). The hormone is excreted in urine and therefore finds its way into human sewage. The active ingredient in contraceptive pills is a synthetic oestrogen known as EE2. Oestrogens are not broken down in all sewage treatment plants so they pass through unchanged to enter rivers and lakes. Male fish and amphibians respond to this oestrogen in the water by gaining certain female characteristics – a condition known as intersex. This 'feminisation', as it is also called, involves producing eggs inside their testes, having some female reproductive structures and making a protein found in egg yolk which is carried in their blood. In addition, their sperm cells do not develop properly.

Drinking water is often taken from rivers, and in the case of long rivers water may be taken downstream of the effluent from sewage works. An increase in oestrogens in drinking water has been linked to the decrease in sperm counts recorded in men in some parts of the world over the past 20 years.

This is only part of the story as there are other sources of oestrogens in the environment and many other synthetic chemicals that have the same effects. These chemicals are in the plastics and in the waste from agriculture and industry, including food processing plants especially those that deal with soya.

There are no easy solutions to this problem as many chemicals are implicated. EE2 is removed from the waste water from some sewage works so the rest could be upgraded. The use of charcoal filters is one of the effective ways to do this. Upgrading sewage works and drinking water treatment works will be expensive.

Fertilisers caused algae to grow and turn this water green.

KEY POINTS

1 Water pollutants include sewage, fertilisers, pesticides and toxic metals.

2 If excess fertilisers are leached into rivers and lakes they provide nutrients causing eutrophication by supplying nutrients for algae.

3 Female contraceptive hormones are causing the feminisation of aquatic organisms.

SUMMARY QUESTIONS

1 a Define the terms *pesticide*, *insecticide* and *herbicide*.

 b Explain why they are used.

 c Explain how pesticides can damage the environment.

2 a Explain the negative effects of female contaceptive hormones in water courses.

 b What steps could be taken to rectify the problem?

The greenhouse effect

Supplement

LEARNING OUTCOMES

- Describe the contribution of the greenhouse gases carbon dioxide and methane to climate change

- Explain how air pollution by greenhouse gases causes climate change

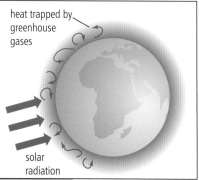

Figure 21.5.1 The greenhouse effect.

Factories, oil refineries and power stations release huge quantities of carbon dioxide into the atmosphere when they burn fossil fuels.

STUDY TIP

There are several air pollutants that you need to know. Look back to the table on page 274 to remind yourself of these.

Greenhouse gases

The atmosphere is like a blanket surrounding the Earth keeping it warm. Some of the gases in the atmosphere act like a greenhouse to keep in heat that otherwise would be radiated into space. These are the so-called **greenhouse gases** which are carbon dioxide, water vapour and methane. Some man-made air pollutants, such as CFCs (chlorofluorocarbons), are greenhouse gases as well as the cause of other problems.

These greenhouse gases allow solar energy to pass through to the Earth's surface. Some energy enters food chains and is eventually lost to the atmosphere as heat, which is radiated away from the Earth's surface.

Some heat energy escapes into space, but much is reflected back towards the Earth. Greenhouse gases keep our atmosphere at the temperatures that allow life to exist.

It is important to understand that the greenhouse effect is a natural process and without it the average temperature on the Earth would be about −17 °C. However, over the last 100 years, there has been a build-up of greenhouse gases.

Power stations, factories, domestic heating and transport use fossil fuels and release huge amounts of carbon dioxide into the atmosphere. As we have seen, deforestation has resulted in large areas of forest being removed. In South America the trees are cleared for farming and they are burned, which releases carbon dioxide into the atmosphere. Roots and other remaining tree parts are decomposed by microbes in the soil, producing even more carbon dioxide.

There has also been a significant increase in methane (another greenhouse gas) due to the expansion of rice cultivation and cattle rearing. Methane is released by cattle and also by bacteria in the anaerobic conditions found in flooded rice fields and natural wetlands. Rotting material in landfill sites and rubbish tips, as well as the extraction of oil and natural gas, are other sources of methane.

Of all the greenhouse gases the largest increase has been in carbon dioxide, which has risen by 10% in the last 30 years. However, methane and CFCs cannot be ignored as they are much more effective greenhouse gases than carbon dioxide.

Climate change

Human activities are causing an increase in the concentration of greenhouse gases so the atmosphere is getting warmer. This is causing the **enhanced greenhouse effect**. The surface of the Earth has warmed by 0.7 °C over the past century. There are fears that this warming is increasing.

If the temperature of the Arctic and Antarctic was to rise above 0 °C then the polar ice would start to melt. This would cause a rise in sea level and flooding of many low-lying areas, for example in Bangladesh. These would include some of the capital cities of the world.

There could also be a change in wind patterns and the distribution of rainfall leading to more extreme weather. Some parts are expected to become very dry. Some of these are important agricultural areas, such as parts of the USA and Asia, so that warming of the climate could mean a massive reduction in the grain crops of Central Asia and North America. The pattern of the world's food distribution could be affected with economic and political consequences.

Measures to reduce the effects of climate change involve reducing carbon emissions. This may be done by encouraging public transport, using energy more efficiently, recycling and changing the diet of cattle to stop them releasing methane.

One method is to use fossil fuels more efficiently as happens in a combined heat and power plant in which the heat is used for domestic heating.

The greenhouse effect

The diagram shows in more detail how the greenhouse effect works.

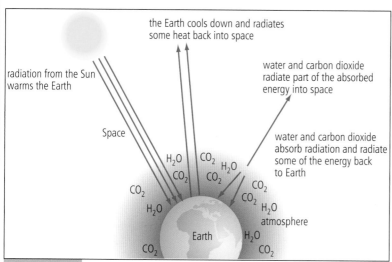

the Earth cools down and radiates some heat back into space

radiation from the Sun warms the Earth

Space

water and carbon dioxide radiate part of the absorbed energy into space

water and carbon dioxide absorb radiation and radiate some of the energy back to Earth

H_2O CO_2 H_2O
CO_2 CO_2
CO_2 CO_2
CO_2 H_2O

atmosphere

Earth H_2O
CO_2 CO_2

Figure 21.5.2 How the greenhouse effect works.

- Solar radiation passes through the atmosphere and warms the Earth's surface.
- The Earth radiates heat energy back into space. This is mainly infra-red radiation.
- Some of this heat energy is absorbed by the greenhouse gases.

This causes the air to warm up. Without carbon dioxide and water vapour the heat energy would pass straight back out into space. Human activity is causing a large increase in the atmosphere of carbon dioxide and other greenhouse gases such as methane, which has a greater impact than carbon dioxide (see question 12 on page 291).

It is the increases in these gases in the atmosphere that are responsible for the enhanced greenhouse effect.

1 The naturally occurring greenhouse gases act as a blanket around the Earth and prevent some heat energy escaping into space.

2 Air pollution by carbon dioxide and methane has increased the greenhouse effect and this is contributing to climate change.

3 The possible consequences of climate change could be a rise in sea level, flooding of low-lying areas, changes in the world's climate and a reduction in crop production in some areas.

SUMMARY QUESTIONS

1 Copy and complete:

warms absorbed air energy space radiates infra-red atmosphere

Solar _____ passes through the _____ and _____ the Earth's surface. The Earth _____ heat energy back into _____.

This is mainly _____ radiation. Some of the heat energy is _____ by the greenhouse gases. This causes the _____ to warm up.

2 List the gases that contribute to the enhanced greenhouse effect and in each case give one human activity that acts as a source of the gas.

3 a Describe the likely consequences of global warming.

  b Describe the ways in which people and their governments can reduce the emission of greenhouse gases and attempt to reduce the effects of global warming.

Acid rain

Sulfur dioxide (SO_2) is released when fossil fuels are burned. Together with nitrogen oxides (NO_x) from exhaust fumes, they cause **acid rain**. Rain is naturally acidic as the water dissolves carbon dioxide. However, when sulfur dioxide and nitrogen oxides dissolve in water in the atmosphere and react with oxygen they produce sulfuric and nitric acids which lower the pH. In some places the pH of the rain may be as low as 3.0.

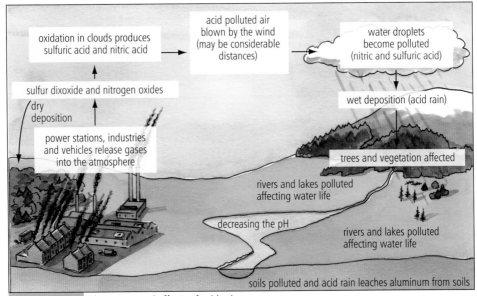

Figure 21.6.1 The causes and effects of acid rain.

These gases can be carried over large distances by winds before being deposited as dry particles or dissolved in the rain. The term 'acid rain' refers to both of these types of deposition.

When acid rain falls on trees it kills the leaves and reduces their ability to resist disease. Many trees in central European forests have been killed by acid rain.

Vegetation is affected by high concentrations of sulfur dioxide. Most sensitive are lichens, which are composed of algae and fungi living together. They absorb all their water and nutrients from the atmosphere. Some species are very sensitive to sulfur dioxide and grow only where the air is free of the gas and some can tolerate high concentrations. Others can tolerate intermediate concentrations. Lichens are useful indicators of pollution. By looking at the species growing in an area, you can deduce the average concentration of sulfur dioxide.

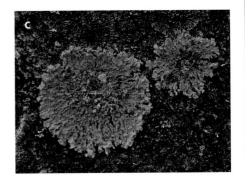

Increasing concentration of sulfur dioxide in the atmosphere ⟶

When acid rain falls on soils on limestone or chalk there is little effect, because the soils are alkaline and this neutralises the acid. When it falls on soils on hard rocks, such as granite, the soil is already acidic and there is nothing to neutralise the acid rain. Acid rain causes plant nutrients, such as potassium and calcium, to become soluble and wash out of soils leaving them infertile.

Aluminium compounds in soils become soluble below pH 5.5. Acidified water that enters lakes, streams and rivers lowers the pH of these bodies of water. The aluminium concentration can increase considerably. Species of invertebrates and fish that live in these ecosystems cannot tolerate water with a low pH, and aluminium ions are toxic. Lakes in Canada, Scotland and Norway have been badly affected by acid rain and some have very few or no animals in them.

The technology exists to combat acid rain such as:

- **low sulfur fuels** could be used. Crushing coal and washing it with a solvent that dissolves the sulfur compounds reducing its sulfur content,

- **flue gas desulfurisation** removes the sulfur from power station chimneys by treating the waste gases with wet powdered limestone, neutralising the acidic gases before they can escape,

- **catalytic converters** can be fitted to reduce the nitrogen oxides in the exhaust fumes of cars.

These solutions are expensive, but are being introduced gradually to stop the damage from acid rain. At a time when Europe and North America are cleaning up some of the damage done by acid rain, China's rapid industrial expansion and use of coal in its many power stations means that the country has problems with acid rain.

Trees in Central Europe killed by acid rain.

A catalytic converter reduces pollutants by oxidising carbon monoxide and hydrocarbons; they also convert oxides of nitrogen (NOx) into nitrogen. The catalysts are platinum or palladium which are recovered and recycled when vehicles are scrapped.

SUMMARY QUESTIONS

1 a Explain why normal rainfall is slightly acidic.

b Name the gases which dissolve in the water in clouds to form acid rain.

c State the sources of these gases.

d Explain the effects of acid rain on coniferous trees, soils and aquatic animals.

e Explain what can be done to reduce the effects of acid rain.

KEY POINTS

1 Sulfur dioxide from the burning of fossil fuels and nitrogen oxides from car exhausts are the main causes of acid rain.

2 Acid rain dissolves useful nutrients out of the soil making it infertile and also causes the release of aluminium ions that are toxic to fish.

3 Acid rain damages leaves and kills trees. Fish and invertebrates die if lakes become acidic.

4 Measures taken to reduce acid rain include removing sulfur from fuels and acidic gases from waste gases.

A flue gas desulfurisation plant which extracts sulfur dioxide from waste gases at a power station that burns coal. The product of this process is gypsum (calcium sulfate) which is used in the construction industry.

LEARNING OUTCOMES

- Define the term *sustainable resource*
- Explain the need to conserve non-renewable resources
- State that forests and fish stocks can be managed sustainably
- Explain how forests and fish stocks are managed sustainably
- Define the term *sustainable development* and explain that it requires management, planning and cooperation at local, national and international levels

A **resource** is any substance, organism or source of energy that we take from the environment.

A **sustainable resource** is a resource which is renewed by the activity of organisms so that there is always enough for us to take from the environment without it running out. Examples of these sustainable or renewable resources, that can be maintained, are timber for building and fish harvested from the sea and lakes.

Non-renewable resources include energy sources such as fossil fuels and minerals, e.g. copper, zinc and lead ores. These resources cannot be replaced: once they are gone, they are gone forever.

The need to conserve fossil fuels

The world's demand for energy is ever increasing and most of this is supplied by fossil fuels – coal, oil and natural gas. We need to conserve these non-renewable resources so that they can be used by future generations until such a time that we have developed renewable sources of energy so that they supply all our energy needs.

There are various ways we can conserve fossil fuels for the future:

- Use fossil fuels more efficiently by burning them to make full use of the energy released on combustion, e.g. combined heat and power plants.
- Reduce wastage of energy, e.g. improving the insulation of houses in cold countries and using alternatives to conventional air conditioning systems that are more energy efficient, such as those that use ice for cooling.
- Reducing the demand for petrol/gasoline, e.g. by providing better public transport and encouraging car pools.
- Recycling materials (see Topic 21.8).

Alternatives to fossil fuels as sources of energy are wind, solar, hydroelectric and nuclear power, and burning rubbish and wood from fast growing willow trees.

Combined Heat and Power plants like this one in Berlin, Germany, generate electricity and distribute heat in the form of hot water or steam to houses and industry in the surrounding area.

In Haiti, the production of charcoal for cooking has contributed to the removal of almost all the tree cover from the country.

Sustainable timber production

Timber is a valuable resource as it used in many industries to make many products. Some trees can be grown in a sustainable way. As soon as trees are felled, they are replanted. The replanting is an example of restocking so that the resource can continue to be harvested at the same rate without becoming depleted. These trees tend to be softwood, coniferous trees. Some governments manage forests by issuing permits to logging companies. They are given quotas so they do not take more trees than can regrow.

Sustainable fishing

Populations of fish are known as fish stocks. Many of these populations are now so severely depleted by overfishing that they are not sustainable. For example, stocks of the Atlantic cod collapsed in the 1990s.

Some species of fish and other sea creatures such as crustaceans and molluscs can be hunted sustainably. To continue to do this various steps need to be taken:

• Certain areas of the sea are declared off limits for fishing during the breeding season. This ensures that a suitable number of young fish are recruited to the stock each year. The Cayman Islands in the Caribbean has a 'no take' zone to help conserve a fish known as the Nassau grouper.

• Quotas are issued by fishing authorities so that boats cannot take too many fish in each season.

• Fishing nets have to be a certain minimum size so that small fish can escape and survive to breed.

• Some stocks of fish can be restocked if it is possible to breed them in captivity and raise them in facilities known as hatcheries. Examples are in Japan where shellfish, prawns and several species of fish including red sea bream and flounder, are released as young stages into the sea each year. Sturgeon to provide caviar are reared and released into the Caspian Sea by facilities in Iran.

Sustainable development

Sustainable development aims to provide the needs of an increasing human population without harming the environment. Many countries signed up to Agenda 21, a global action plan for sustainable development that arose from the 1992 Earth Summit in Rio de Janeiro, Brazil. Agenda 21 promotes the use of environmental impact assessments that assess the risk of developments to the environment. This is one way in which cooperation exists at the international, national and local levels.

Sustainable forestry: the mature trees are ready to be felled for the timber trade. The young trees have another 20 years to grow.

Large factory ships like this one have taken so many fish from the oceans that stocks are not sustainable and some species are being driven to extinction.

KEY POINTS

1 A sustainable resource is one which can be removed from the environment without it running out.

2 There is a need to conserve non-renewable resources, such as fossil fuels, for future generations.

3 Forests and fish stocks can be sustained using education, legal quotas and restocking.

4 Sustainable development provides for the needs of an increasing human population.

5 Cooperation between international, national and local organisations is essential for sustainable development.

SUMMARY QUESTIONS

1 Define the term *sustainable resource*.

2 Explain why forests and fish stocks may be sustainable, but fossil fuels are not.

3 List five ways in which fossil fuels can be conserved.

4 Discuss the role of education in conserving forests and fish stocks.

Activated sludge treatment.

Trickle filters.

Sewage treatment

Raw sewage is made up of a number of unpleasant materials, not just faeces and urine, but industrial waste and run-off from roads and paths. Not only is it unpleasant, but it is also a major health hazard.

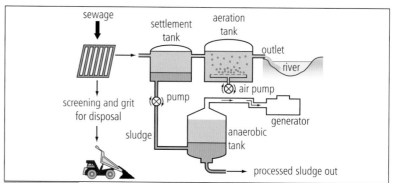

Figure 21.8.1 Components of a sewage treatment plant.

There are three main stages in the treatment of sewage:

- **Primary treatment** involves removing large solids, like paper and twigs, by means of filters or screens. Then the grit and organic solids are allowed to settle out in large tanks.
- **Secondary treatment** involves microbes, mainly bacteria, decomposing material suspended in water. This may happen in two ways.
  - The activated sludge process in which air is pumped through the sewage with a community of bacteria, fungi and protists which respire in aerobic conditions and break down carbohydrates, proteins and fats to carbon dioxide. Urea is broken down to ammonia and then converted to nitrate ions.
  - Trickle filters are beds of gravel covered with microbes. The liquid from the primary treatment is sprayed over these beds from large rotating arms. As the liquid trickles through the gravel, the microbes break down the organic matter.
- **Tertiary treatment** involves a period of settling. The liquid from the secondary process passes into large tanks. Here microbes and any remaining organic material settle out to form sludge. This sludge may be put back into the secondary stage to increase the population of microorganisms or it may go to an anaerobic stage in which bacteria break down waste matter to methane in anaerobic conditions. The methane is burned as a fuel in the sewage works to provide energy for operating the machinery. Sludge may also be dumped at sea or spread on the land as a soil conditioner.

The water leaving the plant will be treated in a variety of ways to make it safe. For example it may be treated with chlorine or ozone to kill all microorganisms. Water leaving the process is now safe to return to the environment or for human use.

Recycling our waste

Another consequence of an increasing population is the production of more waste. Unless this waste is properly handled and disposed of, it may cause pollution which can cause harm to animals, plants and ourselves. Dumping waste in rubbish tips and carefully managed landfill sites also deprives plants and animals of their habitats.

The solution to the problem of conserving useful materials is to **recycle** them. Glass, metals, paper and plastics can all be recycled. This saves raw materials because the recyclable materials do not have to be made in such large quantities, and less energy is used to recycle than to make the material from raw materials. For example, there is a 95% savings in energy if you compare recycled aluminium to the same mass of aluminium extracted from its ore, bauxite.

Paper is a resource in limited supply as it is made from wood. However, paper can be collected and recycled. This means that less paper pulp from trees is required. The waste paper is collected, pulped, the dyes are removed and the pulp is then rolled into sheets. These sheets are dried and converted into newsprint, paper towels, toilet paper and other paper products.

This recycling bin says it all? What do you recycle?

Waste paper is bundled up and shredded. This saves it being otherwise thrown into landfill sites, like the one above right.

SUMMARY QUESTIONS

1 Outline what happens to raw sewage as it passes through a sewage treatment plant.

2 a Make a list of materials that can be recycled.

  b Describe the effects that recycling of these materials have on the following:

    i   the raw materials used in making goods

    ii the energy needed in making these goods

    iii landfill sites and rubbish dumps

3 a Make a list of the different types of packaging that are used by shops.

  b In what ways does this packaging cause problems for society?

  c Explain how the recycling of paper helps the environment.

KEY POINTS

1 Sewage treatment involves screening to remove non-biodegradable material, decomposition by microbes and settling to remove suspended material.

2 Decomposition requires populations of bacteria that digest protein, fats, carbohydrates and cellulose in human wastes.

3 Recycling materials, such as paper, saves money, raw materials, energy and reduces pollution by waste.

Endangered species

LEARNING OUTCOMES

- Explain why organisms become endangered or extinct
- Explain the risks to a species if the size of its population decreases so reducing variation

The last thylacine (Tasmanian wolf) died in Hobart Zoo in 1936.

It is sometimes difficult to know when a species becomes extinct, but the Golden Toad, *Incilius periglenes*, was restricted to the Monteverde Reserve in Costa Rica and has not been seen there since 1989.

Cycads are ancient trees that are little changed since the Jurassic period over 100 million years ago. They are threatened in their natural habitats, but can grow safe in botanic gardens, as here in Florida.

A species becomes **extinct** if there are no individuals left alive anywhere in the world. Some species are kept alive only in captivity and they are described as extinct in the wild. Extinction occurs all the time. However, the number of extinctions that are occurring now are far higher than they have been in the past. This is as a result of all the activities we describe in this Unit.

When the numbers of a species decrease to such a level that it is at risk of becoming extinct the species is described as **endangered**.

We have already considered most of the causes of species becoming endangered and extinct. These are habitat destruction, climate change, pollution and overfishing. Others are:

- hunting for food, sport and to kill animals that are a threat to human life, crops or livestock
- competition with species that humans have moved from one part of the world to another
- predation by introduced or alien species.

Threats to the survival of species

Climate change is occurring as a result of changes to the composition of the atmosphere. Climate change was a feature of the past as the world cooled and heated at different times. During the last ice age most of Europe, North America and much of Asia was covered in ice sheets. As the world warmed, the ice melted and organisms colonised the land that was exposed.

Now the world is warming even more, and plants and animals associated with tropical regions are extending their ranges away from the equator. Plants better able to survive in changing environments will outcompete other species. Examples of this are evident on mountains. Plants adapted to cold conditions are disappearing from lower altitudes and retreating towards mountain peaks. If this continues there will be nowhere left for them to go.

Habitat destruction was covered in Topic 21.2. The Atlantic forests of Brazil have largely been cut down. This has robbed many species, including the golden lion tamarin of their habitats.

Deforestation is not the only cause of habitat loss. Wetlands are very important habitats as they provide ecological services for us and also feeding grounds for migrating birds. Many wetlands across the world have been drained to provide land for farming, housing, industrial development and leisure facilities.

Pollution is occurring on a global scale. As we have seen, coral reefs are endangered ecosystems as a result of coastal pollution. Sea otters, like many marine animals, are at risk of extinction thanks to oil pollution. Oil sticks to their fur and stops providing insulation and they ingest toxic chemicals that cause organ damage (see Topic 21.2).

Many sea otters died following the release of oil from the tanker Exxon Valdez in the Bay of Alaska in 1989.

Introduced species have been moved around the globe by humans. Sometimes we have done this deliberately by introducing animals for sport or to control pests:

- Almost half the small to medium-sized native marsupials in Australia have become extinct due to competition with rabbits and predation by foxes introduced by European colonists.
- The small Indian mongoose was introduced to Jamaica from India in 1872 to control rats that were eating much of the sugar cane crop. They proved so successful that they were introduced to other parts of the Caribbean. However, as with many such introductions, the mongoose fed on other prey as well as the one it was intended to control.
- Plant species can become extinct because of overgrazing by introduced species especially goats, which have done untold damage to many islands.

Hunting has caused the extinction of some species. They have been hunted for food, for sport, for trade and as a deliberate policy to remove dangerous animals or pests. Humans have always hunted animals and harvested plants from the wild for food. With increasing populations many of these populations became unsustainable. About 15 000 years ago people migrated from Asia into the Americas. As they moved southwards they destroyed many of the large mammal species.

People have hunted some animals, such as 'big cats' for sport and collected beautiful and interesting plants for show. Animals have been trapped and shot for their fur in the days before man-made materials became available. In some places, animals are shot as a deliberate policy as is the case with elephants that invade farmland. In Tasmania, people hunted the thylacine, or Tasmanian wolf, to extinction partly because it was believed to kill sheep.

A project to reintroduce the golden lion tamarin to its natural habitat in Brazil has involved captive breeding in many zoos.

Coral reef communities are threatened by pollution, climate change and overexploitation.

Supplement

There is very little genetic variation in small populations, so this makes rare and endangered species at particular risk of becoming extinct. With few individuals left alive, many of the alleles of the genes in that species are no longer left. This reduces the chances of the population evolving in response to changes in the environment (see Topic 18.4).

SUMMARY QUESTIONS

1. Define the following terms: **a** extinction, **b** endangered species.

2. Explain how these threaten the survival of species:
   a hunting   **b** overfishing   **c** hunting
   d habitat destruction   **e** climate change

3. Use examples to explain the possible effects that the introduction of species may have on food webs and ecosystems.

4. When a population decreases in size, variation within the population decreases. What are the implications for the survival of a species when this happens?

KEY POINTS

1. Organisms become extinct when there are no individuals left alive in the wild or in captivity. Endangered species are those that are at risk of extinction.

2. An endangered species is at risk of becoming extinct because its population is small with very little genetic variation that is needed to adapt to changing conditions.

3. Causes of extinction are climate change, habitat destruction, hunting, pollution and predation by introduced species or competition with them.

LEARNING OUTCOMES

- Describe how endangered species can be conserved
- Explain why species should be conserved

Putting a ring onto the legs of birds has given scientists information about how long they live and how far they travel. Some bird species migrate over thousands of miles.

Horton Plains National Park, Sri Lanka. Areas of wilderness like this should be conserved as the habitats of rare plants and animals, and also for the enjoyment of future generations.

The Arabian oryx, *Oryx leucoryx*, has been bred in zoos and reintroduced into the wild, sometimes successfully.

Endangered species can be conserved. This is never achieved simply by putting a barrier around an organism's habitat and leaving the area alone. Human activities affect everywhere on the planet so it is important to manage the organism's habitat and closely monitor the changes in population.

Monitoring can involve labelling, radio tagging and counting flocks of birds and herds of animals. Researchers in East Africa use aeroplanes to find herds of elephants and count them. Other techniques have to be used for forest elephants.

No species lives in isolation so we have a duty to conserve ecosystems and habitats. Here are some ways in which this is done:

- National Parks – large tracts of land set aside for wildlife but which may be occupied by people and which are patrolled by wardens, e.g. the game parks of East Africa such as Masai Mara in Kenya and Serengeti in Tanzania.
- Marine parks to protect areas of the sea from damage by fishing and pollution, e.g. Goat Island Marine Reserve in New Zealand.
- Rescuing endangered animals and breeding them in captivity and then returning them to the wild. Many species of *Partula* snails became extinct on Pacific islands during the 20th century. Some are now bred in captivity ready to be released back into their habitats. The Arabian oryx was bred in zoos and reintroduced to Oman where it had become nearly extinct.
- Growing endangered plants in botanical gardens and re-establishing them in the wild.
- Reducing habitat destruction, e.g. issuing licences for logging in forests to prevent deforestation and protecting wetlands to prevent them being drained.
- Re-establishment of ecosystems where land has become degraded, e.g. establishing dry forest in Guanacaste National Park in Costa Rica which may take up to 300 years to achieve!
- Preventing trade in endangered species. CITES, the Convention of International Trade in Endangered Species, imposed a worldwide ban on the ivory trade in 1989. This led to an increase in elephant populations in countries like Kenya.
- Encouraging sustainable management of ecosystems. Trees removed from forests should be replaced by planting or allowing time for natural replacement from seeds.
- Seed banks are cold stores that conserve seeds of endangered or valuable species. Seed banks around the world hope to collect and store seeds from many species in case they become extinct in the wild. Their genes may be useful for crop improvement in the future or to produce valuable products such as medicinal drugs. Seeds are collected from plants in the wild and are put into long-term storage. Seeds of many species are stored by dehydration so they contain only 5% water and can thus survive being kept at $-20\,°C$.

Removing water from seeds slows down their metabolism so that they remain viable for many years. However, seeds do not remain viable in seed banks forever. Some seeds from each sample are removed from storage, thawed and tested to see if they will germinate. This takes place every five years. Collections continue to be made if possible to 'top up' the bank for each species.

Increasing people's knowledge and understanding of the way in which their actions affect wildlife will help to ensure the success of conservation programmes.

These wardens in Papua New Guinea are conserving the eggs of leatherback turtles.

Reasons for conservation

We should conserve ecosystems, habitats and species because:

- Ecosystems provide us with services such as treating waste, providing food and fuels and giving us areas for recreation. They provide us with useful substances such as medicines.
- Ecosystems help to maintain the balance of life on the planet, e.g. nutrient cycles.
- Habitats support a wide variety of organisms that interact in ways we do not fully understand, often to continue life on this planet, for example by keeping pests and diseases in check.
- There are few foods, apart from some fish species, that we take directly from the wild in large quantities. But there are many that we take in smaller quantities, such as Brazil nuts. There may be many more plants and animals that we could utilise as food sources.
- Fuels – fossil fuels will not last forever. We still need timber to provide fuels, such as biomass fuels (see Topic 21.7).
- Drugs – drugs for treating cancer have been discovered in plants.
- Genes – as a result of selective breeding there is very little genetic diversity in our three main staples – rice, wheat and maize. It is important to conserve any locally adapted varieties that exist and also any wild relatives that have genes we could use in the future. These plants can be kept in botanic gardens and their seeds in seed banks.
- Prevent species becoming extinct, especially those threatened directly by human activities.

The Svalbard seed bank in northern Norway is cut into the Arctic permafrost.

KEY POINTS

1 Endangered species can be conserved by monitoring the sizes of their populations, protecting habitats, using captive breeding programmes and putting seeds into long term storage in seed banks.

2 Educating people of all ages about the importance and practices of conservation is essential for the survival of endangered species.

3 There are many reasons for conservation, such as reducing extinction, protecting vulnerable environments and maintaining ecosystem functions, such as nutrient cycling and providing resources such as food, drugs, fuel and genes.

SUMMARY QUESTIONS

1 Explain the meaning of each of the following:
  a endangered species   b species monitoring
  c captive breeding

2 State two species that are endangered in your region or country. Describe the steps being taken to conserve them.

3 Explain the part played by each of the following in conservation:
  a national parks   b zoos
  c botanic gardens   d seed banks

4 Explain why it is necessary to conserve habitats and ecosystems.

Supplement

1 Competition between crop plants and weeds for resources is one of the biggest potential causes of reduced yields. Which reduces competition between crop plants and weeds?

 A chemical fertilisers

 B herbicides

 C pesticides

 D selective breeding

(Paper 1)   *[1]*

2 The boll weevil shown below is a notorious pest of cotton plants. What should be used to control the boll weevil?

 A fungicide

 B herbicide

 C insecticide

 D molluscicide

(Paper 1)   *[1]*

3 An ecosystem is all the:

 A organisms in an area

 B organisms in an area and the physical factors that influence them

 C physical factors that influence an organism in an area

 D plants and animals in an area and the interactions between them

(Paper 1)   *[1]*

4 What is the negative effect of the over-use of fertilisers on farmland?

 A acid rain

 B eutrophication

 C global warming

 D more photosynthesis

(Paper 1)   *[1]*

5 Which pair of gases contributes to the enhanced greenhouse effect?

 A carbon dioxide and methane

 B oxygen and carbon dioxide

 C sulfur dioxide and water vapour

 D water vapour and oxygen

(Paper 2)   *[1]*

6 Fish stocks are in serious decline in many major fishing areas. Which is *not* a method that could be used to make stocks sustainable?

 A decreasing the mesh size of nets

 B setting up exclusion zones and 'no-take' zones

 C reducing the time that fishing boats can stay at sea

 D issuing quotas for the maximum numbers of fish that can be caught.

(Paper 2)   *[1]*

7 When a population becomes endangered the genetic variation within the population decreases. This means that within the population there are fewer:

 A alleles

 B chromosomes

 C DNA base sequences

 D genes

(Paper 2)   *[1]*

8 Eutrophication may happen when sewage or fertilisers enter bodies of water. The oxygen concentration in the water may decrease because:

 A decomposers are respiring aerobically

 B less oxygen dissolves in water from the air

 C more carbon dioxide is released by bacteria

 D there are more animals in the water

(Paper 2)   *[1]*

9 Human wastes enter sewage treatment works which use microorganisms to breakdown complex compounds in urine and faeces.

 (a) Outline what happens in the primary, secondary and tertiary stages of sewage treatment.   *[6]*

 (b) State three reasons why sewage should be treated for the safety of humans and the environment.   *[3]*

 (c) Microorganisms in sewage treatment works recycle carbon as part of the carbon cycle. Explain how they do this.   *[2]*

(Paper 3)

10 (a) Define the term *pollution*. [2]

When populations reach a certain size they are declared to be endangered.

(b) State what is meant by the term *endangered species*. [2]

(c) The tusked weta is a large carnivorous insect that used to live throughout New Zealand. It did not survive the introduction of alien species and was rescued by captive breeding and a reintroduction programme. Suggest:

(i) why the tusked weta did not survive the introduction of alien species to New Zealand [2]

(ii) how a captive breeding programme for such an insect would be carried out. [2]

(Paper 3)

11 The Asian elephant, *Elephas maximus*, is a large herbivorous animal. It is classified by the International Union for Conservation of Nature as an endangered animal. In January 2009, researchers published the first estimate of the population in the Taman Negara National Park in the centre of Peninsular Malaysia. The researchers counted piles of dung to estimate the population. Their estimate was 631.

(a) Suggest why the researchers could not count the elephants, but had to count piles of dung instead. [3]

(b) Suggest why it is important for conservationists to have reliable estimates of the population sizes of endangered species. [3]

(c) It has been estimated that Asian elephant populations can increase at a rate of 2% per year if there are no limiting factors.

(i) List four factors that are likely to limit the population growth of Asian elephants in a National Park such as Taman Negara. [4]

(ii) Explain why it is important to conserve large herbivorous animals, such as the Asian elephant, but not allow the population in a National Park to increase too much. [5]

(Paper 4)

12 The table shows how the concentrations of three gases have increased between 1890 and 1990. It also shows how effective they are as greenhouse gases compared with carbon dioxide.

gas	concentration in the atmosphere in certain years / parts per million			relative effect as greenhouse gas compared with CO_2
	1890	**1990**	**2030**	
carbon dioxide (CO_2)	290	354	400–500	1
methane (CH_4)	0.9	1.7	2.2–2.5	30
nitrous oxide (N_2O)	0.28	0.3	0.33–0.35	160

(a) Calculate the percentage increase in the concentration of carbon dioxide in the atmosphere between 1890 and 1990. Show your working. [2]

(b) The main sources of nitrous oxide are motor vehicles and fertilisers.

Explain why the concentrations of **(i)** carbon dioxide, and **(ii)** methane increased in the twentieth century. [4]

(c) Explain how the gases shown in the table may cause global warming. [5]

(d) Suggest why it is important to know the relative effects of these gases as shown in the table. [3]

(Paper 4)

Alternative to practical paper

Experimental skills and abilities for assessment

1 Use techniques, apparatus, and materials (including following a sequence of instructions, where appropriate).

2 Make and record observations and measurements.

3 Interpret and evaluate experimental observations and data.

4 Plan and carry out investigations, evaluate methods and suggest possible improvements (including the selection of techniques, apparatus and materials).

These skills are tested together in an alternative to practical paper or practical test such as the one below.

QUESTION 1

Pancreatin is a mixture of enzymes, including trypsin, that will digest proteins. Pancreatin will make a sample of skimmed milk turn from white to clear. Two students, A and B, are given dried milk powder.

They are given the following instructions:

1 Dissolve the milk powder in water and stir until there is no powder left.

2 Mix the dissolved milk and a pancreatin solution together.

3 Record the time taken for the milk to go clear.

The table on the left shows their results when they carried out their experiment at five different temperatures.

a Plot the results obtained for student B in the form of a suitable graph. [4]

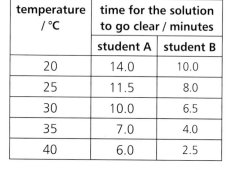

temperature / °C	time for the solution to go clear / minutes	
	student A	student B
20	14.0	10.0
25	11.5	8.0
30	10.0	6.5
35	7.0	4.0
40	6.0	2.5

EXPERIMENTAL SKILLS

• Following a sequence of instructions

• Interpret experimental data:
  • Select suitable scales and axes from graphs
  • Draw graphs from given data

Tips for drawing the graph:
• Scale chosen to make the graph as large as possible and to be easy to read.
• Plots are clear but not too large. Line is clear but not too thick.
• Scale labelled.
• Plots joined with ruled straight lines.
• Axis labelled with units. Note that the axis labels are the same as the table headings.

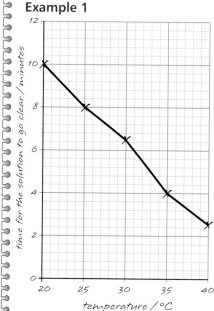

Example 1

time for the solution to go clear / minutes vs *temperature / °C*

STUDY TIP

Examination questions requiring a graph include the grid printed on the examination paper.
This graph gains full marks. Note that the variable that we change in this experiment is the temperature. This is the independent variable and is used as the left-hand column of the table and the x-axis of the graph.
The 'time to go clear' is the dependent variable which goes into the right hand column of the table and the y-axis of the graph.

Example 2

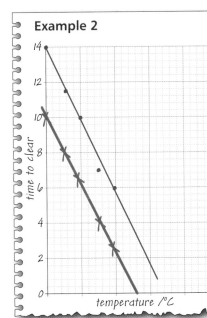

Answer

STUDY TIP

The plots in this graph are perfect but there are seven errors. What are they?*
In biology graphs points are joined with ruled straight lines, not with freehand lines. Lines of best fit can be used if requested in the question, but normally they are not used in biology. Uncertainty is common in biology where there is genetic variation and it can be difficult to control all other variables. This uncertainty is shown by joining points with ruled lines. In physics, chemistry and mathematics there is less uncertainty and then graphs are drawn with lines of best fit. Example 2 has used different plot markers for the two curves. This is good practice when two curves are required on the same grid.

* Answer
1. Curves for both students were plotted, even though only one was asked for.
2. The plot markers for student A shown in red are too large.
3. The red line is too thick.
4. There are no units on the y-axis and the x-axis scale has no numbers on it.
5. The x-axis scale has not been chosen well and the graph is not as large as it could be.
6. Lines of best fit have been used.
7. The lines are extrapolated.

b From your graph, describe and explain the effect of temperature on this enzyme between 20°C and 35°C. [3]

Example 1

The greater the temperature the less time is needed for the milk to go clear. The reaction time at 20°C is 10 minutes and 4 minutes at 35°C. This means that the reaction is faster by 6 minutes at the higher temperature. At 35°C the molecules have more energy so are moving faster and the milk molecules are more likely to collide with the enzyme.

STUDY TIP

A good description of a graph trend should say how the dependent variable changes as we changed the independent variable. In this case 'The greater the temperature the less time is needed for the milk to go clear'. The answer also includes at least two figures from the graph to illustrate the change as well as a calculation. In a question such as this, it is good practice to carry out a calculation of the rate of change without being directly asked to do so.
The two ideas of molecules moving faster at higher temperatures and of collisions being more likely gives 2 marks (of the maximum 3 marks).

Example 2

The reaction time at 20 is 10 minutes and 2.5 minutes at 40. This means that the reaction rate falls at the higher temperature. At very high temperatures enzymes will denature so the rate of reaction falls.

STUDY TIP

Example 2 makes two mistakes when trying to describe the effect. First, we do not know what 20 and 40 refer to because units are not given. This is a very common mistake. Units should always be given in an answer. Second, the question asks for the effect between 20°C and 35°C but the answer describes the effect between '20' and '40', so the example has not answered the question.
'This means that the reaction rate falls at the higher temperature' seems to be confusing time with rate of reaction.
The discussion of enzymes denaturing is true but it is not relevant to this graph.   No marks.

EXPERIMENTAL SKILLS

- Interpret and evaluate experimental observations and data

- Recall simple physiological experiments

TOP TIPS

Join points on a graph with ruled straight lines.

c Suggest reasons for differences between the results of the two students. [3]

Example

Student B may have used more milk powder or less water to dissolve the milk in.

Student B may have used a greater volume of the enzyme solution.

They may have had a different pH.

STUDY TIP

This answer would gain 2 out of the 3 marks.

The first two sentences correctly identify that other variables, such as milk concentration and enzyme volume, have not been controlled. Temperature and pH can change the rate of enzyme reactions so the suggestion that the reactions were carried out at a different pH is good. Take care to write pH and not PH or ph.

In these results student B has the smaller times so the larger rate of reaction. It must be clear from the answer whether concentrations are greater or smaller.

d Suggest and explain what would happen if student A carried out the experiment at 80°C. [2]

Example 1

The enzyme will be denatured.

Example 2

The milk would not go clear at all because the enzyme will be denatured. This means that the active sites of the enzymes do not work anymore.

STUDY TIP

Example 1 only gives an explanation so only scores 1 mark. Example 2 makes a correct suggestion and then goes on to give a correct, more detailed explanation so gains 2 marks.

QUESTION 2

The banded snail, *Cepaea hortensis*, exists as two forms within a single population. These forms can breed with each other and lay eggs. A pair of banded snails was allowed to breed together and their eggs placed in a glass tank with food. Two months after the eggs hatched the number of snails was counted.

The banded form of *Cepaea hortensis*.

The non-banded form of *C. hortensis*

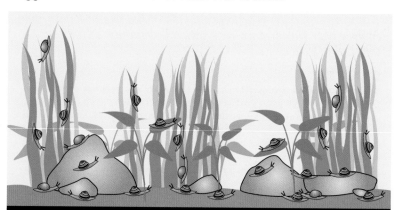

Figure 2.1

a i Count the number of banded snails and non-banded snails in the tank shown in Figure 2.1 and complete the banded and non-banded columns in the table below. *[2]*

	banded	non-banded	total number of snails
number of snails in Figure 2.1			

EXPERIMENTAL SKILLS

- Make and record observations
- Perform simple mathematical calculations

ii Calculate the total number of snails and complete the table above. *[1]*

Example 1

	banded	non-banded	total number of snails
number of snails in Figure 2.1	*18*	*6*	*24*

STUDY TIP

Tip: cross out each snail in pencil as you count so that you do not count one twice or miss one.

TOP TIPS

Make your handwriting clear, especially for numbers.

iii Calculate the fraction of banded snails in the sample. *[2]*

Fraction of banded snails =
Answer =

Example 1

Fraction of banded snails = $\dfrac{number\ of\ banded\ snails}{total\ number\ of\ snails}$

Answer = $\dfrac{2}{3}$

Example 2

Fraction of banded snails = $\dfrac{number\ of\ banded\ snails}{total\ number\ of\ snails}$

Answer = $\dfrac{18}{24}$

STUDY TIP

Example 1 gains 2 marks for showing how the fraction was calculated and the correct answer. Fractions do not have units but other calculations in the alternative to practical paper may require units. Example 2 gains only 1 mark because the fraction given is not simplified.
Care is needed that handwritten numbers are clear.

b Suggest and explain what these results in the table above indicate about the inheritance of the banding pattern in snails. *[3]*

EXPERIMENTAL SKILLS

- Interpret and evaluate experimental observations and data

STUDY TIP

This is another example of a question that tests interpretation of experimental data. Marks would be given for:
1. Recognising a 3:1 ratio, banded : non-banded.
2. Banded is dominant.
3. Non-banded is recessive.
4. The original parents were heterozygous, or the non-banded snails were homozygous.
5. Correct use of the term 'allele'; e.g. the banded allele is dominant.

To a maximum of 3 marks.

c Figure 2.2 shows a Cape Rock Thrush, *Monticola rupestris*. Make a labelled drawing of the head of the thrush. Ensure that your drawing has a ×2 magnification compared with the photograph. *[3]*

Figure 2.2

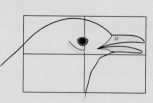

Draw a grid with a ruler over the relevant part of the photograph and draw the same grid lightly onto the answer space. This will help you get the proportions and magnification correct.

Magnification

You could be asked to calculate the magnification of your drawing:

$$Magnification = \frac{drawing\ size}{actual\ size}$$

If the photograph already has a magnification, do not forget to include that in your final answer.

Example 1

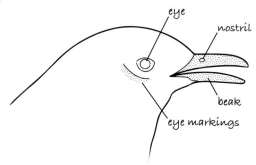

STUDY TIP

- The drawing is of the head only at a x2 magnification.
- The beak is in perfect proportion and the eye and nostril are in the correct place.
- There is minimum shading.
- The lines are smooth giving a clear outline.
- The label lines are clear and go precisely to the part labelled. Tip: avoid arrows.
- The drawing is accurately labelled. 3 marks

Example 2

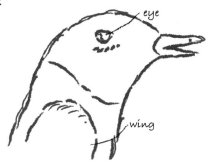

STUDY TIP

- The drawing includes some of the body and the head part is not x2 magnification.
- The beak is out of proportion and the eye and nostril are in the wrong places.
- There is too much shading.
- The lines are not smooth; they do not give a clear outline.
- The label lines do not go precisely to the part labelled.
- The line for the eye goes to the feathers above the eye and the wing is labelled but the question asks for a drawing of the head. No marks

Use smooth lines without too much shading for drawings.

Describe an investigation to compare the rate of water loss between plants in the light and plants in the dark, using the apparatus in Figure 3.1. In your answer describe how you will make the results as reliable and valid as possible. [5]

TOP TIPS

Keep controlled variables constant in an investigation.

Good investigations always involve taking repeated readings to improve the reliability of the data collected.

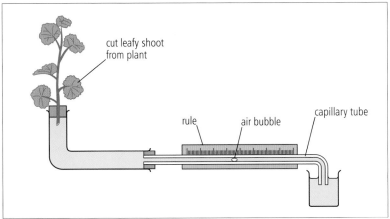

Figure 3.1 Shows a potometer that is used to measure the rate of water uptake by leafy shoots.

Example 1

You collect your results by measuring the distance travelled by the bubble in 2 minutes. Divide the distance by 2 and you will get the rate in mm per minute.
Make sure your apparatus has no leaks and that the temperature and wind speed are kept the same. Use the same piece of plant.
Do the experiment in the light three times, check for anomalies and take an average.
Do the same in the dark. Draw a bar chart to compare the averages.
If you know the diameter of the capillary tube you can work out the rate as mm³ of water lost per minute.

STUDY TIP

✓ Measuring the movement of the bubble (dependent variable) is explained.
✓ Gives 2 minutes as the standard time and explains how rate is measured.
✓ Shows the need to prevent leaks. This student has used the apparatus.
✓ Three control variables, temperature, wind speed and the piece of plant are controlled. Humidity could also have been mentioned.
✓ The need to carry out repeat readings, check for anomalies and calculation of the average is stated.
✓ A bar chart is suggested because there are only two values of the independent variable.
✓ Further detail about rates is given. Candidates should be careful when writing superscript numbers by hand.

A maximum of 5 marks is given.

Example 2

You have one set of apparatus with the plant in the dark and another with the plant in the light.
Measure the distance the bubble moves. When we did this experiment the bubble did not move at all because our apparatus was broken.
It is important to keep everything the same.
Measure the time with a stop watch.
Do the whole experiment three times and take an average.

STUDY TIP

- The first sentence simply repeats the question.
✓ How to measure the dependent variable by measuring the movement of the bubble is explained.
✓ Time measured with stopwatch.
- Variables kept the same need to be named.
✓ 'Do the experiment 3 times' is acceptable as referring to repeat readings.
- 'Take an average' is too vague, which figures are averaged?

3 marks awarded out of 5.

Glossary

This glossary provides definitions of some of the important biological terms used in this book. The underlined words are key terms as defined in the syllabus, which you are required to know for the examinations. You may, however, be expected to explain terms other than these.

A

Absorption movement of digested food molecules through the wall of the intestine into the blood or lymph.

Accommodation the adjustment of the shape of the lens so as to focus light onto the retina.

Acid rain rain or snow which is acidic. Polluting gases such as sulfur dioxide and the oxides of nitrogen react with water in the atmosphere to form acid rain.

Active immunity the body produces its own antibodies in response to an antigen, either on the surface of a pathogen or in a vaccine.

Active site part of the surface of an enzyme molecule into which the substrate fits.

Active transport the movement of molecules or ions in or out of a cell through the cell membrane from a region of their lower concentration to a region of their higher concentration against a concentration gradient, using energy released from respiration.

Adaptation a feature that helps an organism to survive in its environment.

Adaptive feature an inherited feature that increases an organism's fitness, i.e. by helping it to survive and reproduce in its environment.

Addiction when a person has taken a drug so regularly that they cannot do without it.

Adrenaline a hormone produced by the adrenal glands that prepares the body for emergencies and stress, for example by increasing the glucose concentration of the blood.

Aerobic respiration the process that happens in cells to release a relatively large amount of energy from food substances, such as glucose, in the presence of oxygen.

AIDS Acquired immunodeficiency syndrome – a collection of diseases that result from a weakening of the body's immune system following infection by HIV.

Allele any of two or more alternative forms of a gene.

Alveolus a tiny air sac in the lungs where exchange of gases between the air and blood occurs.

Amino acid a molecule made up of carbon, hydrogen, oxygen and nitrogen. Amino acids link together by chemical bonds to form protein molecules. Some amino acids also contain sulfur.

Amnion the membrane that surrounds the developing fetus in the uterus.

Amniotic fluid the liquid that is contained within the amnion and which protects the fetus from mechanical damage.

Amylase the enzyme that breaks down starch.

Anabolic steroid a hormone-like substance that increases the production of protein to build up muscle.

Anaerobic respiration the process that happens in cells to release a relatively small amount of energy from glucose in the absence of oxygen.

Anaemia a disorder of haemoglobin in the red blood cells which means that the blood cannot carry sufficient oxygen.

Antagonistic muscles a pair of muscles which brings about movement at a joint. When one contracts, the other relaxes, e.g. the biceps and triceps which move the lower arm.

Anther the part of a flower where pollen grains are produced.

Antibiotic a drug that is taken to kill or stop the growth of bacteria.

Antibody a protein released by lymphocytes to protect against pathogens.

Antitoxin a chemical released by lymphocytes that neutralises the poisonous waste products (toxins) produced by bacteria.

Arteriole type of blood vessel between artery and capillaries. Contraction and relaxation of muscle in the wall controls flow of blood into capillaries. See vasoconstriction and vasodilation.

Artery a blood vessel through which blood travels away from the heart.

Arthropods the phylum that consists of animals that have jointed legs and a protective exoskeleton.

Artificial insemination (AI) placing semen into the uterus or oviduct as a means of treating infertility

Artificial selection the selection of plants and animals for breeding because of their useful characteristics, e.g. high crop yield. Also called selective breeding.

Asexual reproduction reproduction without the formation of gametes. New individuals are genetically identical (clones) of the parent.

Assimilation the movement of digested food molecules into the cells of the body where they are used, becoming part of the cells.

Atherosclerosis a narrowing of the arteries caused by deposits of cholesterol in the internal walls of arteries which slows down the rate of blood flow.

Atrium a chamber of the heart that receives blood from veins and pumps it to a ventricle (plural atria).

Auxin a plant growth hormone which controls cell elongation.

B

Bacterium a type of microorganism consisting of single cells. Each bacterial cell has cytoplasm surrounded by a cell membrane and cell wall, but has no nucleus.

Balanced diet a diet that provides sufficient energy for a person's needs and all the food nutrients in the correct proportions.

Bile an alkaline fluid made in the liver and stored in the gall bladder. It is released into the small intestine through the bile duct to help with the digestion and absorption of fats.

Binary fission a type of asexual reproduction in which one cell divides into two. Bacteria reproduce by binary fission.

Binomial system a method for naming organisms in which each is given two names: a genus name and a trivial name, e.g. *Homo sapiens* for humans.

Biodegradable something that can be broken down by biological processes. Some plastics are biodegradable.

Biomass the mass of living material in a particular area.

Biotechnology the use of organisms to make useful products, such as foods and medicines, or to carry out useful services, such as making wastes harmless.

Birth control prevention of a birth either by contraception or abortion.

Bladder a muscular sac that stores urine and passes it out through the urethra.

Blind spot part of the retina where there are no receptors.

Brain the part of the central nervous system that coordinates most activities of the body.

Bronchiole a small branch of a bronchus which ends in alveoli (air sacs).

Bronchus one of the two tubes that branch off the trachea and pass into the lungs (plural: bronchi).

C

Capillary the smallest blood vessel with walls only one cell thick. Substances are exchanged through capillary walls between blood and tissue fluid.

Cancer a disease resulting from the uncontrolled division of cells in one or more parts of the body.

Carbohydrase an enzyme that digests carbohydrates to simple sugars, e.g. amylase digests starch to sugars.

Carbohydrate a class of food substance that provides energy, e.g. starch and glucose. Composed of carbon, hydrogen and oxygen.

Carbon cycle the flow of carbon compounds through plants, animals, decomposers and their environment.

Carbon dioxide the gas produced as a waste product during respiration and absorbed by plants and used during photosynthesis to make simple sugars.

Carnivore an animal that eats other animals – a meat-eater.

Carrier (1) an individual who has a recessive allele for a genetic disease and can transmit it to the next generation; (2) a person who is infected by the pathogen for an infectious disease, does not have the symptoms of the disease, but is able to transmit it to others.

Catalyst a substance that speeds up a chemical reaction and is not changed by the reaction.

Cell structural and functional unit of living organisms. All organisms are composed of cells. Viruses are not.

Cell membrane the boundary of the cell which controls the materials that pass into and out of it.

Cell sap the liquid that fills the vacuole of a plant cell.

Cell wall the outer layer of a plant cell made of cellulose, which supports the cell and gives it shape.

Cellulose a complex carbohydrate that makes up the cell walls of plant cells.

Central nervous system brain and spinal cord.

Chemical digestion the breakdown of large, insoluble molecules into small, soluble molecules.

Chlorophyll the green pigment found in many plant cells that absorbs light for photosynthesis.

Chloroplast small structure containing chlorophyll that is found in plant cells. Carries out photosynthesis.

Cholera an intestinal disease caused by a bacterium.

Cholesterol a lipid-based chemical made in the liver and found in the blood. High levels of cholesterol in the blood are linked to an increased risk of atherosclerosis and heart disease.

Chromosome a thread-like structure made up of genes. They are found inside the nucleus and are visible only when a cell is dividing.

Cilium a tiny process found on cells that line some tubular organs, e.g. airways and oviducts (plural: cilia). In the airways, they beat to move dust and microbes out of the lungs and up to the throat.

Ciliary muscle a muscle in the eye that controls the shape of the lens during focussing.

Circulatory system the organ system made up of blood vessels and the heart that transports blood. Mammals have a double circulation with blood passing through the heart twice in one circuit of the body.

Classify to sort living organisms into groups according to features they have in common.

Clone an organism that is genetically identical to its parent.

Clotting a series of chemical reactions that cause blood cells to stick together. At a wound this stops the loss of blood and results in the formation of a scab.

Codominance the existence of two alleles for a particular characteristic where neither is dominant over the other and both are expressed in heterozygous individuals.

Colon the part of the alimentary canal between the small intestine and the rectum where the absorption of water occurs.

Glossary

Community all the animals, plants and microorganisms that are found in a particular habitat.

Competition contest between organisms for resources such as food, water and mates.

Compost decaying plant remains used as a source of nutrients in gardens.

Concentration gradient the difference in concentration of a substance between two places, e.g. either side of a cell membrane, between air in the alveolus and blood in the lungs.

Cone a sensory cell in the retina of the eye that responds to light of high intensity and detects colour.

Constipation a condition where compacted faeces are difficult to pass out of the body.

Consumer an organism that gains its energy by feeding on other organisms.

Continuous variation variation in a feature that shows a range of phenotypes between two extremes with many intermediates, e.g. human height.

Contraception any method of birth control that prevents fertilisation.

Contraceptive any device or substance that prevents fertilisation.

Coronary arteries arteries that branch from the aorta to supply oxygenated blood to heart muscle.

Cornea the transparent layer at the front of the eye which helps to refract light rays onto the retina.

Coronary heart disease heart disease caused by blockage of coronary arteries that supply heart muscle with blood.

Cotyledon part of the embryo of a flowering plant – a seed leaf. In many plants cotyledons are food stores for the embryo.

Cross-pollination the transfer of pollen grains from the anther of a flower to the stigma of a flower on a different plant of the same species.

Cuticle the waxy covering of the epidermis in plant stems and leaves that reduces the loss of water by transpiration.

Cystic fibrosis an inherited disease affecting the lungs and the digestive system, caused by a faulty, recessive allele.

Cytoplasm jelly-like contents of the cell not including the nucleus.

D

Deamination the process, which takes place in the liver, where the nitrogen-containing part of amino acids is removed to form ammonia. Ammonia is then converted into urea.

Decay the breakdown of dead organisms and waste material by decomposers.

Decomposers microorganisms, mainly bacteria and fungi that gain their energy by breaking down dead organisms and waste material.

Deficiency disease a condition when an important nutrient, such as a vitamin or mineral, is missing and results in a disease.

Deforestation the removal of trees by humans in order to exploit the land.

Denitrification conversion of nitrate ions into nitrogen gas by bacteria.

Denitrifying bacteria bacteria which live in anaerobic conditions, such as water-logged soil, and convert nitrate ions into nitrogen gas.

Development an increase in complexity as an embryo grows and gains new tissues, organs and organ systems.

Diabetes a medical condition in which the blood glucose concentration is not controlled. One cause of diabetes is the failure of the pancreas to secrete insulin.

Dialysis the use of a partially permeable membrane to separate substances.

Diaphragm a sheet of muscular and fibrous tissue that separates the thorax from the abdomen. Its movements cause air to flow in and out of the lungs.

Diarrhoea a condition where the faeces released are loose and watery. A symptom of diseases, such as cholera.

Dicotyledon a type of flowering plant with an embryo that has two cotyledons and a net-like arrangement of veins in its leaves. (Also known as eudicotyledon.)

Diffusion the net movement of molecules or ions, from a place with a higher concentration to a place with a lower concentration down a concentration gradient as a result of random movement.

Digestion the breakdown of large, insoluble food molecules into small, water-soluble molecules using mechanical and chemical processes.

Digestive system the organ system that breaks down food and absorbs it into the blood.

Diploid nucleus a nucleus containing two sets of chromosomes (e.g. in body cells).

Discontinuous variation limited number of phenotypes for a feature with no intermediates, e.g. human blood groups.

DNA the molecule that forms the genetic material. The sequence of bases in DNA codes for the sequence of amino acids in proteins.

Dominant an allele that is expressed if it is present (e.g. T or G).

Drug any substance taken into the body that modifies or affects chemical reactions in the body.

Duodenum the first part of the small intestine.

E

Ecosystem all the living organisms (the community) in a place, and the interactions between them and their physical environment.

Egestion the passing out of food that has not been digested or absorbed, as faeces, through the anus.

Embryo the early stage of an animal or plant as it develops from a fertilised egg.

Emphysema a medical condition in which the walls of the alveoli are broken down. This reduces the surface area for gas exchange and means that the person cannot get enough oxygen into their blood.

Emulsification the breakdown of large fat globules into many tiny globules.

Energy transfer the transfer of energy from one trophic (feeding) level to the next.

Enhanced greenhouse effect increase in concentration of greenhouse gases in the atmosphere leading to global warming.

Enzymes biological catalysts that speed up the rate of chemical reactions in the body.

Eutrophication the enrichment of waters with plant nutrients that can stimulate growth of algae and plants.

Evolution the process in which inherited features change in populations of organisms over time.

Excretion the removal from organisms of toxic materials, the waste products of metabolism and substances in excess of requirements.

Extinct when a species no longer exists on the Earth.

Fats lipid molecules that consist of glycerol and three fatty acids. Fats are energy rich as they contain a high proportion of carbon and hydrogen atoms.

Fatty acids molecules that combine with glycerol to form fats.

Fermentation (1) anaerobic respiration in which glucose is converted into ethanol and carbon dioxide; (2) an industrial process in which microorganisms are used to make a useful product.

Fermenter a large container in which fungi or bacteria are grown under sterile, controlled conditions.

Fertilisation the fusion of the male and female sex cells, e.g. sperm and egg, to form a zygote (fertilised egg) that grows into a new individual.

Fertilisers chemicals that provide plant nutrients and are put on the land to increase the growth of a crop and produce a higher yield.

Fertility drug a drug that is used to increase the chances of a woman becoming pregnant.

Fetus a stage during development of a mammal when all the major organs are recognisable.

Fibre indigestible plant material, mainly cellulose and lignin that provides bulk to assist the passage of food through the gut by peristalsis.

Fitness probability of an organism surviving and reproducing in the environment in which it is found.

Flaccid a plant cell that has lost water by osmosis so that the cell contents no longer push outwards against the cell wall is described as flaccid.

Food chain this shows the feeding relationships in a community beginning with a producer. Each organism is fed on by the next organism in the chain. Food chains show the flow of energy and nutrients.

Food web a network of interconnected food chains showing the energy flow through an ecosystem.

Gall bladder a sac in the liver which stores bile before it is released down the bile duct into the small intestine.

Gametes sex cells with the haploid number of chromosomes.

Gas exchange system organ system comprising trachea, bronchi and lungs that involves moving air into the alveoli where oxygen and carbon dioxide are exchanged with the blood.

Gene a length of DNA, found on a chromosome, that codes for a particular characteristic.

Gene mutation a change in the base sequence of DNA that gives rise to new alleles of genes.

Genetic engineering changing the genetic material of an organism by removing, changing or inserting individual genes.

Genotype the genetic makeup of an organism in terms of the alleles it possesses (e.g. Tt or GG).

Genus a group of species with similar characteristics, e.g. the lion, *Panthera leo* and the tiger, *Panthera tigris* are in the same genus.

Glucagon a hormone produced by the pancreas that stimulates the liver to convert glycogen to glucose and so increase the concentration of glucose in the blood.

Glycogen a complex carbohydrate found as an energy store in the liver and muscles.

Goblet cells mucus-secreting cells.

Gravitropism a growth response by a plant to the stimulus of gravity.

Greenhouse effect the Earth is kept warm because carbon dioxide and other greenhouse gases in the atmosphere reduce the escape of heat energy into space. Greenhouse gases radiate heat towards the Earth and maintain the temperature higher than it would be otherwise.

Growth a permanent increase in size and dry mass by an increase in cell number or cell size or both.

Haemoglobin the red pigment in red blood cells that combines with oxygen to form oxyhaemoglobin.

Haploid nucleus a nucleus containing a single set of unpaired chromosomes (e.g. sperm and egg).

Hepatic portal vein the vein through which absorbed food travels from the small intestine to the liver.

Glossary

Herbicide a chemical that kills weeds.

Herbivore an animal that feeds only on plants.

Heroin a highly addictive, depressant drug derived from opium extracted from poppies.

Heterozygous a genotype where the two alleles of a gene are different.

HIV (Human immunodeficiency virus) the virus that causes AIDS. HIV attacks and destroys lymphocytes reducing the body's ability to defend itself against disease.

Homeostasis maintenance of a constant internal environment. This involves controlling factors, such as temperature and the concentration of glucose in the blood.

Homologous chromosomes a pair of matching chromosomes that carry genes for the same characteristics in the same positions.

Homozygous a genotype where both alleles of a gene are identical.

Hormone a chemical messenger produced by an endocrine gland that is transported in the blood and alters the activity of one or more specific target organs. Hormones are destroyed by the liver.

Hydrophyte a plant that is adapted to living in water.

Hypha (plural: hyphae) thin thread-like structure that is part of the body (mycelium) of a mould fungus.

I

Ileum the part of the small intestine between the duodenum and the colon, the major function of which is the absorption of digested food.

Immune system tissues, cells and chemicals that act together to give a defence against pathogens.

Immunity protection against disease provided by the immune system, including lymphocytes and antibodies.

Implantation the embedding of an embryo into the lining of the uterus.

Ingestion the process of taking in food.

Inheritance the transmission of genetic information from one generation to the next.

Inherited disorders disorders caused by dominant or recessive alleles that are passed from one generation to the next.

Insulin a hormone produced by the pancreas that stimulates the liver and muscles to store glucose as glycogen so causing a reduction in the concentration of glucose in the blood.

Intercostal muscles muscles between the ribs; external intercostal muscles contract during inspiration; internal intercostal muscles contract during expiration.

in vitro **fertilisation (IVF)** fertilisation occurs outside the body by mixing sperm and eggs in a laboratory dish. The resulting embryo is placed into the uterus.

Involuntary action an action that we do not have to think about.

Iris the coloured part of the eye around the pupil. It alters the size of the pupil and so controls the amount of light entering the eye.

K

Kidney the organ that filters waste chemicals out of the blood and controls the water and salt levels in the body.

L

Lactase the enzyme that breaks down lactose (milk sugar).

Lacteal a lymph capillary found inside a villus, which contains absorbed fats.

Lactic acid the chemical produced in muscles when glucose is respired anaerobically. Some bacteria also produce lactic acid during anaerobic respiration.

Large intestine the final parts of the alimentary canal that absorb water and store faeces.

Leaf the plant organ that absorbs light energy to convert into chemical energy during photosynthesis.

Lens part of the eye that focuses light rays onto the retina.

Limiting factor the factor that is in the shortest supply and restricts processes, such as the rate of photosynthesis or the rate of growth.

Lipase the enzyme that digests fats to fatty acids and glycerol.

Liver the organ in the abdomen that produces bile, breaks down amino acids into ammonia, converts ammonia into urea, stores glucose in the form of glycogen and carries out detoxification of toxic chemicals, e.g. alcohol.

Lymph the fluid formed when tissue fluid drains into lymph vessels.

Lymphatic system system of thin-walled vessels that transport lymph and lymph nodes where white blood cells are found.

M

Magnesium the element needed for the synthesis of chlorophyll. It is absorbed by the roots in the form of magnesium ions.

Malnutrition the condition caused by eating an unbalanced diet. Undernutrition occurs when a diet is deficient in one or more food types. Overnutrition can lead to obesity and coronary heart disease.

Maltase the enzyme that breaks down maltose to glucose.

Mechanical digestion the breakdown of food into smaller pieces without chemical change to the food molecules.

Meiosis a type of division of the nucleus to reduce the chromosome number by half. Diploid nuclei give rise to haploid nuclei. Also called a reduction division.

Memory cells lymphocytes produced during the first invasion of a pathogen or to a vaccine; these cells provide a fast immune response to infection.

Menstrual cycle the sequence of events, that occurs in a woman, where the lining of the uterus thickens in order to receive a fertilised egg after ovulation. The cycle is controlled by hormones from the pituitary gland and the ovaries.

Menstruation the breakdown of the soft lining of the uterus, discharging blood and cells through the vagina.

Metabolism all the chemical reactions that take place inside a living organism.

Microorganism (microbe) microscopic organisms, such as bacteria, fungi and viruses.

Mineral salts inorganic nutrients such as iron and calcium that are needed for a balanced diet. Also required by plants to make compounds, such as amino acids.

Mitochondrion cell structure where aerobic respiration occurs.

Mitosis a type of division of the nucleus that gives rise to genetically identical cells in which the chromosome number is maintained by the exact duplication of chromosomes.

Monocotyledon a type of flowering plant with an embryo with one cotyledon and parallel veins in its leaves.

Monohybrid inheritance the inheritance of one gene with two or more alleles.

Motor (effector) neurone a neurone that transmits impulses away from the brain or spinal cord to effector organs, e.g. muscles and glands.

Movement an action by an organism or part of an organism causing a change of position or place.

Mucus a slimy, sticky substance that traps dust and some microbes in the air pathways. It also acts as a lubricant to help the passage of food along the gut.

Muscle tissue that is capable of contracting and relaxing to bring about movement of bones at joints. Muscle tissue also moves food along our gut and keeps our heart beating.

Mutation a change in a gene or in a chromosome.

Mycelium a network of thin threads (hyphae) that make up the body of a fungus.

N

Natural selection factors such as competition and predation affect the survival of a species. As a result, only those individuals adapted to survive have the greater chance to pass on their genes to the next generation.

Negative feedback the mechanism used in homeostasis for maintaining near constant conditions in the body.

Nerve a bundle of nerve cells (neurones) which pass from the central nervous system to a certain part of the body.

Neurone a nerve cell.

Nitrification conversion of ammonium ions to nitrate ions by bacteria.

Nitrifying bacteria bacteria found in the soil that convert ammonium ions into nitrite ions and nitrite ions into nitrate ions.

Nitrogen-fixing bacteria bacteria found both in the soil and in the roots of legume plants that convert atmospheric nitrogen into nitrogen-containing compounds such as amino acids.

Nitrogen an element needed for healthy plant growth, taken up by plant roots in the form of nitrate ions. Chemical fertilisers often supply nitrate ions.

Non-biodegradable materials that will not break down in the environment, e.g. many plastics.

Nucleus the part of the cell which contains genetic information in the form of chromosomes. The nucleus controls the activities of the cell. Nuclei may be haploid or diploid.

Nutrition taking in of nutrients which are organic substances and mineral ions, containing raw materials or energy for growth and tissue repair, absorbing and assimilating them.

O

Obesity when a person is extremely overweight (see page 75 for two ways in which this is determined).

Oesophagus the muscular tube connecting the mouth with the stomach. Food passes down the oesophagus by a wave of muscular contraction called peristalsis.

Oestrogen a hormone secreted by the ovaries that stimulates the development of secondary sexual characteristics in females and helps to control the menstrual cycle.

Optic nerve Nerve from the retina in the eye to the brain.

Organ a number of tissues working together to carry out a function in the body.

Organ system a number of different organs that work together to carry out functions for the body.

Osmoregulation the control of the water content of the body.

Osmosis the diffusion of water molecules through a partially permeable membrane, from a region of higher concentration of water molecules to a region of lower concentration of water molecules (or from a region of higher water potential to a region of lower water potential).

Ovary female sex organ where ova or eggs are produced.

Ovulation the release of an ovum (egg) from the ovary.

Ovule the structure inside the ovary of a plant that contains the female gamete. After fertilisation an ovule develops into a seed.

Ovum the female sex cell (gamete) produced in the ovary. Also called the egg.

Glossary

Oxygen debt the extra oxygen that is needed by the body to respire lactic acid produced during anaerobic respiration.

P

Palisade mesophyll tissue in leaves that contain many chloroplasts and are the main site of photosynthesis.

Pancreas the organ in the abdomen that produces digestive enzymes and makes insulin and glucagon that regulate the concentration of glucose in the blood.

Partially permeable membrane a membrane that allows small molecules to pass through, but does not allow large molecules to pass through.

Passive immunity providing antibodies from another person or an animal; the person receiving the antibodies does not make the antibodies himself or herself.

Pathogen an organism that causes a disease.

Pectinase an enzyme that breaks down cell walls of plants.

Peripheral nervous system nerves that arise from the brain and spinal cord and go to all the organs of the body.

Peristalsis a wave of muscular contraction that squeezes food down the oesophagus to the stomach.

Pesticide a chemical that kills pests.

Phagocyte a type of white blood cell that ingests and destroys pathogens.

Phagocytosis the ingestion of food particles into a cell.

Phenotype the physical or other features of an organism due to both its genotype and its environment.

Phloem the plant tissue that transports sugars and amino acids.

Photosynthesis process by which plants manufacture carbohydrates from raw materials using energy from light.

Phototropism a growth response of a plant to the direction of light.

Pituitary gland a gland at the base of the brain that secretes hormones to control the activity of other organs, e.g. kidneys, testes and ovaries.

Placenta the organ that connects a mammalian embryo to its mother and through which it receives food and oxygen and removes carbon dioxide and chemical waste.

Plaque (1) a mixture of food remains, saliva and bacteria which can build up on the surface of teeth; (2) fatty deposit in the lining of an artery.

Plasma the liquid part of the blood which transports dissolved foods, urea, carbon dioxide and hormones.

Plasmid small piece of circular DNA used as a vector in genetic engineering.

Plasmolysis the separation of the cell membrane from the cell wall of a plant cell when water leaves the cell by osmosis.

Platelets small pieces of cells that release substances that cause blood to clot.

Pollination the transfer of pollen from the anther of a flower to the stigma.

Pollution the release by humans of materials or energy that will harm the environment.

Population a group of individuals of the same species living in the same habitat at the same time.

Population growth growth in numbers of a population. In absence of limiting factors growth is usually exponential.

Predators animals that hunt and kill other animals for their food.

Prey animals that are hunted and killed for food by predators.

Producer an organism that makes its own organic nutrients, usually using energy from sunlight through photosynthesis.

Progesterone a hormone secreted by the ovaries and by the placenta that maintains the lining of the uterus during the second half of the menstrual cycle and during pregnancy.

Protease an enzyme that breaks down proteins to amino acids.

Protein-energy malnutrition form of malnutrition in which energy and protein are limited in the diet, e.g. marasmus and kwashiorkor.

Proteins compounds made up of amino acids, which are needed for growth and repair of tissues in the body. They contain the elements carbon, hydrogen, oxygen, nitrogen and sulfur.

Puberty the age at which secondary sexual characteristics appear in boys and girls.

Pulse when the left ventricle of the heart contracts, it forces blood out of the heart along the arteries. The arteries swell and this can be felt as a pulse in various parts of the body, such as the wrist.

Pupil the hole through which light enters the eye. The size of the pupil is controlled by the iris.

Pyramid of biomass a way to show biomass at each trophic level in an ecosystem. The area of each horizontal bar is proportional to the mass of living material at each trophic level.

Pyramid of numbers a way to show numbers of organisms at each trophic level in an ecosystem. The area of each horizontal bar is proportional to the number of individuals at each trophic level.

 Q

Quota a limited number or quantity of a resource that can be taken from the environment, e.g. timber and fish.

 R

Receptor cells or organs that are sensitive to a stimulus.

Recessive an allele that is expressed only when there is no dominant allele of the gene present (e.g. t or g).

Rectum the last part of the large intestine where faeces are stored before passing out through the anus.

Recycle to convert material back into a form that can be useful to humans.

Reflex action an automatic, rapid response to a stimulus which is often protective.

Reflex arc the arrangement of neurones that controls a reflex action.

Relay (connector) neurone a neurone in the central nervous system that transmits impulses from sensory to motor neurones.

Renal dialysis the use of a machine to remove the body's waste chemicals when the kidneys fail.

Reproduction processes that make more of the same kind of organism.

Respiration a complex series of reactions, taking place in all living cells, that break down nutrient molecules to release energy from food.

Response the reaction of an organism to a particular stimulus.

Restriction enzyme an enzyme used in genetic engineering to cut DNA at specific places.

Retina the part of the eye that contains light sensitive cells.

Ribosome small cell structure where proteins are made.

Rod a sensory cell in the retina of the eye that responds to light of low intensity.

Root the organ that anchors plants into the ground and absorbs water and mineral salts.

Root hairs specialised cells in a root that provide a large surface area for the absorption of water and mineral salts.

Rough endoplasmic reticulum system of membranes inside cells where ribosomes are attached and proteins are made.

S

Sclerotic the tough, white outer layer of the eye which keeps it in shape.

Secretion the release of useful substances by cells.

Selective breeding see artificial selection.

Self-pollination the transfer of pollen grains from the anther of a flower to the stigma of the same flower or different flower on the same plant.

Sense organ a receptor organ that is sensitive to a particular stimulus or stimuli, e.g. touch, light, sound, temperature and chemicals.

Sensitivity the ability to detect changes in the environment (stimuli) and make responses.

Sensory neurone a neurone that transmits impulses from receptors to the brain or spinal cord.

Sex chromosomes the pair of chromosomes that determine a person's sex. In humans, XX is female and XY is male.

Sex linkage the location of genes on a sex chromosome, usually the X chromosome.

Sexual reproduction the process involving fusion of haploid nuclei to form a diploid zygote and the production of genetically dissimilar offspring.

Sexually transmitted infection a disease that is transmitted through body fluids during sexual activity, e.g. HIV.

Sickle cell anaemia an inherited disease affecting haemoglobin, which alters the shape of the red blood cells and results in severe anaemia.

Small intestine the region of alimentary canal that completes digestion and absorbs digested food.

Solute a substance that dissolves in a solvent.

Solvent a liquid that dissolves substances.

Species a group of organisms with similar characteristics which are capable of interbreeding and producing fertile offspring.

Sperm the male sex cell or gamete.

Starch an insoluble carbohydrate made from glucose molecules. An energy store in plant cells and an important component of the human diet.

Stem the organ that supports the leaves, flowers and fruits of a plant. It contains xylem tissue to transport water and mineral salts and phloem tissue to transport sugars and amino acids.

Stem cells unspecialised cells that divide by mitosis to produce daughter cells that can become specialised for specific functions.

Stimulus (plural: stimuli) a change in the environment that is detected by a sense organ.

Stoma (plural: stomata) a small hole in the epidermis of leaves that allows gases to diffuse in and out. The size of the hole is controlled by guard cells.

Stomach a muscular sac at the end of the oesophagus that mixes food with gastric juice. The chemical digestion of protein begins in the stomach.

Suspensory ligaments fibres that hold the lens in the eye in place and alter its shape when the ciliary muscles contract and relax.

Sustainable development development that is able to conserve natural resources so that they are available in the future.

Sustainable resource a resource, e.g. timber or fish, which is produced as rapidly as it is removed from the environment so that it does not run out.

Sweat glands coiled glands found in the dermis of the skin which secrete sweat to lose heat by evaporation.

Synapse a gap between two neurones across which a chemical transmitter is released to stimulate an impulse in the second neurone.

T

Test cross a genetic cross used to find out the genotype of an organism.

Testis (plural: testes) the male sex organs where sperm cells and testosterone are produced.

Testosterone the hormone, produced by the testes, that stimulates the development of secondary sexual characteristics in males and the development of sperm cells.

Glossary

Tissue a group of similar cells that act together to perform the same function.

Tissue fluid the fluid that bathes the cells when plasma and some white blood cells pass out of the capillaries.

Toxin a poisonous substance. Bacteria release toxins as waste products which bring about the symptoms of a disease, e.g. high fever.

Translocation the movement of sucrose and amino acids in the phloem.

Transmissible disease a disease in which a pathogen can be passed from one host to another.

Transpiration the evaporation of water from the surfaces of mesophyll cells in leaves followed by diffusion of water vapour through stomata into the atmosphere.

Trophic level the position of an organism in a food chain, food web, pyramids of numbers, biomass or energy. For example, herbivores are primary consumers and form the second trophic level.

Tropism a growth response by part of a plant to a stimulus, e.g. light or gravity.

Thrombosis a blood clot that occurs in a vein or an artery.

Tuber a swollen stem of a potato for food storage and asexual reproduction.

Turgor the pressure exerted by a plant cell onto its cell wall. Turgid cells are firm because they are full of water; this can support plant organs such as stems.

U

Umbilical cord the cord connecting the placenta to the fetus. It contains two umbilical arteries and a vein.

Urea a waste product formed from excess amino acids in the liver. It is filtered out of the blood in the kidneys and passed out in the urine.

Ureter a tube through which urine passes from the kidney to the bladder.

Urethra a tube that passes urine from the bladder to the outside at intervals.

Urinary system organ system that filters blood, produces and stores urine.

Urine an excretory fluid produced in the kidneys and stored in the bladder. It contains excess water, urea, and excess salts.

Uterus the womb; a muscular chamber with a soft lining in which the fetus develops.

V

Vaccine a preparation containing mild or dead pathogenic microorganisms to stimulate the production of antibodies and give immunity.

Vaccination giving a vaccine to provide immunity.

Vacuole fluid-filled sac containing cell sap present in most plant cells.

Vagina a tube leading from the uterus to the outside. It receives the erect penis during sexual intercourse and is the birth canal.

Variation differences between individuals in features that may show continuous variation or discontinuous variation.

Vasoconstriction contraction of muscles in arterioles to reduce blood flow through capillaries.

Vasodilation relaxation of muscles in arterioles that increases blood flow through capillaries.

Vein a blood vessel through which blood travels towards the heart.

Ventricle the lower, more muscular chamber of the heart. In mammals, the right ventricle pumps blood to the lungs and the left ventricle pumps blood to the rest of the body.

Vertebrate an animal with a backbone.

Vesicles small cell structures surrounded by membrane that contain substances made by cells, such as enzymes or neurotransmitters.

Villus (plural: villi) tiny, finger-like projection from the wall of the small intestine. Villi increase the surface area of the small intestine for absorption of digested food.

Virus a microorganism that cannot be seen under the light microscope. Smaller than bacteria, they consist of nucleic acid (e.g. DNA) surrounded by a protein coat. They can reproduce only inside living cells.

Vitamin a micronutrient, needed in small amounts in the diet. If not enough is provided in the diet a deficiency disease can result.

Voluntary action an action that we decide to make.

W

Water potential the tendency for water molecules to move by diffusion. Water diffuses from an area of higher water potential to an area of lower water potential.

White blood cells they form part of the body's defence system. Lymphocytes release antibodies and antitoxins to combat pathogens. Phagocytes ingest and kill pathogens.

Wilting loss of turgor causes plant leaves to droop if water is lacking.

Withdrawal symptoms side effects that a person feels when they give up an addictive drug.

X

Xerophyte a plant adapted to dry conditions.

Xylem the plant tissue that transports water and mineral salts from roots to leaves, flowers and fruits.

Y

Yeast a single-celled fungus that ferments sugar to produce ethanol and carbon dioxide.

Yield the quantity of product from a crop or from a fermentation process.

Z

Zygote a fertilised egg.

Index

Index

Index